Lecture Notes in Computer Science 9994

Commenced Publication in 1973
Founding and Former Series Editors:
Gerhard Goos, Juris Hartmanis, and Jan van Leeuwen

More information about this series at http://www.springer.com/series/7409

Shaoping Ma · Ji-Rong Wen
Yiqun Liu · Zhicheng Dou
Min Zhang · Yi Chang
Xin Zhao (Eds.)

Information Retrieval Technology

12th Asia Information Retrieval Societies Conference, AIRS 2016
Beijing, China, November 30 – December 2, 2016
Proceedings

 Springer

Editors
Shaoping Ma
Tsinghua University
Beijing
China

Ji-Rong Wen
Renmin University of China
Beijing
China

Yiqun Liu
Tsinghua University
Beijing
China

Zhicheng Dou
Renmin University of China
Beijing
China

Min Zhang
Tsinghua University
Beijing
China

Yi Chang
Yahoo Labs
Sunnyvale, CA
USA

Xin Zhao
Renmin University of China
Beijing
China

ISSN 0302-9743   ISSN 1611-3349   (electronic)
Lecture Notes in Computer Science
ISBN 978-3-319-48050-3   ISBN 978-3-319-48051-0   (eBook)
DOI 10.1007/978-3-319-48051-0

Library of Congress Control Number: 2016954933

LNCS Sublibrary: SL3 – Information Systems and Applications, incl. Internet/Web, and HCI

This Springer imprint is published by Springer Nature
The registered company is Springer International Publishing AG
The registered company address is: Gewerbestrasse 11, 6330 Cham, Switzerland

Preface

The 2016 Asian Information Retrieval Societies Conference (AIRS 2016) was the 12th instalment of the conference series, initiated from the Information Retrieval with Asian Languages (IRAL) workshop series back in 1996 in Korea. The conference was held from November 30 to December 2, 2016, at Tsinghua University, Beijing, China.

The annual AIRS conference is the main information retrieval forum for the Asia-Pacific region and aims to bring together academic and industry researchers, along with developers, interested in sharing new ideas and the latest achievements in the broad area of information retrieval. AIRS 2016 enjoyed contributions spanning the theory and application of information retrieval, both in text and multimedia.

This year we received 74 submissions form all over the world, among which 57 were full-paper submissions. Submissions were peer reviewed in a double-blind process by at least three international experts and one session chair. The final program of AIRS 2016 featured 21 full papers divided into seven tracks: "Machine Learning and Data Mining for IR," "IR Models and Theories" (two tracks), "IR Applications and User Modeling," "Personalization and Recommendation" (two tracks) and "IR Evaluation." The program also featured 11 short or demonstration papers.

AIRS 2016 featured three keynote speeches: "New Ways of Thinking About Search with New Devices" from Emine Yilmaz (University College London); "Will Question Answering Become the Main Theme of IR Research?" from Hang Li (Huawei Noah's Ark Lab); and "NLP for Microblog Summarization" from Kam-Fai Wong (The Chinese University of Hong Kong).

The conference and program chairs of AIRS 2016 extend our sincere gratitude to all authors and contributors to this year's conference. We are also grateful to the Program Committee for the great reviewing effort that guaranteed AIRS 2016 could feature a quality program of original and innovative research in information retrieval. Special thanks go to our sponsors for their generosity: GridSum Incorporation, Sogou.com Incorporation, Alibaba Group, and Airbnb. We also thank Springer for supporting the best paper award of AIRS 2016 and the Special Interest Group in Information Retrieval (SIGIR) for supporting AIRS by granting it in-cooperation status and sponsoring the student travel grant.

November 2016

Yiqun Liu
Zhicheng Dou
Yi Chang
Min Zhang
Ke Zhou
Wayne Xin Zhao

Organization

Honorary Conference Co-chairs

Shaoping Ma   Tsinghua University, China
Ji-Rong Wen   Renmin University of China, China

Conference Co-chairs

Yiqun Liu   Tsinghua University, China
Zhicheng Dou   Renmin University of China, China

Program Co-chairs

Yi Chang   Yahoo! Research, USA
Min Zhang   Tsinghua University, China

Poster and Demos Program Chair

Ke Zhou   Yahoo! Research, UK

Publication Chair

Wayne Xin Zhao   Renmin University of China, China

Program Committee

Giambattista Amati   Fondazione Ugo Bordoni, Italy
Bin Bi   Microsoft Research, USA
Shiyu Chang   University of Illinois at Urbana-Champaign, USA
Yi Chang   Yahoo! Research, USA
Jianhui Chen   Yahoo! Research, USA
Zhumin Chen   Shandong University, China
Bruce Croft   University of Massachusetts Amherst, USA
Ronan Cummins   University of Cambridge, UK
Hongbo Deng   Google, USA
Zhicheng Dou   Renmin University of China, China
Koji Eguchi   Kobe University, Japan
Carsten Eickhoff   ETH Zurich, Switzerland
Hui Fang   University of Delaware, USA
Yi Fang   Santa Clara University, USA
Geli Fei   University of Illinois at Chicago, USA

Kavita Ganesan	University of Illinois at Urbana-Champaign, USA
Bin Gao	Microsoft Research, China
Suriya Gunasekar	University of Texas at Austin, USA
Martin Halvey	University of Strathclyde, UK
Jialong Han	Nanyang Technological University, Singapore
Xianpei Han	Chinese Academy of Sciences, China
Satoshi Hara	National Institute of Informatics, Japan
Claudia Hauff	Delft University of Technology, The Netherlands
Kohei Hayashi	National Institute of Informatics, Japan
Ben He	University of Chinese Academy of Sciences, China
Jiyin He	Centrum Wiskunde & Informatica, The Netherlands
Yunlong He	Yahoo! Research, USA
Qinmin Hu	East China Normal University, China
Jimmy Huang	York University, Canada
Xuanjing Huang	Fudan University, China
Mark Hughes	IBM, USA
Adam Jatowt	Kyoto University, Japan
Jiepu Jiang	University of Massachusetts Amherst, USA
Jing Jiang	Singapore Management University, Singapore
Gareth Jones	Dublin City University, Ireland
Yannis Kalantidis	Yahoo! Research, USA
Jaap Kamps	University of Amsterdam, The Netherlands
Noriko Kando	National Institute of Informatics, Japan
Makoto P. Kato	Kyoto University, Japan
Hayato Kobayashi	Yahoo Japan Corporation, Japan
Udo Kruschwitz	University of Essex, UK
Lun-Wei Ku	Institute of Information Science, Academia Sinica, Taiwan, China
Hady Lauw	Singapore Management University, Singapore
Chenliang Li	Wuhan University, China
Fangtao Li	Google, USA
Jingyuan Li	Chinese Academy of Sciences, China
Liangda Li	Yahoo! Research, USA
Zhe Li	The University of Iowa, USA
Zhuo Li	University of Delaware, USA
Nut Limsopatham	University of Cambridge, UK
Jialu Liu	University of Illinois at Urbana-Champaign, USA
Kang Liu	Chinese Academy of Sciences, China
Xitong Liu	Google Inc., USA
Yiqun Liu	Tsinghua University, China
Zongyang Ma	Nanyang Technological University, Singapore
Kevin McGuinness	Dublin City University, Ireland
Qiaozhu Mei	University of Michigan, USA
Alistair Moffat	The University of Melbourne, Australia
Richi Nayak	Queensland University of Technology, Australia
Canh Hao Nguyen	Kyoto University, Japan

Dong Nguyen	University of Twente, The Netherlands
Jian-Yun Nie	Université de Montréal, Canada
Neil O'Hare	Yahoo! Research, USA
Hiroaki Ohshima	Kyoto University, Japan
Dae Hoon Park	University of Illinois at Urbana-Champaign, USA
Mingjie Qian	Yahoo! Research, USA
Miriam Redi	Yahoo! Research, USA
Zhaochun Ren	University of Amsterdam, The Netherlands
Tetsuya Sakai	Waseda University, Japan
Rodrygo Santos	Universidade Federal de Minas Gerais, Brazil
Huawei Shen	Chinese Academy of Sciences, China
Mahito Sugiyama	Osaka University, Japan
Aixin Sun	Nanyang Technological University, Singapore
James A. Thom	RMIT University, Australia
Bart Thomee	Yahoo! Research, USA
Andrew Trotman	University of Otago, New Zealand
Xiaojun Wan	Peking University, China
Bin Wang	Chinese Academy of Sciences, China
Hongning Wang	University of Virginia, USA
Jingjing Wang	University of Illinois at Urbana-Champaign, USA
Sheng Wang	Peking University, China
Yilin Wang	Arizona State University, USA
Yue Wang	University of Michigan, USA
Ingmar Weber	Qatar Computing Research Institute, Qatar
Dayong Wu	Chinese Academy of Sciences, China
Hao Wu	University of Delaware, USA
Weiran Xu	Beijing University of Posts and Telecommunications, China
Yao Wu	Simon Fraser University, Canada
Hongbo Xu	Chinese Academy of Sciences, China
Linli Xu	University of Science and Technology of China, China
Makoto Yamada	Kyoto University, Japan
Takehiro Yamamoto	Kyoto University, Japan
Yuya Yoshikawa	Chiba Institute of Technology, Japan
Haitao Yu	University of Tokushima, Japan
Mo Yu	Johns Hopkins University, USA
Aston Zhang	University of Illinois at Urbana-Champaign, USA
Jiawei Zhang	University of Illinois at Chicago, USA
Jin Zhang	Chinese Academy of Sciences, China
Min Zhang	Tsinghua University, China
Mingyang Zhang	Google Inc., USA
Xin Zhao	Renmin University of China, China
Yu Zhao	Beijing University of Posts and Telecommunications, China
Wei Zheng	University of Delaware, USA
Ke Zhou	Yahoo research, UK

Mianwei Zhou   Yahoo! Research, USA
Guido Zuccon   Queensland University of Technology, Australia

Steering Committee

Hsin-Hsi Chen   National Taiwan University, Taiwan, China
Youngjoong Ko   Dong-A University, Korea
Wai Lam   The Chinese University of Hong Kong, Hong Kong,
                                    SAR China
Alistair Moffat   University of Melbourne, Australia
Hwee Tou Ng   National University of Singapore, Singapore
Zhicheng Dou   Renmin University of China, China
Dawei Song   Tianjin University, China
Masaharu Yoshioka   Hokkaido University, Japan

Sponsors

Contents

Personalization and Recommendation

IR Evaluation

Short Paper

IR Models and Theories

Modeling Relevance as a Function of Retrieval Rank

Xiaolu Lu[1(✉)], Alistair Moffat[2], and J. Shane Culpepper[1]

[1] RMIT University, Melbourne, Australia
{xiaolu.lu,shane.culpepper}@rmit.edu.au
[2] The University of Melbourne, Melbourne, Australia
ammoffat@unimelb.edu.au

Abstract. Batched evaluations in IR experiments are commonly built using relevance judgments formed over a sampled pool of documents. However, judgment coverage tends to be incomplete relative to the metrics being used to compute effectiveness, since collection size often makes it financially impractical to judge every document. As a result, a considerable body of work has arisen exploring the question of how to fairly compare systems in the face of unjudged documents. Here we consider the same problem from another perspective, and investigate the relationship between relevance likelihood and retrieval rank, seeking to identify plausible methods for estimating document relevance and hence computing an inferred gain. A range of models are fitted against two typical TREC datasets, and evaluated both in terms of their goodness of fit relative to the full set of known relevance judgments, and also in terms of their predictive ability when shallower initial pools are presumed, and extrapolated metric scores are computed based on models developed from those shallow pools.

1 Introduction

A comprehensive set of judged documents derived from human relevance assessments is a key component in the successful evaluation of IR systems. However, growing collection sizes make it prohibitively expensive to judge all of the documents that are potentially relevant, and sampling methods such as *pooling* [15] are now commonly used to select a subset of documents to be judged. Partial judgments present an interesting challenge in carrying out reliable evaluation, and can result in subtle problems when comparing the quality of two or more systems.

The main issue arising from partial judgments is how to handle unjudged documents during evaluation. One simple rule – and the one often used in practice – is to assume that all unjudged documents are non-relevant. Although an evaluation score can be obtained using this assumption, any conclusions drawn may be a biased view of a system's relative performance. Two approaches to handling these issues have been proposed: metric-based solutions [1,3,5,9,11,17,18], and score adjustment [7,10,16]. Metric-based solutions can be further categorized

© Springer International Publishing AG 2016
S. Ma et al. (Eds.): AIRS 2016, LNCS 9994, pp. 3–15, 2016.
DOI: 10.1007/978-3-319-48051-0_1

as those that ignore the unjudged documents, and work only with the known documents; and those that attempt to infer the total relevance gain achieved by the system, or, at least, to quantify the extent of the uncertainty in the measured scores. Score adjustment approaches require a different type of collection pooling process, which can greatly impact the reusability of the test collection. They also seek to minimize the bias between the pooled and unpooled systems, which is different than the pooling depth bias. Pooling depth bias can occur in contributing systems as well as new systems since using a pooling depth less than the evaluation depth can result in unjudged documents occurring in any system ranking.

Here we consider traditionally pooled collections, and consider the problem from a fresh angle: *does the rank position of a previously unseen document influence the likelihood of it being relevant, and if so, can that relationship be exploited to allow more accurate system scores to be computed?* Our estimations of gain based on rank fit well with weighted-precision metrics, and allow both types of bias to be incorporated when performing evaluations. In particular, we measure the aptness of several possible models that build on existing judgments, from which we obtain an observed likelihood of relevance at different ranks. The benefit of assessing relevance as a function of rank is that the model can be applied both within the original pooling depth and also beyond it. A further advantage of the proposed approach is that in making the model topic-specific, it automatically adapts to differing numbers of relevant documents and to query difficulty, both of which can vary greatly across topics.

As a specific example of how our techniques might be employed, we consider the rank-biased precision (RBP) metric [9], which computes a *residual* as a quantification of the net metric weight associated with the unjudged documents in a ranking. Using an estimator, a value within that identified residual range can also be computed, and given as a proposed "best guess" score. To demonstrate the validity of our proposal, empirical studies are conducted on two representative TREC datasets: those associated with the 2004 Robust Track; and with the 2006 Terabyte Track. The first collection is believed to be relatively complete [13], while the second is understood to be less comprehensive [8,12]. The proposed models are fitted using topics in the two datasets and compared using a standard goodness-of-fit criterion at different nominal pooling depths. We then explore the predictive power of those models, by comparing extrapolated system scores generated from shallow-depth pools with the corresponding scores computed using deeper pools.

2   Background

Batch IR evaluations require a set of judgments for each included topic. *Pooling* [15] is often used to generate those judgments, but has limitations, since there is no guarantee that all relevant documents for a topic are identified. The usual way of handling that problem during evaluations is to assume that unjudged documents are not relevant. Incomplete judgments have been shown to have little effect in the NewsWire collections [19], but the evaluation results in larger

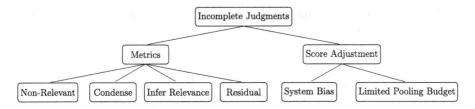

Fig. 1. A taxonomy of approaches for minimizing the effects of unjudged documents on system evaluation.

web collections can be biased [2]. As a result, several strategies for dealing with unknown documents have been developed [1,3,5,7,9–11,16–18]. Broadly speaking, these strategies can be categorized into two types – metrics that deal in some way with the missing judgments, and methods for adjusting the bias. Figure 1 provides a taxonomy of approaches, which we now explore.

Metrics for Incomplete Judgments. Widely used metrics such as AP and NDCG [6] were developed on the assumption that the judgments were complete. When they are used with incomplete judgments, unjudged documents are typically assumed to be non-relevant during the calculation process, an assumption that can result in underestimating the effectiveness of a system if it returns many unjudged documents, or overestimating the effectiveness of all systems if there are many undetected relevant documents. Alternative approaches have been proposed that use only the documents which are judged, including condensed scoring [11,17], and BPref [3]. Sakai [11] compared different condensed metrics with BPref and concluded that condensed Q-measure and NDCG work well in practice, and have a higher discriminative power than BPref.

In a quest to make better use of both judged and unjudged documents, metrics using inference [17,18] have also been proposed. For example, InfAP [17] estimates the precision at ranks where relevant documents occur, and assumes that relevant documents are distributed uniformly between identified ranks. A drawback is that inferred metrics depend on pools being constructed using a predefined sampling method. A recent study by Voorhees [14] concluded that a two-strata sampling is a suitable method for constructing collections for inferred metrics.

The metric StatAP [1] embeds another approach to sampling based estimation by deploying importance sampling when judgment pools are created in order to minimize the likelihood of missing relevant documents. StatAP estimates precision based on a joint distribution derived from the relevance probability of a pair of ranks. The total number of relevant documents is estimated via a uniform sampling process over a depth 100 pool. Combining both estimates produces the final StatAP score. Both InfAP and StatAP have been shown to be highly correlated with AP when the judgments are incomplete, using a range of collections [17,18]. However, inferred metrics and StatAP are reliant on specific sampling strategies being followed when pool construction occurs, meaning that applying these methods on unpooled systems may not be appropriate.

The final metric-based approach is to provide both the minimum and the maximum effectiveness score for a system, using the notion of a *residual* that was introduced alongside Rank-Biased Precision (RBP) [9]. Instead of generating a point effectiveness score, RBP provides a lower and an upper bound, with the gap between them representing the extent of score uncertainty associated with the unjudged documents. RBP supports traditional score-based system comparisons, and also provides quantitative evidence of the potential impact the unjudged documents may have on that comparison.

Score Adjustments Based on Estimated Relevance. The alternative is to try and adjust for the bias. The first option is to compensate for system bias – the difference between pooled and unpooled systems when using a fixed pool depth – using either a metric-based approach [7,10,16] or a metric-independent approach [5]. Based on RBP@10, Webber and Park [16] propose adjustment methods to deal with the inference from systems and from topics. In separate work, Ravana and Moffat [10] propose estimation schemes for picking a point within the RBP residual range: a background method; an interpolation method; and a smoothing method that blends the first two. Although Ravana and Moffat primarily focus on system bias, their results also indicate that the same approaches could be applied to adjust the bias resulting from a limited pooling budget.

Recent work by Lipani et al. [7] views the problem from another perspective, proposing an "anti-precision" measure in order to determine when to correct the pooling bias. By using a Monte Carlo method to estimate the adjustment score to be added to a run, Lipani et al. empirically obtain better results than previous work. Lastly, Büttcher et al. [5] consider the problem independent of the evaluation metric. By transforming bias adjustment into a document classification problem, the relevance of a document can be predicted to minimize rank variance when a leave-one-out experiment is applied.

Most of this prior work has focused on adjusting the bias between pooled and unpooled systems. When the pooling budget is limited, condensed runs and BPref may be vulnerable to relatively high score variance. Residual-enabled metrics such as RBP at least allow this variance to be quantified, but do not necessarily provide any way of drawing useful conclusions. Sampling methods and inferred metrics may be of some benefit in this regard, but give rise to different issues when systems not contributing to the original pool are to be scored. It is this set of trade-offs that motivates us to revisit the question of system comparisons in the face of a limited pooling budget.

3   Models and Analysis

We now describe methods for modeling relevance as a function of ranking depth.

Gain Models. Consider a weighted-precision metric such as RBP, which is computed as $\sum_{i=1}^{\infty} W(i) \cdot r_i$, where $W(i)$ is the ranking-independent weight attached to the item at rank i according to the metric definition, and r_i is the gain associated with that ith item in the ranking generated for the topic in question. When the judgments are incomplete, and the value r_j is not known for one or more ranks j, we propose that an estimated gain $\hat{r}_j$ be used, where $\hat{r}_j$ is computed via a model of relevance in which topic and retrieval rank j are the inputs.

Focusing on a single topic, we let $\langle r_{k,n} \rangle$ be a *gain matrix* spanning n systems that have contributed to a pooled evaluation to a maximum run length (or evaluation depth) of $k = d$, so that $r_{i,s}$ is the gain attributed to system s by the document it placed at rank i. The *empirical gain* vector $\mathbf{g} = \langle g_1, g_2, \ldots, g_k \rangle$ is then:

$$g_i = \frac{1}{n} \sum_{j=1}^{n} r_{i,j}. \tag{1}$$

A *gain model* is a function $G(\mathbf{g}, k)$ that generates a value $\hat{g}_k$ as an approximation for g_k, the empirical gain at rank k. For example, one simple gain model is to assert that if a document is unjudged its predicted gain is minimal, that is, $G_0(\mathbf{g}, k) = mingain$, where $mingain$ is the lower limit to the gain range and is usually zero. This is the pessimal approach to dealing with unjudged documents that was discussed in Sect. 2. Similarly, the residuals associated with RBP combine $G_0()$ at one extreme, and $G_1(\mathbf{g}, k) = maxgain$ at the other, where $maxgain$ is the upper limit to the gain range, and is often (but not necessarily always) one.

Increasingly Flexible Models. We are interested in gain models that lie between the extremes of $G_0()$ and $G_1()$, and consider five different interpolation functions in our evaluation, embodying different assumptions as to how gain varies according to rank. Table 1 lists the five options. The first model listed, $G_s()$, assumes that the gain is static and both topic and rank invariant. For early ranks this is perhaps more realistic than using G_0 or G_1, but is intuitively implausible for large ranks, since the goal of any retrieval system is to bring the relevant documents to the top of the ranking.

The second model is a truncated constant model, G_c, which is predicated on the assumption that all relevant documents appear in a random manner

Table 1. Five possible gain models, where $k \geq 1$ is the rank, and "Parameters" lists the free parameters in the estimated model.

Model	Description	Parameters	Assumptions
G_s	$(maxgain - mingain)/2$	–	Static, constant across all ranks
G_c	$\begin{cases} \lambda_0 & 1 \leq k \leq m \\ 0 & k > m \end{cases}$	λ_0, m	Constant until rank m, zero thereafter
G_ℓ	$\max\{-\lambda_0 \cdot k + c, 0\}$	$\lambda_0 \geq 0, c$	Linear, decreasing until rank m, zero thereafter
G_z	$\lambda_0/(k^c \cdot H_{n,c})$	$\lambda_0, c \geq 0$	Zipfian, monotonic decreasing, never zero
G_w	$\lambda_0 \cdot \left((1-\lambda_1)^{(k-1)^c} - (1-\lambda_1)^{k^c} \right)$	$\lambda_1 \in [0, 1], c > 0, \lambda_0$	Weibull, might increase before decreasing, never zero

at the early ranks of each run, and that beyond some cutoff rank m, no further relevance gain occurs. This model is rank-sensitive in a binary sense, and because m is a parameter that is selected in the context of a particular topic, it is also topic-sensitive. That is, the constant model G_c adds a level of flexibility to the static $G_s()$, and while it may also be implausible to assert that average gain is a two-valued phenomena determined by rank for any individual topic, in aggregate over a set of topics, each with a fitted value of m, the desired overall behavior might emerge.

The third step in this evolution is the model G_ℓ. The constant model G_c allows an abrupt change in predicted gain as a function of rank, at the topic-dependent cutoff value m. If we add further flexibility and suppose that average relevance gain decreases linearly as ranks increase, rather than abruptly, we get G_ℓ. This model also has cutoff rank m beyond which the expected gain from an unjudged document is presumed to be zero, given by $m = \lceil c/\lambda_0 \rceil$. A fourth option is to allow a tapered decrease, and this is what G_z achieves, via the Zipfian distribution, in which $H_{n,c}$ is a normalizing constant determined by the controlling parameter c and the ranking length n. The expected gain rate decreases at deeper pooling depths but remains non-zero throughout, due to the long-thin tailed property of the Zipfian distribution.

Another possibility is that the gain may initially increase or be constant, and then decrease in the longer term. To achieve this option, the monotonicity expectation is relaxed, a possibility captured by the discrete Weibull distribution, model G_w. Note that this function allows the possibility of an initial increase, but does not make that mandatory. In particular, when $c = 1$, the underlying distribution becomes a simple decreasing geometric distribution. Since this model is derived from a discrete Weibull distribution, the gain rate decreases faster than G_z when the distribution of relevance by rank is similar.

Given a model G that has been determined in response to a empirical gain vector $\mathbf{g}$, we take $\hat{r}_j = G(\mathbf{g}, j)$ for unjudged documents when r_j is unavailable, and then compute a weighted-precision metric such as RBP in exactly the same manner as before. That is, the estimated gain for that topic is used whenever the actual gain is unknown.

Measuring Model Fit. With a choice of ways in which relevance might be modeled, an obvious question is how to compare them and identify which ones provide the most accurate matches to actual ranking data. To measure goodness-of-fit we use root-mean-squared-error, or RMSE. That is, given a model G fitted to an empirical gain vector $\mathbf{g} = \langle g_j \rangle$ by choosing values for the controlling parameters (Table 1), we compute

$$\text{RMSE} = \sqrt{\frac{1}{n} \sum_{j=1}^{n} (G(\mathbf{g}, j) - g_j)^2}$$

as an indicator of how well that model and those parameters fit the underlying distribution. Small values of this measure – ideally, close to zero – will indicate that the corresponding model is a good estimator of the underlying observed behavior.

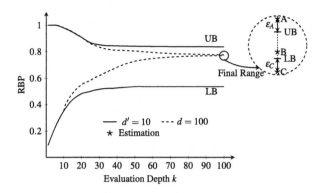

Fig. 2. An example of score bound convergence for a single system and a single topic. The two pairs of lines indicate the RBP score range at different evaluation depths k, based on two different pooling depths $d' = 10$ and $d = 100$. The "Final Range" is the metric score range at $k = 100$ using a $d = 100$ judgment pool; A, B, and C indicate three possible outcomes of a predictive model starting with the $d' = 10$ judgment pool.

Measuring Model Predictive Power. A second important attribute of any model is its ability to be predictive over unseen data, that is, its ability to be used as a basis for extrapolation. In particular, we wish to know if a model fitted to an empirical gain vector computed using judgments to some depth d' (training data) can then be used to predict system scores in an evaluation to some greater depth $d > d'$. Figure 2 illustrates this notion. Suppose that pooled relevance judgments to depth $d' = 10$ are available. If a weighted-precision metric such as RBP is used at an evaluation depth $k = 10$, all required judgments are available, but even so, there is a still a non-zero score range, or residual. That $d' = 10$ score range is illustrated in Fig. 2 by the solid lines, plotted as a function of k, the evaluation depth. Note that as the evaluation depth k is increased beyond 10 there is still some convergence in the metric, because documents beyond depth $d' = 10$ in this system's run might have appeared in the top-10 for some other system, and thus have judgments. The endpoints of those lines, at an evaluation depth of $k = 100$, are marked LB and UB. The dotted lines in the figure show the bounds on the score range that would arise if evaluation to k was supported by pooling to $d = 100$. The final $d = 100$ LB-UB range – a subset of the wider $d' = 10$ LB-UB range – is still non-empty, because the residual at depth k accounts for all documents beyond depth k, even if full or partial judgments beyond that depth are available.

Now consider an evaluation to depth $k = d$, but based on a model $G()$ derived from a pooling process to depth d'. If the model has strong predictive power, then the extended-evaluation using the predicted $\hat{r}_j$ values should give rise to a metric score that falls close to – or even within – the dotted-line LB-UB range that would have been computed using the deeper $d = 100$ judgment pool. That is, a metric score based on a predictive extrapolation will give rise to one of the three situations shown within the dotted circle: it will either overshoot the

$d = 100$ range by an amount ϵ_A; or it will undershoot the $d = 100$ range by an amount ϵ_C; or it will fall within that range, as suggested by the point labeled B. In the latter case, we take $\epsilon_B = 0$.

The overall process followed is that for each topic we use the set of system runs for that topic, together with the depth-d' pooled judgments, and compute the parameters for an estimated gain function. We then use that gain function to extrapolate the depth-d metric scores for that topic for each system, using $\hat{r}_j$ values generated by the model in place of r_j values whenever the corresponding document does not appear within the depth-d' pool. So, for each combination of topic and system an ϵ difference is computed relative to the score range generated by a pooled-to-d evaluation.

4   Experiments

Test Collections. We employ two different test collections, the 2004 Robust task (Rob04, topics 651–700) and the Terabyte06 task (TB06, topics 801–850), considering only the runs that contributed to the judgment pool. The first dataset has a pooling depth of $d = 100$ and a set of 42 contributing runs [13]; the second a pooling depth of $d = 50$, and 39 contributing runs [4]. Figure 3 provides a breakdown of document relevance in the two collections. Although the TB06 dataset uses shallower pooling, on average it contains more relevant documents per topic than Rob04 (left pane); and the percentage of relevant documents decreases more slowly as a function of pool depth (right pane). For example, approximately 8% of the TB06 documents that first enter the pool as it is extended from $d = 40$ to $d = 50$ are found to be relevant.

Goodness-of-Fit Evaluation. Regression was used to compute the two or three parameters for each model (Table 1), fitting them on a per-topic-basis, and using a range of nominal pooling depths d. In the static model $G_s()$ the predicted gain was set to 0.5 at all ranks; and in the constant model $G_c()$ the

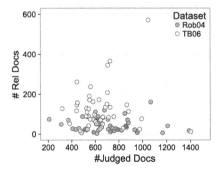
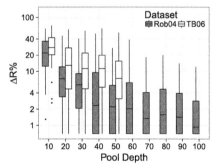

Fig. 3. Datasets used, showing the balance between judged documents and relevant documents on a per-topic basis (left); and the rate at which relevant documents are discovered by increasing pool depth bands (right, with a logarithmic vertical scale).

Table 2. RMSE of models, evaluated to depth d, averaged across topics and systems, using parameters computed using pooling data to depth d. Model $G_H()$ is a hybrid that selects the best of the other models on a per-topic basis. Daggers indicate values not significantly worse than the hybrid model at $p = 0.05$, using a two-tail paired t-test.

d	Rob04						TB06					
	G_s	G_c	G_ℓ	G_z	G_w	G_H	G_s	G_c	G_ℓ	G_z	G_w	G_H
10	0.237	0.106	0.080	0.086	0.071	0.070	0.190	0.040	0.020	0.018	0.011†	0.011
20	0.256	0.113	0.085	0.088	0.072†	0.071	0.186	0.049	0.020	0.022	0.011†	0.011
30	0.275	0.114	0.088	0.087	0.071†	0.070	0.186	0.056	0.022	0.024	0.011†	0.011
40	0.292	0.114	0.090	0.087	0.069†	0.069	0.187	0.060	0.024	0.023	0.012	0.011
50	0.307	0.112	0.090	0.087	0.068†	0.068	0.189	0.063	0.025	0.025	0.012†	0.011

cutoff parameter m was capped at the pooling depth. All of the judgments to the specified test depth d were used, in order to gauge the suitability of the various models. Note that the large volume of input data used per topic and the small number of parameters being determined means that there is only modest risk of over-fitting, even when d is small. Predictive experiments that bypass even this low risk are described shortly.

Table 2 lists average RMSE scores, categorized by dataset, by model, and by pooling depth d. The two columns labeled $G_H()$ are discussed shortly. Two-tail paired t-tests over topics were used to compare the RMSE values associated with the five models. When all available judged documents are used, G_w has the smallest RMSE on both datasets compared to the other four models, at a significance level $p \leq 0.05$ in all cases, and is a demonstrably better fit to the observed data than are the other four approaches.

We also explored a hybrid model, denoted $G_H()$, which selects the smallest RMSE over the available fitting data for the five primary approaches on a topic-by-topic basis. Two-tail paired t-tests were also conducted between model G_H and each of the others, and in Table 2 superscript daggers indicate the RMSE measurements that were not found to be significantly inferior to the hybrid approach, again using $p \leq 0.05$. The Weibull model is a very close match to the hybrid approach, and of the per-topic selections embedded in the hybrid, the Weibull was preferred around 85% of the time.

Looking in detail at Table 2, we also conclude that the Rob04 judgments are harder to fit a curve to, with overall higher RMSE values for each corresponding depth and model compared to the TB06 judgments. It is also apparent that little separates the Zipfian $G_z()$ and linear $G_\ell()$ approaches, and that either could be used as a second-choice to the Weibull mechanism. Finally in connection with Table 2, the consistency of values down each column as data points are added confirms the earlier claim that there is only a modest risk of over-fitting affecting the results of this experiment.

Figure 4 illustrates the five fitted curves for two topics, and their approximation of the empirical gain, which is shown in the graphs as a sequence of

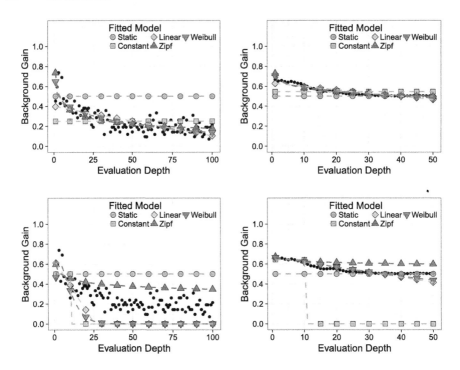

Fig. 4. Topic 683 for Rob04 (left column), and Topic 819 for TB06 (right column), with models fitted using all available judgments (top row) and using a depth $d' = 10$ pool (bottom row). The black dots show the empirical gain, and are the same in both rows.

black dots. One topic from each of the two datasets is plotted, with two different pooling depths – one graph in each vertical pair using all of the available judgments ($d = 100$ for Rob04, and $d = 50$ for TB06, in the top row), and one graph showing the models that were fitted when pooling was reduced to a nominal $d' = 10$ (bottom row). One observation is immediately apparent, and that is that empirical gain does indeed decrease with rank; moreover, in the case of TB06 Topic 819, it does so surprisingly smoothly. Also worth noting is that the empirical gain for the Rob04 topic decreases more quickly than it does for the TB06 topic as the evaluation depth k increases, which both fits with the overall data plotted in the right pane of Fig. 3, and helps explain the better TB06 scores for the static model in Table 2. Comparing the top two graphs with the lower two, it is clear that the more volatile nature of the empirical gain in the Rob04 topic has meant that when only $d' = 10$ judgments are available, the models all diverge markedly from the actual g_k values when they are extrapolated beyond the fitted range. The smoother nature of the TB06 empirical gain function means that the extrapolated models based on $d' = 10$ continue to provide reasonable projections.

Predictive Strength Evaluation. The most important test of the various models is whether they can be used to generate reliable estimates of metric scores when extrapolated beyond the pooling depth, the process that was illustrated in Fig. 2. Table 3 lists the results of such an experiment, using RBP0.95 throughout, a relatively deep metric (at an evaluation depth of 50, the inherent RBP0.95 tail-residual is 0.07, and at an evaluation depth of 100, it is 0.006), and with $G_s()$ omitted for brevity. To generate each of the table's entries, a pool to depth d' is constructed, and the corresponding model fitted to the empirical gain values associated with that pool. Each run is then evaluated to depth $k = 100$ (Rob04) or $k = 50$ (TB06) using pooled-to-d' judgments, if they are available, or using estimated gain values $\hat{r}_j$ generated by the model for that topic. The RBP score estimate that results is then compared to the score and residual range generated using the full pool, $d = 100$ for Rob04 and $d = 50$ for TB06. If the extrapolated RBP score falls within that pooled-to-d range, an ϵ of zero is registered for that system-topic combination; if it falls outside the range, a non-zero ϵ is registered, as described in Sect. 3. Each value in the table is then the average over systems of the root-mean-square of that system's topic ϵ's; with the parenthesized number beside it recording the percentage of the ϵ values that are zero, corresponding to predictions that fell within the final RBP score range. We also measured the "interpolative" method of estimating a final RBP score that was described Ravana and Moffat [10], denoted as "RM" in the table. It predicts RBP scores assuming that the residual can be assigned a gain at the same weighted rate as is indicated by the judged documents for that run.

All of the models are sensitive to the pooling depth d', and it is only when sufficient initial observations are available that it is appropriate to extrapolate.

Table 3. Root-mean-square of ϵ prediction errors using different pooling depths d', compared to an evaluation and pooling depth of $k = d = 100$ (Rob04) and $k = d = 50$ (TB06). The method labeled RM is the "interpolation" method of Ravana and Moffat [10]. Bold values are the best in that row, and the numbers in parentheses are the percentage of the system-topic combinations for which $\epsilon = 0$ (point B in Fig. 2).

d'	G_c	G_ℓ	G_z	G_w	G_H	RM
Robust04						
20	0.020 **(33)**	**0.015 (33)**	0.027 (15)	0.018 (31)	0.018 (31)	0.040 (9)
40	0.004 (60)	**0.003 (65)**	0.005 (56)	0.004 **(65)**	0.004 **(65)**	0.010 (31)
60	**0.001** (78)	**0.001** (84)	**0.001 (89)**	**0.001** (88)	**0.001** (88)	0.002 (71)
80	**0.000** (92)	**0.000** (95)	**0.000 (99)**	**0.000** (97)	**0.000** (97)	**0.000** (98)
Terabyte06						
10	0.089 (24)	**0.061 (45)**	0.082 (39)	0.065 **(45)**	0.067 **(45)**	0.065 (42)
20	0.033 (40)	**0.022** (71)	0.031 (68)	0.023 **(73)**	0.024 **(73)**	0.023 (68)
30	0.013 (58)	0.006 (88)	0.008 (87)	**0.005 (90)**	0.006 **(90)**	0.008 (87)
40	0.004 (78)	0.001 (98)	0.001 (98)	**0.000 (99)**	**0.000 (99)**	0.001 (97)

Also interesting in Table 3 is that the linear model, $G_\ell()$, provides score predictions that are as reliable as those of the Weibull model. As a broad guidance, based on Table 3, we would suggest that if an evaluation is to be carried out to depth k, then pooled judgments to depth $d' \geq k/2$ are desirable, and that application of either the Weibull model $G_w()$ or the simpler linear model G_ℓ to infer any missing gain values between d' and k will lead to reliable final score outcomes. Both outperformed the previous RM approach [10].

That then leaves the choice of k, the evaluation depth to be used; as noted by Moffat and Zobel [9], k is in part determined by the properties of the user model that is embedded in the metric. In the RBP model used in Table 3, the persistence parameter $p = 0.95$ indicates a deep evaluation. When p is smaller and the user is considered to be less patient, the fact that the tail residual is given by p^k means that smaller values of k can be adopted to yield that same level of tail residual. Note that it is not possible to analyze AP in the same way, hence our reliance on RBP in these experiments.

5   Conclusions and Future Work

We have investigated a range of options for modeling the relationships between relevance and retrieval rank, calculating the probability of a document being relevant conditioned on a set of systems and the evaluation depths. Our experiments show that it is possible to use the models to estimate final scores in weighted-precision metrics with a reasonable degree of accuracy, and hence that pooling costs might be usefully reduced for this type of metric. To date the predictive score models have not been conditioned on the document itself, and the fact that it might be unjudged in multiple runs at different depths. We plan to extend this work to incorporate the latter, hoping to develop more refined estimation techniques. We also plan to explore the implications of stratified pooling, whereby only a subset of documents within the pool depth are judged.

Acknowledgment. This work was supported by the Australian Research Council's *Discovery Projects* Scheme (DP140101587). Shane Culpepper is the recipient of an Australian Research Council DECRA Research Fellowship (DE140100275).

References

1. Aslam, J.A., Pavlu, V., Yilmaz, E.: A statistical method for system evaluation using incomplete judgments. In: Proceedings of SIGIR, pp. 541–548 (2006)
2. Buckley, C., Dimmick, D., Soboroff, I., Voorhees, E.M.: Bias and the limits of pooling for large collections. Inf. Retr. **10**(6), 491–508 (2007)
3. Buckley, C., Voorhees, E.M.: Retrieval evaluation with incomplete information. In: Proceedings of SIGIR, pp. 25–32 (2004)
4. Büttcher, S., Clarke, C.L.A., Soboroff, I.: The TREC 2006 Terabyte Track. In: Proceedings of TREC, pp. 39–53 (2006)
5. Büttcher, S., Clarke, C.L.A., Yeung, P.C.K., Soboroff, I.: Reliable information retrieval evaluation with incomplete and biased judgements. In: Proceedings of SIGIR, pp. 63–70 (2007)

6. Järvelin, K., Kekäläinen, J.: Cumulated gain-based evaluation of IR techniques. ACM Trans. Inf. Syst. **20**(4), 422–446 (2002)
7. Lipani, A., Lupu, M., Hanbury, A.: Splitting water: Precision and anti-precision to reduce pool bias. In: Proceedings of SIGIR, pp. 103–112 (2015)
8. Lu, X., Moffat, A., Culpepper, J.S.: The effect of pooling and evaluation depth on IR metrics. Inf. Retr. **19**(4), 416–445 (2016)
9. Moffat, A., Zobel, J.: Rank-biased precision for measurement of retrieval effectiveness. ACM Trans. Inf. Syst. **27**(1), 2 (2008)
10. Ravana, S.D., Moffat, A.: Score estimation, incomplete judgments, and significance testing in IR evaluation. In: Cheng, P.-J., Kan, M.-Y., Lam, W., Nakov, P. (eds.) AIRS 2010. LNCS, vol. 6458, pp. 97–109. Springer, Heidelberg (2010)
11. Sakai, T.: Alternatives to BPref. In: Proceedings of SIGIR, pp. 71–78 (2007)
12. Soboroff, I.: A comparison of pooled and sampled relevance judgments in the TREC 2006 Terabyte Track. In: Proceedings of EVIA (2007)
13. Voorhees, E.M.: Overview of the TREC 2004 robust retrieval track. In: Proceedings of TREC, pp. 69–77 (2004)
14. Voorhees, E.M.: The effect of sampling strategy on inferred measures. In: Proceedings of SIGIR, pp. 1119–1122 (2014)
15. Voorhees, E.M., Harman, D.K. (eds.): TREC: Experiment and Evaluation in Information Retrieval. The MIT Press, Cambridge (2005)
16. Webber, W., Park, L.A.F.: Score adjustment for correction of pooling bias. In: Proceedings of SIGIR, pp. 444–451 (2009)
17. Yilmaz, E., Aslam, J.A.: Estimating average precision when judgments are incomplete. Knowl. Inf. Syst. **16**(2), 173–211 (2008)
18. Yilmaz, E., Kanoulas, E., Aslam, J.A.: A simple and efficient sampling method for estimating AP and NDCG. In: Proceedings of SIGIR, pp. 603–610 (2008)
19. Zobel, J.: How reliable are the results of large-scale information retrieval experiments? In: Proceedings of SIGIR, pp. 307–314 (1998)

The Effect of Score Standardisation on Topic Set Size Design

Tetsuya Sakai[(⊠)]

Waseda University, Tokyo, Japan
tetsuyasakai@acm.org

Abstract. Given a topic-by-run score matrix from past data, *topic set size design* methods can help test collection builders determine the number of topics to create for a new test collection from a statistical viewpoint. In this study, we apply a recently-proposed *score standardisation* method called **std-AB** to score matrices before applying topic set size design, and demonstrate its advantages. For topic set size design, **std-AB** suppresses score variances and thereby enables test collection builders to consider realistic choices of topic set sizes, and to handle unnormalised measures in the same way as normalised measures. In addition, even discrete measures that clearly violate normality assumptions look more continuous after applying **std-AB**, which may make them more suitable for statistically motivated topic set size design. Our experiments cover a variety of tasks and evaluation measures from NTCIR-12.

1 Introduction

Given a topic-by-run score matrix from past data, *topic set size design* methods can help test collection builders determine the number of topics for a new test collection from a statistical viewpoint [8]. These methods enable test collection builders such as the organisers of evaluation conferences such as TREC, CLEF and NTCIR to improve the test collection design across multiple rounds of the tracks/tasks, through accumulation of topic-by-run score matrices and computation of better variance estimates.

In this study, we apply a recently-proposed *score standardisation* method called **std-AB** [7] to score matrices before applying topic set size design, and demonstrate its advantages. A standardised score for a particular topic means how different the system is from an "average" system in standard deviation units, and therefore enables cross-collection comparisons [14]. For topic set size design, **std-AB** suppresses score variances and thereby enables test collection builders to consider realistic choices of topic set sizes, and to handle unnormalised measures in the same way as normalised measures. In addition, even discrete measures that clearly violate normality assumptions look more continuous after applying **std-AB**, which may make them more suitable for statistically motivated topic set size design. Our experiments cover four different tasks from the recent NTCIR-12 conference[1]: MedNLP [1], MobileClick-2 [4], STC (Short Text Conversation) [11]

---

[1] http://research.nii.ac.jp/ntcir/workshop/OnlineProceedings12/index.html.

© Springer International Publishing AG 2016
S. Ma et al. (Eds.): AIRS 2016, LNCS 9994, pp. 16–28, 2016.
DOI: 10.1007/978-3-319-48051-0_2

and QALab-2 [12], and some of the official evaluation measure scores from these tasks kindly provided by the task organisers.

2 Prior Art and Methods Applied

The present study demonstrates the advantages of our score standardisation method called **std-AB** [7] in the context of topic set size design, which determines the number of topics to be created for a new test collection [8]. This section situates these methods in the context of related work.

2.1 Power Analysis and Topic Set Size Design

Webber/Moffat/Zobel, CIKM 2008. Webber, Moffat and Zobel [15] proposed procedures for building a test collection based on power analysis. They recommend adding topics and conducting relevance assessments incrementally while examining the achieved *statistical power* (i.e., the probability of detecting a between-system difference that is real) and re-estimating the standard deviation σ_t of the between-system differences. They considered the comparison of two systems only and therefore adopted the t-test; they did not address the problem of the *family-wise error rate* [2,3]. Their experiments focused on Average Precision (AP), a binary-relevance evaluation measure. In order to estimate σ_t (or equivalently, the variance σ_t^2), they relied on empirical methods such as 95 %-percentile computation.

Sakai's Topic Set Size Design. Unlike the incremental approach of Webber *et al.* [15], Sakai's topic set size design methods seek to provide a straightforward answer to the following question: "I have a topic-by-run score matrix from past data and I want to build a new and statistically reliable test collection. How many topics should I create?" [8]. His methods cover not only the paired t-test but also one-way ANOVA for comparing more than two systems at the same time, as well as confidence interval widths. The present study focusses on the ANOVA-based approach, as it has been shown that the topic set sizes based on the other two methods can be deduced from ANOVA-based results. His ANOVA-based topic set size design tool[2] requires the following as input:

α, β: Probability of Type I error α and that of Type II error β.
m: Number of systems that will be compared ($m \geq 2$).
minD: *Minimum detectable range* [8]. That is, whenever the performance difference between the best and the worst systems is *minD* or larger, we want to ensure $100(1 - \beta)\%$ power given the significance level α.
$\hat{\sigma}^2$: Estimated variance of a system's performance, under the *homoscedasticity* (i.e., equal variance) assumption [2,8]. That is, as per ANOVA, it is assumed that the scores of the i-th system obey $N(\mu_i, \sigma^2)$ where σ^2 is common to all systems. This variance is heavily dependent on the evaluation measure.

---

[2] http://www.f.waseda.jp/tetsuya/CIKM2014/samplesizeANOVA.xlsx.

Sakai recommends estimating within-system variances σ^2 for topic set size design using the sample residual variance V_E which can easily be obtained as a by-product of one-way ANOVA; it is known that V_E is an unbiased estimate of σ^2. Let x_{ij} denote the performance score for the i-th system with topic j ($i = 1, \ldots, m'$ and $j = 1, \ldots, n'$); let $\bar{x}_{i\bullet} = \frac{1}{n'} \sum_{j=1}^{n'} x_{ij}$ (sample system mean) and $\bar{x} = \frac{1}{m'n'} \sum_{i=1}^{m'} \sum_{j=1}^{n'} x_{ij}$ (sample grand mean). Then:

$$\hat{\sigma}^2 = V_E = \frac{\sum_{i=1}^{m'} \sum_{j=1}^{n'} (x_{ij} - \bar{x}_{i\bullet})^2}{m'(n'-1)}. \tag{1}$$

If there are more than one topic-by-run matrices available from past data, a pooled variance may be calculated to improve the accuracy of the variance estimate [8]. However, this is beyond the scope of the present study, as we are interested in obtaining a future topic set size based on a single matrix from NTCIR-12 for each measure in each task.

The present study uses the above method with *existing* NTCIR test collections and propose topic set sizes for the next NTCIR rounds. Sakai and Shang [9] considered the problem of topic set size design for a *new* task, where we can only assume the availability of a small pilot topic-by-run matrix rather than a complete test collection. Based on reduced versions of the NTCIR-12 STC official Chinese subtask topic-by-run matrices, they conclude that accurate variance estimates for topic set size design can be obtained if there are about $n' = 25$ topics and runs from only a few different teams.

2.2   Score Standardisation

Webber/Moffat/Zobel, SIGIR 2008. Webber, Moffat and Zobel [14] proposed *score standardization* for information retrieval evaluation with multiple test collections. Given m' runs and n' topics, a topic-by-run raw score matrix $\{raw_{ij}\}$ ($i = 1, \ldots, m', j = 1, \ldots, n'$) is computed for a given evaluation measure. For each topic, let the sample mean be $mean_{\bullet j} = \frac{1}{m'} \sum_i raw_{ij}$, and the sample standard deviation be $sd_{\bullet j} = \sqrt{\frac{1}{m'-1} \sum_i (raw_{ij} - mean_{\bullet j})^2}$. The standardised score is then given by

$$std_{ij} = \frac{raw_{ij} - mean_{\bullet j}}{sd_{\bullet j}}, \tag{2}$$

which quantifies how different a system is from the "average" system in standard deviation units. Using standardised scores, researchers can compare systems across different test collections without worrying about topic hardness (since, for every j, the mean $mean_{\bullet j}$ across runs is subtracted from the raw score) or normalisation (since the standardised scores, which are in the $[-\infty, \infty]$ range, are later mapped to the $[0, 1]$ range as described below). In practice, runs that participated in the pooling process for relevance assessments (*pooled systems*) can also serve as the runs for computing the *standardisation factors* ($mean_{\bullet j}, sd_{\bullet j}$) for each topic (*standardising systems*) [14]. The same standardisation factors are then used also for evaluating new runs.

In order to map the standardised scores into the $[0, 1]$ range, Webber *et al.* chose to employ the cumulative density function (CDF) of the standard normal distribution. The main reason appears to be that, after this transformation, a score of 0.5 means exactly "average" and that outlier data points are suppressed.

Our Method: Std-AB. Recently, we proposed to replace the aforementioned CDF transformation of Webber *et al.* [14] by a simple linear transformation [7]:

$$lin_{ij} = A * std_{ij} + B = A * \frac{raw_{ij} - mean_{\bullet j}}{sd_{\bullet j}} + B, \tag{3}$$

where A and B are constants. By construction, the sample mean and the standard deviation of std_{ij} over the known systems are 0 and 1, respectively $(j = 1, \ldots, n')$. It then follows that the sample mean and the standard deviation of lin_{ij} are B and A, respectively $(j = 1, \ldots, n')$. Regardless of what distribution raw_{ij} follows, Chebyshev's inequality guarantees that at least 89 % of the transformed scores lin_{ij} fall within $[-3A, 3A]$. In the present study, we let $B = 0.5$ as we want to assign a score of 0.5 to "average" systems, and let $A = 0.15$ so that the 89 % score range will be $[0.05, 0.95]$. Furthermore, in order to make sure that even outliers fall into the $[0, 1]$ range, we apply the following *clipping* step:

    if $lin_{ij} > 1$ **then** $lin_{ij} = 1$

    else if $lin_{ij} < 0$ **then** $lin_{ij} = 0$;

This means that *extremely* good (bad) systems relative to others are all given a score of 1 (0). Note that if A is too small, the achieved range of **std-AB** scores would be narrower than the desired $[0, 1]$; if it is too large, the above clipping would be applied to too many systems and we would not be able to distinguish among them. The above approach of using A and B with standardisation is quite common for comparing students' scores in educational research: for example, SAT (Scholastic Assessment Test) and GRE (Graduate Record Examinations) have used $A = 100, B = 500$ [5]; the Japanese *hensachi* ("standard score") uses $A = 10, B = 50$.

In our previous work [7], we demonstrated the advantages **std-AB** over the CDF-based method of Webber *et al.*: **std-AB** ensures pairwise system comparisons that are more consistent across different data sets, and is arguably more convenient for designing a new test collection from a statistical viewpoint. More specifically, using a small value of A ensures that the variance estimates $\hat{\sigma}^2$ will be small, which facilitates test collection design, as we shall demonstrate later. Moreover, as score *normalisation* becomes redundant if we apply standardisation [14], we can handle unnormalised measures (i.e., those that do not lie between 0 and 1). Furthermore, even discrete measures (i.e., those that only have a few possible values), which clearly violate the normality assumptions, look more continuous after applying **std-AB**. While our previous work was limited to the discussion of TREC robust track data and normalised ad hoc IR evaluation measures, the present study extends the work substantially by experimenting with four different NTCIR tasks with a variety of evaluation measures, including unnormalised and discrete ones for the first time.

3   NTCIR-12 Tasks Considered in the Present Study

The core subtask of the *MedNLPDoc* task is *phenotyping*: given a medical record, systems are expected to identify possible disease names by means of ICD (International Classification of Diseases) codes [1]. Systems are evaluated based on recall and precision of ICDs. MedNLPDoc provided us with a *precision* matrix with $n' = 78$ topics (i.e., medical records) and $m' = 14$ runs, as well as a *recall* matrix with $n' = 76$ topics and $m' = 14$ runs.

The *MobileClick-2* task evaluates search engines for smartphones. Systems are expected to output a two-layered textual summary in response to a query [4]. The basic evaluation unit is called *iUnit*, which is an atomic piece of factual information that is relevant to a given query. In the *iUnit ranking* subtask, systems are required to rank given iUnits by importance, and are evaluated by *nDCG* (normalised discounted cumulative gain) and *Q-measure*. In the *iUnit summarisation* subtask, systems are required to construct a two-layered summary from a given set of iUnits. The systems are expected to minimise the reading effort of users with different search intents; for this purpose the subtask employs a variant of the *intent-aware U-measure* [6], called *M-measure* [4], which is an unnormalised measure. MobileClick-2 provided us with 12 topic-by-run matrices in total: six from the English results and six from the Japanese results. While the variances of the unnormalised M-measure are too large for the topic set size design tool to handle, we demonstrate that the problem can be solved by applying **std-AB**.

The *STC* (Short Text Conversation) task requires systems to return a human-like response given a tweet (a Chinese Weibo post or a Japanese twitter post) [11]. Rather than requiring systems to generate natural language responses, however, STC makes them search a repository of past responses (posted in response to some other tweet in the past) and rank them. The STC Chinese subtask provided us with three matrices, representing the official results in *nG@1* (normalised gain at 1), *P+* (a variant of Q-measure), and *nERR@10* (normalised expected reciprocal rank at 10), all of which are navigational intent measures [10].

The *QALab-2* task tackles the problem of making machines solve university entrance exam questions. From the task organisers, we received two matrices based on National Center Test multiple choice questions, one for Phase-1 (where question types are provided to the system) and one for Phase-3 (where question types are not provided). As each topic is a multiple choice question, the evaluation measure is "Boolean" (either 0 or 1).

nG@1 for STC takes only three values: 0, 1/3 or 1 [10], and Boolean for QAlab-2 takes only two values: 0 or 1. These clearly violate the normality assumptions behind ANOVA: $x_{ij} \sim N(\mu_i, \sigma^2)$ for each system i. Thus, it should be noted that, when we apply topic set size design using the variances of these *raw* measures, what we get are topic set sizes for some normally distributed measure M that happens to have the same variance as that discrete measure, rather than topic set sizes for that measure per se. Whereas, if we apply **std-AB**, these measures behave more like continuous measures, as we shall demonstrate later.

Table 1. Columns (e) and (f) show within-system variance estimates $\hat{\sigma}^2$ based on the NTCIR-12 topic-by-run matrices and their **std-AB** versions. The values in bold are those plugged into the topic set design tool in this study. Column (g) compares the system rankings before and after applying **std-AB** in terms of Kendall's τ, with 95 % confidence intervals.

(a) Task/subtask	(b) Measure	(c) m'	(d) n'	(e) $\hat{\sigma}^2$ (raw scores)	(f) $\hat{\sigma}^2$ (**std-AB**)	(g) τ [95 %CI]
MedNLPDoc	precision	14	78	.0597	.0139	.978 [.585, 1.371]
	recall	14	76	**.0601**	**.0127**	.956 [.563, 1.349]
MobileClick	Q-measure	25	100	.0023	.0211	.867 [.587, 1.147]
iUnit ranking	nDCG@3	25	100	**.0259**	**.0215**	.720 [.440, 1.000]
(English)	nDCG@5	25	100	.0198	.0214	.713 [.433, .993]
	nDCG@10	25	100	.0141	.0212	.773 [.493, 1.053]
	nDCG@20	25	100	.0077	.0211	.853 [.573, 1.133]
(Japanese)	Q-measure	12	100	.0189	.0155	.970 [.537, 1.403]
	nDCG@3	12	100	**.0570**	**.0176**	.970 [.537, 1.403]
	nDCG@5	12	100	.0466	.0173	.909 [.476, 1.342]
	nDCG@10	12	100	.0355	.0163	.970 [.537, 1.403]
	nDCG@20	12	100	.0276	.0159	1 [.567,1.433]
MobileClick						
iUnit summarisation						
(English)	M-measure	16	100	44.3783	**.0072**	.983 [.620, 1.346]
(Japanese)	M-measure	13	100	93.5109	**.0077**	.949 [.537, 1.361]
STC (Chinese)	nG@1	44	100	**.1144**	**.0193**	.884 [.679, 1.089]
	P+	44	100	.0943	.0186	.962 [.757, 1.167]
	nERR@10	44	100	.0867	.0182	.947 [.742, 1.152]
QALab Phase-1	Boolean	27	41	.2124	.0191	.892 [.624, 1.160]
Phase-3	Boolean	34	36	**.2130**	**.0204**	.964 [.728, 1.200]

4   Results and Discussions

4.1   Results Overview

Table 1 Columns (e) and (f) show the variance estimates obtained by applying Eq. 1 to the aforementioned topic-by-run matrices, before and after performing **std-AB** as defined by Eq. 3. It can be observed that the variances are substantially smaller after applying **std-AB**. This means that the required topic set sizes will be smaller, provided that the tasks take up the habit of using **std-AB** measures. For each subtask (and language), we selected the *largest* raw score variance, shown in bold in Column (e), and plugged into the topic set size design tool (except for the unnormalised M-measure, whose variances were too large for the tool to handle); that is, we focus on the least stable measures to obtain topic set sizes that are reliable enough for all evaluation measures. We then used the variances of the corresponding **std-AB** measures, shown in bold in Column (f).

Currently, there is no task at NTCIR that employs score standardisation. Now, how would **std-AB** actually affect the official results? Table 1 Column (g) compares the run rankings before and after applying **std-AB** in terms of Kendall's τ for each evaluation measure in each subtask. The 95 % confidence intervals show that the two rankings are statistically equivalent for all cases, except for nDCG@5 in MobileClick English iUnit ranking whose 95 % CI is [.433, .993]. These results suggest that, by and large, **std-AB** enables cross-collection comparisons without affecting within-collection comparisons.

Table 2. Recommended topic set sizes for four NTCIR-12 Tasks ($\alpha = 0.05, \beta = 0.80$).

(I) MedNLPDoc	(a) raw recall ($\hat{\sigma}^2 = .0601$)				(b) **std-AB** recall ($\hat{\sigma}^2 = .0127$)			
$m \downarrow minD \rightarrow$	0.02	0.05	0.10	0.20	0.02	0.05	0.10	0.20
2	2301	369	93	24	487	79	20	6
10	4680	750	188	48	990	159	40	11
20	6159	986	247	62	1302	209	53	14
30	7262	1163	291	73	1535	246	62	16
50	8986	1438	360	91	1899	305	77	20

(II) MobileClick English	iUnit Ranking								iUnit Summarisation			
	(a) raw nDCG@3 ($\hat{\sigma}^2 = .0259$)				(b) **std-AB** nDCG@3 ($\hat{\sigma}^2 = .0215$)				(c) **std-AB** M-measure ($\hat{\sigma}^2 = .0072$)			
$m \downarrow minD \rightarrow$	0.02	0.05	0.10	0.20	0.02	0.05	0.10	0.20	0.02	0.05	0.10	0.20
2	992	159	41	11	824	133	34	9	276	45	12	4
10	2017	323	82	21	1675	269	68	18	561	91	23	6
20	2655	425	107	27	2204	353	89	23	739	119	30	8
30	3130	501	126	32	2598	416	105	27	871	140	36	9
50	3873	620	156	39	3215	515	129	33	1077	173	44	12

(III) MobileClick Japanese	(a) raw nDCG@3 ($\hat{\sigma}^2 = .0570$)				(b) **std-AB** nDCG@3 ($\hat{\sigma}^2 = .0176$)				(c) **std-AB** M-measure ($\hat{\sigma}^2 = .0077$)			
$m \downarrow minD \rightarrow$	0.02	0.05	0.10	0.20	0.02	0.05	0.10	0.20	0.02	0.05	0.10	0.20
2	2182	350	88	23	674	109	28	8	296	48	13	4
10	4439	711	178	45	1371	220	56	15	600	97	25	7
20	5842	935	234	59	1804	289	73	19	790	127	32	9
30	6887	1103	276	70	2127	341	86	22	931	150	38	10
50	8522	1364	342	86	2632	422	106	27	1152	185	47	12

(IV) STC	(a) raw nG@1 ($\hat{\sigma}^2 = .1144$)				(b) **std-AB** nG@1 ($\hat{\sigma}^2 = .0193$)			
$m \downarrow minD \rightarrow$	0.02	0.05	0.10	0.20	0.02	0.05	0.10	0.20
2	4379	701	176	45	739	119	30	8
10	8908	1426	357	90	1504	241	61	16
20	11724	1876	470	118	1979	317	80	21
30	13822	2212	554	139	2333	374	94	24
50	17104	2737	685	172	2886	462	116	30

(V) QALab	(a) raw Boolean ($\hat{\sigma}^2 = .2130$)				(b) **std-AB** Boolean ($\hat{\sigma}^2 = .0204$)			
$m \downarrow minD \rightarrow$	0.02	0.05	0.10	0.20	0.02	0.05	0.10	0.20
2	8152	1305	327	82	782	126	32	9
10	16585	2654	664	167	1589	255	64	17
20	21828	3493	874	219	2091	335	84	22
40	28992	4639	1160	291	2777	445	112	29
50	31845	5096	1275	319	3051	489	123	31

Table 2 shows the recommended topic set sizes with $\alpha = 0.05, \beta = 0.20$ (Cohen's five-eighty convention [3]), for several values of m (i.e., number of systems to be compared) and $minD$ (i.e., minimum detectable range), based on the variances shown in bold in Table 1. It should be noted first, that the values of $minD$ are not comparable across Parts (a) and (b). For example, a $minD$ of 0.02 with raw scores and a $minD$ of 0.02 with **std-AB** scores are not equivalent, because **std-AB** applies score standardisation (Eq. 2) followed by a linear transformation (Eq. 3). Nevertheless, it can be observed that, after applying **std-AB**, the choices of topic set sizes look more realistic. For example, let us consider the $m = 2$ row in Table 2(I). If we want to guarantee 80 % power whenever the difference between the two systems is $minD = 0.05$ (i.e., 5 % of the score range) or larger in raw recall, we would require 369 topics. Whereas, if we want to guarantee 80 % power whenever the difference between the two systems is $minD = 0.05$ (i.e., 5 % of the score range) or larger in **std-AB** recall, we would require only 79 topics. Although the above two settings of $minD$ mean different things, the latter is much more practical. In other words, while ensuring 80 % power for a $minD$ of 0.05 in raw recall is not realistic, ensuring the same power for a $minD$ of 0.05 in **std-AB** is.

Figure 1 visualises the per-topic scores before and after applying **std-AB** for some of our data. Below, we discuss the effect of **std-AB** on recommended topic set sizes for each task in turn.

4.2 Recommendations for MedNLPDoc

The effect of **std-AB** on the recall scores from MedNLPDoc can be observed by comparing Fig. 1(a) and (a'). Note that while many of the raw recall values are 0's, all values are positive after applying **std-AB**. Moreover, there are fewer 1's after applying **std-AB**.

From Table 2(I), a few recommendations for a future MedNLPDoc test collection would be as follows. If the task is continuing to use raw recall, then:

– Create 100 topics: this guarantees 80 % power for comparing any $m = 2$ systems with a $minD$ of 0.10 (93 topics are sufficient), and for comparing any $m = 50$ systems with a $minD$ of 0.20 (91 topics are sufficient);
– Create 50 topics: this guarantees 80 % power for comparing $m = 10$ systems with a $minD$ of 0.20 (48 topics are sufficient).

Whereas, if the task adopts **std-AB** recall, then:

– Create 80 topics: this guarantees 80 % power for comparing $m = 2$ systems with a $minD$ of 0.05 (79 topics are sufficient), and for comparing $m = 50$ systems with a $minD$ of 0.10 (77 topics are sufficient).

Note that MedNLPDoc actually had 76–78 topics (Table 1(d)), and therefore that the above recommendation is quite practical.

4.3 Recommendations for MobileClick-2

The effect of **std-AB** on the nDCG@3 scores from MobileClick-2 iUnit ranking (English) can be observed by comparing Fig. 1(b) and (b'). It can be observed that, after applying **std-AB**, the scores are more evenly distributed within the [0, 1] range. Similarly, the effect of **std-AB** on the unnormalised M-measure from MobileClick-2 iUnit summarisation (English) can be observed by comparing Fig. 1(c) and (c'). Note that the scale of the y-axis for Fig. 1(c) is very different

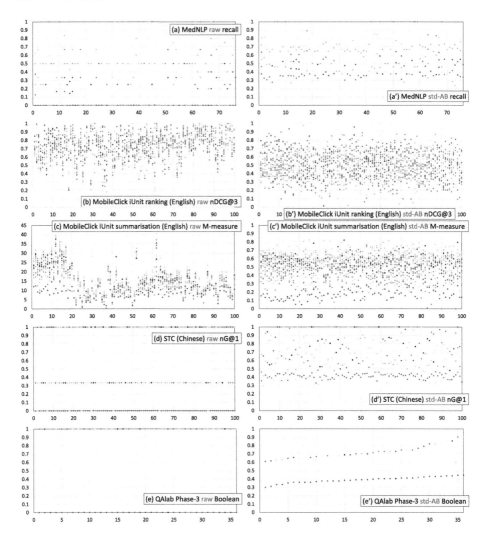

Fig. 1. Per-topic raw and **std-AB** scores for selected NTCIR-12 tasks. The horizontal axes represent topics. Different colours represent different runs (best viewed in colour).

from others. Despite this, Fig. 1(c') shows that **std-AB** transforms the scores into the [0, 1] range without any problems. In this way, **std-AB** can handle any unnormalised measure. Put another way, if we take up the habit of using **std-AB** scores, normalisation becomes no longer necessary.

Since MobileClick-2 is a multilingual task, let us discuss topic set sizes that work for both English and Japanese. Moreover, since the topic set is shared across the iUnit ranking and summarisation subtasks, we want topic set sizes that work across these two subtasks. From Table 2(II) and (III), a few recommendations for future MobileClick test collections would be as follows. If the task is continuing to use raw nDCG@3, then:

- Create 90 topics: this guarantees 80 % power for comparing any $m = 10$ English iUnit ranking systems with a $minD$ of 0.10 (82 topics are sufficient), and for comparing any $m = 2$ Japanese iUnit ranking systems with a $minD$ of 0.10 (88 topics are sufficient).

However, the above setting cannot guarantee anything for the iUnit summarisation task, due to the use of the unnormalised M-measure. In contrast, if the tasks adopts **std-AB** nDCG@3 and **std-AB** M-measure, then:

- Create 100 topics: this guarantees 80 % power for comparing any $m = 20$ English iUnit ranking systems with a $minD$ of 0.10 (89 topics are sufficient), and for comparing any $m = 30$ Japanese iUnit ranking systems with a $minD$ of 0.10 (86 topics are sufficient), and for comparing any $m = 10$ English iUnit summarisation systems with a $minD$ of 0.05 (91 topics are sufficient), and for comparing any $m = 10$ Japanese iUnit summarisation systems with a $minD$ of 0.05 (97 topics are sufficient).

Thus being able to handle unnormalised measures just like normalised measures seems highly convenient. Also, recall that MobileClick-2 actually had 100 topics.

4.4   Recommendations for STC

The effect of **std-AB** on the nG@1 scores from STC (Chinese) can be observed by comparing Fig. 1(d) and (d'). It can be verified from Fig. 1(d) that nG@1 indeed take only three values: 0, 1/3 and 1. Whereas, Fig. 1(d') shows that **std-AB** nG@1 is more continuous, and that there are fewer 1's, and no 0's.

From Table 2(IV), a few recommendations for a future STC test collection would be as follows. If the task is continuing to use raw nG@1, then:

- Create 120 topics: this guarantees 80 % power for comparing any $m = 20$ systems with a $minD$ of 0.20 (118 topics are sufficient);
- Create 90 topics: this guarantees 80 % power for comparing any $m = 10$ systems with a $minD$ of 0.20 (exactly 90 topics are needed).

But note that, strictly speaking, the above recommendations are for normally distributed measures that have a variance similar to that of nG@1, since nG@1 takes only three values. Whereas, if the tasks adopts **std-AB** nG@1, then:

- Create 100 topics: this guarantees 80 % power for comparing any $m = 30$ systems with a $minD$ of 0.10 (94 topics are sufficient).

The STC task actually had 100 topics; this was actually a decision based on topic set size design with raw evaluation measures and pilot data [10].

4.5   Recommendations for QALab

The effect of **std-AB** on the Boolean scores from QALab Phase-3 can be observed by comparing Fig. 1(e) and (e'). It can be observed that **std-AB** transforms the raw Boolean scores (0's and 1's) into something a little more continuous, but that the resultant scores still fall into two distinct score ranges; hence our topic set size design results for QALab should be taken with a large grain of salt even after applying **std-AB** as the scores are clearly not normally distributed. The reason why the **std-AB** scores are monotonically increasing from left to right is just that the QALab organisers sorted the topics by the number of systems that correctly answered them before providing the matrices to the present author. This is equivalent to sorting the topics by $mean_{\bullet j}$ (in decreasing order, i.e., easy topics first).

From Table 2(V), a few recommendations for a future STC test collection would be as follows. If the task is continuing to use raw Boolean, then:

– Create 90 topics: this guarantees 80 % power for comparing any $m = 2$ systems with a $minD$ of 0.20 (82 topics are sufficient).

Whereas, if the tasks adopts **std-AB** Boolean, then:

– Create 40 topics: this guarantees 80 % power for comparing any $m = 2$ systems with a $minD$ of 0.10 (32 topics are sufficient), or any $m = 50$ systems with a $minD$ of 0.20 (31 topics are sufficient).

But recall that the above recommendations are for normally distributed measures whose variances happen to be similar to those of the Boolean measures.

QALab-2 Phase-3 actually had 36 topics only. Note that $n = 36$ is not satisfactory in any of the settings shown in Table 2(V)(a); $n = 36$ does not even satisfy the suggested setting shown above for (a normally distributed equivalent of) **std-AB** Boolean. These results suggest that the QALab task should have more topics to ensure high statistical power.

5   Conclusions and Future Work

Using topic-by-run score matrices from the recent NTCIR-12 MedNLPDoc, MobileClick-2, STC and QALab tasks, we conducted topic set design experiments with and without score standardisation and demonstrated the advantages of employing **std-AB** in this context. It is clear from our results that **std-AB** suppresses score variances and thereby enables test collection builders to consider realistic choices of topic set sizes, and that it can easily handle even unnormalised measures such as M-measure. Other unnormalised measures such as Time-Biased Gain [13], U-measure [6] and those designed for diversified search may be handled similarly. Furthermore, we have demonstrated that discrete measures such as nG@1, which clearly violate the normality assumptions, can be "smoothed" to some extent by applying **std-AB**. Recall that topic set size design assumes that the scores are indepent and identically distributed: that the scores for system i obey $N(\mu_i, \sigma^2)$. While this is clearly a crude assumption especially for unnormalised and discrete measures, **std-AB** makes it a little more believable at least, as shown in the right half of Fig. 1.

In our previous work [7], we performed a preliminary investigation into the robustness of standardisation factors $mean_{\bullet j}, sd_{\bullet j}$ for handling unknown runs (i.e., those that contributed to neither pooling nor the computation of standardising factors). However, our experiments were limited to handling unknown runs from the *same* round of TREC. Hence, to examine the longevity of standardisation factors over technological advances, we have launched a new web search task at NTCIR, which we plan to run for several years[3]. The standardisation factors obtained from the first round of this task will be compared to those obtained from the last round: will the initial standardisation factors hold up against the latest, more advanced systems?

Acknowledgement. We thank the organisers of the NTCIR-12 MedNLPDoc, QALab-2, MobileClick-2, and STC tasks, in particular, Eiji Aramaki, Hideyuki Shibuki, and Makoto P. Kato, for providing us with their topic-by-run matrices of the official results prior to the NTCIR-12 conference.

References

1. Aramaki, E., Morita, M., Kano, Y., Ohkuma, T.: Overview of the NTCIR-12 MedNLPDoc task. In: Proceedings of NTCIR-12, pp. 71–75 (2016)
2. Carterette, B.: Multiple testing in statistical analysis of systems-based information retrieval experiments. ACM TOIS **30**(1) (2012). Article No. 4
3. Ellis, P.D.: The Essential Guide to Effect Sizes. Cambridge University Press, Cambridge (2010)
4. Kato, M.P., Sakai, T., Yamamoto, T., Pavlu, V., Morita, H., Fujita, S.: Overview of the NTCIR-12 MobileClick task, pp. 104–114 (2016)
5. Lodico, M.G., Spaulding, D.T., Voegtle, K.H.: Methods in Educational Research, 2nd edn. Jossey-Bass, San Francisco (2010)
6. Sakai, T.: How intuitive are diversified search metrics? Concordance test results for the diversity U-measures. In: Banchs, R.E., Silvestri, F., Liu, T.-Y., Zhang, M., Gao, S., Lang, J. (eds.) AIRS 2013. LNCS, vol. 8281, pp. 13–24. Springer, Heidelberg (2013)
7. Sakai, T.: A simple and effective approach to score standardisation. In: Proceedings of ACM ICTIR 2016 (2016)
8. Sakai, T.: Topic set size design. Inf. Retr. **19**(3), 256–283 (2016)
9. Sakai, T., Shang, L.: On estimating variances for topic set size design. In: Proceedings of EVIA 2016 (2016)
10. Sakai, T., Shang, L., Lu, Z., Li, H.: Topic set size design with the evaluation measures for short text conversation. In: Zuccon, G., Geva, S., Joho, H., Scholer, F., Sun, A., Zhang, P. (eds.) AIRS 2015. LNCS, vol. 9460, pp. 319–331. Springer, Heidelberg (2015). doi:10.1007/978-3-319-28940-3_25

---

[3] http://www.thuir.cn/ntcirwww/.

11. Shang, L., Sakai, T., Lu, Z., Li, H., Higashinaka, R., Miyao, Y.: Overview of the NTCIR-12 short text conversation task. In: Proceedings of NTCIR-12, pp. 473–484 (2016)
12. Shibuki, H., Sakamoto, K., Ishioroshi, M., Fujita, A., Kano, Y., Mitamura, T., Mori, T., Kando, N.: Overview of the NTCIR-12 QA Lab-2 task. In: Proceedings of NTCIR-12, pp. 392–408 (2016)
13. Smucker, M.D., Clarke, C.L.A.: Time-based calibration of effectiveness measures. In: Proceedings of ACM SIGIR 2012, pp. 95–104 (2012)
14. Webber, W., Moffat, A., Zobel, J.: Score standardization for inter-collection comparison of retrieval systems. In: Proceedings of ACM SIGIR 2008, pp. 51–58 (2008)
15. Webber, W., Moffat, A., Zobel, J.: Statistical power in retrieval experimentation. In: Proceedings of ACM CIKM 2008, pp. 571–580 (2008)

Incorporating Semantic Knowledge into Latent Matching Model in Search

Shuxin Wang[1]([✉]), Xin Jiang[2], Hang Li[2], Jun Xu[1], and Bin Wang[3]

[1] Institute of Computing Technology, Chinese Academy of Sciences, Beijing, China
wangshuxin@ict.ac.cn
[2] Huawei Noah's Ark Lab, Hong Kong, China
[3] Institute of Information Engineering, Chinese Academy of Sciences, Beijing, China

Abstract. The relevance between a query and a document in search can be represented as matching degree between the two objects. Latent space models have been proven to be effective for the task, which are often trained with click-through data. One technical challenge with the approach is that it is hard to train a model for tail queries and tail documents for which there are not enough clicks. In this paper, we propose to address the challenge by learning a latent matching model, using not only click-through data but also semantic knowledge. The semantic knowledge can be categories of queries and documents as well as synonyms of words, manually or automatically created. Specifically, we incorporate semantic knowledge into the objective function by including regularization terms. We develop two methods to solve the learning task on the basis of coordinate descent and gradient descent respectively, which can be employed in different settings. Experimental results on two datasets from an app search engine demonstrate that our model can make effective use of semantic knowledge, and thus can significantly enhance the accuracies of latent matching models, particularly for tail queries.

Keywords: Latent Matching Model · Semantic knowledge · Learning to match · Regularized mapping to latent structures

1 Introduction

In search, given a query documents are retrieved and ranked according to their relevance, which can be represented by the matching score between the query and each of the documents, referred to as semantic matching in [9]. Traditional IR models, including Vector Space Model (VSM), BM25, and Language Models for Information Retrieval (LMIR) can be viewed as matching models for search, created without using machine learning. The models work well to some extent, but they sometimes suffer from mismatch between queries and documents.

Recently significant effort has been made on automatic construction of matching models in search, using machine learning and click-through data. The learned models can effectively deal with mismatch and outperform traditional IR models [9]. Among the proposed approaches, learning a latent space model for

© Springer International Publishing AG 2016
S. Ma et al. (Eds.): AIRS 2016, LNCS 9994, pp. 29–41, 2016.
DOI: 10.1007/978-3-319-48051-0_3

matching in search becomes the state-of-the-art. The class of semantic matching models, called latent matching models in this paper, map the queries and documents from their original spaces to a lower dimensional latent space, in which the matching scores are calculated as inner products of the mapped vectors.

Despite the empirical success of latent matching models, the problem of query document mismatch in search is still not completely solved. Specifically, it remains hard to effectively train a matching model which works well not only for frequent queries and documents, but also for rare queries and documents, because there is not sufficient click data for rare queries and documents. This in fact belongs to the long tail phenomenon, which also exists in many different tasks in web search and data mining. One way to conquer the challenge would be to incorporate additional semantic knowledge into the latent matching models. Specifically, semantic knowledge about synonyms and categories of queries and documents can make the latent space better represent similarity between queries and documents. Suppose that "Sacramento" and "capital of California" are synonyms and it would be difficult to observe their association directly from click information(e.g., a query and the title of clicked document), because both rarely occur in the data. If we can embed the knowledge into the learned latent space, then it will help to make judgment on the matching degrees between queries and documents containing the synonyms. The technical question which we want to address in this paper is how to incorporate semantic knowledge in the learning of latent space model in a theoretically sound and empirically effective way.

In this paper, as the first step of the work, we propose a novel method for learning a linear latent matching model for search, leveraging not only click-through data, but also semantics knowledge such as synonym dictionary and semantic categories. The semantic knowledge can either be automatically mined or manually created. Specifically, we reformulate the learning of latent space model by adding regularization, in which way semantic knowledge can be naturally embedded into the latent space and be utilized in matching. The learning problem becomes maximization of the matching degrees of relevant query document pairs as well as the agreement with the given semantic knowledge. Regularization is also imposed on the linear mapping matrices as well as their product in order to increase the generalization ability of the learned model.

Without loss of generality, we take Regularized Mapping in Latent Space (RMLS) [13], one of the state-of-the-art methods for query document matching, as the basic latent matching model and augment it with semantic knowledge. We improve the optimization procedure of RMLS by introducing a new regularization term. We further develop a coordinate descent algorithm and a gradient descent algorithm to solve the optimization problem. The algorithms can be employed in different settings and thus the learning can be generally carried out in an efficient and scalable way. We conduct experiments on two large-scale datasets from a mobile app search engine. The experimental results demonstrate that our model can make effective use of the semantic knowledge, and significantly outperform existing matching models.

2   Related Work

Matching between queries and documents is of central importance to search [9]. Traditional information retrieval models based on term matching may suffer from term mismatch.

Topic modeling techniques aim to discover the topics as well as the topic representations of documents in the document collection, and can be used to deal with query document mismatch. Latent semantic indexing (LSI) [3] is one typical non-probabilistic topic model. Regularized Latent Semantic Indexing (RLSI) [12] formalizes topic modeling as matrix factorization with regularization of ℓ_1/ℓ_2-norm on topic vectors and document representation vectors. Probabilistic Latent Semantic Indexing (PLSI) [5] and Latent Dirichlet Allocation (LDA) [2] are two widely used probabilistic topic models. By employing one of the topic models, one can project queries and documents into the topic space and calculate their similarities in the space. However, topic modeling does not directly learn the query document matching relation, and thus its ability of dealing with query document mismatch is limited.

In a latent matching model, queries and documents are deemed as objects in two different spaces and are mapped into the same latent space for matching degree calculation (e.g., inner product). The learning of the mapping functions is performed by using training data such as click-through log in a supervised fashion, and thus is more effective to deal with mismatch. Partial Least Square (PLS) [11] is a method developed in statistics and can be utilized to model the matching relations between queries and documents. PLS is formalized as learning of two linear projection functions represented by orthonormal matrices and can be solved by Singular Value Decomposition (SVD). Canonical Correspondence Analysis (CCA) [4] is an alternative method to PLS. The difference between CCA and PLS is that CCA takes cosine as the similarity measure and PLS takes inner product as the similarity measure. Bai et al. [1] propose Supervised Semantic Indexing(SSI), which makes use of a pairwise loss function and learns a low-rank model for matching and ranking. Wu et al. [13] propose a general framework for learning to match heterogeneous objects, and a matching model called Regularized Mapping to Latent Structures (RMLS) is specified. RMLS extends PLS by replacing its orthonormal constraints with ℓ_1 and ℓ_2 regularization. RMLS is superior to PLS in terms of computation efficiency and scalability.

Recently, non-linear matching models have also been studied. For example, Huang et al. [6] propose a model referred to as Deep Structured Semantic Model (DSSM), which performs semantic matching with deep learning techniques. Specifically, the model maps the input term vectors into output vectors of lower dimensions through a multi-layer neural network, and takes cosine similarities between the output vectors as the matching scores. Lu and Li [10] propose a deep architecture for matching short texts, which can also be queries and documents. Their method learns matching relations between words in the two short texts as a hierarchy of topics and takes the topic hierarchy as a deep neural network.

3   Incorporating Semantic Knowledge into Latent Matching Model

3.1   Latent Matching Model

Let $\mathcal{X} \subset \mathbb{R}^{d_x}$ and $\mathcal{Y} \subset \mathbb{R}^{d_y}$ denote the two spaces for matching, and $x \in \mathcal{X}$ and $y \in \mathcal{Y}$ denote the objects in the spaces. In search, x and y are a query vector and a document vector, respectively. Suppose that there is a latent space $\mathcal{L} \subset \mathbb{R}^d$. We intend to find two mapping functions that can map the objects in both $\mathcal{X}$ and $\mathcal{Y}$ into $\mathcal{L}$. When the two mapping functions are linear, they can be represented as matrices: $L_x \in \mathbb{R}^{d \times d_x}$ and $L_y \in \mathbb{R}^{d \times d_y}$. The degree of matching between objects x and y is then defined as inner product of $L_x x$ and $L_y y$, which is $x^T L_x^T L_y y$.

To learn the linear mappings, we need training data that indicates the matching relations between the objects from the two spaces. In search, click-through logs are often used as training data, because they provide information about matching between queries and documents. Following the framework by Wu et al. [13], given a training dataset of positive matching pairs $\{(x_i, y_i)\}_{i=1}^n$, the learning problem is formalized as

$$\arg \max_{L_x, L_y} \frac{1}{n} \sum_{i=1}^n x_i^T L_x^T L_y y_i, \text{subject to } L_x \in \mathcal{H}_x, L_y \in \mathcal{H}_y. \tag{1}$$

where $\mathcal{H}_x$ and $\mathcal{H}_y$ denote the hypothesis spaces for the linear mappings L_x and L_y, respectively. This framework subsumes Partial Least Square (PLS) and Regularized Mapping to Latent Structure (RMLS) as special cases. For PLS, the hypothesis spaces are confined to matrices with orthonormal rows. RMLS replaces the orthonormal assumption with sparsity constraints on L_x and L_y. More specifically, the hypothesis spaces in RMLS become:

$$\mathcal{H}_x = \{L_x \mid \|l_{xu}\|_p \leqslant \tau_{x,p}, p = 1, 2, u = 1, \ldots, d_x\},$$
$$\mathcal{H}_y = \{L_y \mid \|l_{yv}\|_p \leqslant \tau_{y,p}, p = 1, 2, v = 1, \ldots, d_y\},$$

where l_{xu} is the u-th column vector of L_x and l_{yv} is the v-th column vector of L_y. The column vectors are actually latent representations of the elements in the original spaces, for instance, the terms in queries and documents. $\| \cdot \|_p$ denotes ℓ_p norm, and both ℓ_1 and ℓ_2 are used in RMLS. $\tau_{x,p}$ and $\tau_{y,p}$ are thresholds on the norms.

We point out that RMLS is not very robust, both theoretically and empirically. Wu et al. [13] prove that RMLS gives a degenerate solution with ℓ_1 regularization only. Specifically, the solution of L_x and L_y will be matrices of rank one and all the column vectors l_{xu} and l_{yv} will be proportional to each other. Wu et al. [13] propose addressing the problem with further ℓ_2 regularization on l_{xu} and l_{yv}. However, this does not solve the problem, which we will explain later in this Section. Our experiments also show that RMLS tends to create degenerate solutions.

We notice that RMLS does not penalize the case in which any x in one space matches any y in the other space, which may happen even when L_x and L_y are sparse. To cope with the problem, we introduce additional constraints on the matching matrix $L_x^T L_y$, whose (u, v)-th element corresponds to the matching score between the u-th basis vector from $\mathcal{X}$ and the v-th basis vector from $\mathcal{Y}$. Specifically, we add ℓ_1 and ℓ_2 norms on $L_x^T L_y$ as follows, which can limit the overall degree of matching any two objects.

The regularizations $\|L_x^T L_y\|_1 = \sum_{u,v} |l_{xu}^T l_{yv}|$, $\|L_x^T L_y\|_2^2 = \sum_{u,v} (l_{xu}^T l_{yv})^2$ can prevent the model from becoming a degenerate solution, and thus make the model more robust. With all of the constraints the hypothesis spaces of L_x and L_y become:

$$\mathcal{H}_x = \{L_x \mid \|l_{xu}\|_p \leqslant \tau_{x,p}, \|l_{xu}^T l_{yv}\|_p \leqslant \sigma_p, p = 1, 2, \forall u, v\},$$
$$\mathcal{H}_y = \{L_y \mid \|l_{yv}\|_p \leqslant \tau_{y,p}, \|l_{xu}^T l_{yv}\|_p \leqslant \sigma_p, p = 1, 2, \forall u, v\}.$$

Note that $\mathcal{H}_x$ and $\mathcal{H}_y$ are now related to each other because of the constraints on the interaction of the two mappings.

We then reformalize the learning of latent matching model, referred to as LMM for short, as the following optimization problem:

$$\arg\min_{L_x, L_y} -\frac{1}{n} \sum_{i=1}^n x_i^T L_x^T L_y y_i + \sum_{p=1,2} \frac{\theta_p}{2} \|L_x^T L_y\|_p^p + \sum_{p=1,2} \frac{\lambda_p}{2} \|L_x\|_p^p + \sum_{p=1,2} \frac{\rho_p}{2} \|L_y\|_p^p, \quad (2)$$

where θ_p, λ_p and ρ_p are the hyper-parameters for regularization.

In general, there is no guarantee that a global optimal solution of (2) exists, and thus we employ a greedy algorithm to conduct the optimization. Let F denote the corresponding objective function. The matching term in F can be reformulated as $\frac{1}{n} \sum_{i=1}^n x_i^T L_x^T L_y y_i = \sum_{u,v} c_{u,v} l_{xu}^T l_{yv}$, where $c_{u,v}$ is the (u, v)-th element of the empirical cross-covariance matrix $C = \frac{1}{n} \sum_{i=1}^n x_i y_i^T$.

For simplicity, in the following derivation, let us only consider the use of ℓ_2 regularization, i.e., set $\theta_1 = \lambda_1 = \rho_1 = 0$.

By setting the derivatives to zeros, the optimal values of l_{xu} and l_{yv} can be solved as:

$$l_{xu}^* = (\theta_2 \sum_v l_{yv} l_{yv}^T + \lambda_2 I)^{-1} (\sum_v c_{u,v} l_{yv}), l_{yv}^* = (\theta_2 \sum_u l_{xu} l_{xu}^T + \rho_2 I)^{-1} (\sum_u c_{u,v} l_{xu}). \quad (3)$$

The parameters of L_x and L_y are updated alternatively until convergence.

It should be noted that since the parameters are directly calculated, the convergence rate is fast for the coordinate descent algorithm. However, the calculations at each step in Eq. 3 involve inversion of two d-dimension matrices, which could become a computation bottleneck when the dimension of latent space is high. Therefore, we can obtained a gradient descent algorithm for LMM as an alternative[1], specifically for the case of high-dimensional latent space. The gradient descent algorithm has less computation at each step but generally needs more iterations to converge. Therefore, one always needs to consider selecting a more suitable optimization method in a specific situation.

---

[1] $l_{xu}' = l_{xu} + \gamma(\sum_v c_{u,v} l_{yv} - \theta_2 \sum_v l_{yv} l_{yv}^T l_{xu} - \lambda_2 l_{xu}), l_{yv}' = l_{yv} + \gamma(\sum_u c_{u,v} l_{xu} - \theta_2 \sum_u l_{xu} l_{xu}^T l_{yv} - \rho_2 l_{yv})$, where γ is the learning rate.

When Eq. 3 is applied to RMLS (by letting $\theta_2 = 0$), the updates of parameters in each iteration become $L_x^{(t+1)} = L_x^{(t)}(\lambda_2\rho_2)^{-1}CC^T$ and $L_y^{(t+1)} = L_y^{(t)}(\lambda_2\rho_2)^{-1}C^TC$. They are equivalent to conducting power iteration on each row of L_x and L_y independently. Consequently, all rows of L_x will converge to the eigenvector (with the largest eigenvalue) of the matrix $(\lambda_2\rho_2)^{-1}CC^T$, and so will be all rows of L_y. Thus, the optimal parameters L_x^* and L_y^* are both matrices of rank one. This justifies the necessity of regularization on the matching matrix $L_x^T L_y$.

3.2 Incorporating Semantic Knowledge to Latent Matching Model

A latent matching model trained with the method described in the previous section can perform well for head queries and documents, since it can capture the matching information from click-through data. However, for tail queries and documents, there is not enough click-through data, and it is almost impossible to accurately learn the matching relations between them. To alleviate this problem, we propose incorporating semantic knowledge of synonyms and semantic categories into the learning of the latent matching model.

Without loss of generality, we assume that in one space the semantic knowledge is represented as a set of pairs of similar objects (e.g., words or tags), denoted as $\{w_i^{(1)}, w_i^{(2)}, s_i\}_{i=1}^m$, where $w_i^{(1)}$ and $w_i^{(2)}$ represent the term vectors of the objects, and s_i is a scalar representing their weight. Therefore, the matching degrees of the pairs become $\sum_{i=1}^m s_i (w_i^{(1)})^T L^T L w_i^{(2)}$.

We extend the latent matching model (2) by incorporating the above 'regularization' term, for the two spaces $\mathcal{X}$ and $\mathcal{Y}$ respectively, into the objective function of learning:

$$\arg\min_{L_x, L_y} -\frac{1}{n}\sum_{i=1}^n x_i^T L_x^T L_y y_i + \sum_{p=1,2}\frac{\theta_p}{2}\|L_x^T L_y\|_p^p + \sum_{p=1,2}\frac{\lambda_p}{2}\|L_x\|_p^p +$$

$$\sum_{p=1,2}\frac{\rho_p}{2}\|L_y\|_p^p - \frac{\alpha}{m_x}\sum_{i=1}^{m_x} s_{x,i} (w_{x,i}^{(1)})^T L_x^T L_x w_{x,i}^{(2)} - \frac{\beta}{m_y}\sum_{i=1}^{m_y} s_{y,i} (w_{y,i}^{(1)})^T L_y^T L_y w_{y,i}^{(2)}.$$

$$(4)$$

The hyper-parameters α and β control the importance of semantic knowledge from the two spaces.

Similarly, coordinate descent can be employed to solve the optimization (4). The optimal values of l_{xu} and l_{yv} are then given by

$$l_{xu}^* = (\theta_2 \sum_v l_{yv} l_{yv}^T + \lambda_2 I)^{-1}(\sum_v c_{u,v} l_{yv} + \alpha \sum_v r_{x,u,v} l_{xv}),$$

$$l_{yv}^* = (\theta_2 \sum_u l_{xu} L_{xu}^T + \rho_2 I)^{-1}(\sum_u c_{u,v} l_{xu} + \beta \sum_u r_{y,u,v} l_{yu}),$$

$$(5)$$

where $r_{x,u,v}$ and $r_{y,u,v}$ denote the (u,v)-th elements of the empirical covariance matrices R_x and R_y respectively, where $R_x = \frac{1}{m_x}\sum_{i=1}^{m_x} s_{x,i} w_{x,i}^{(1)}(w_{x,i}^{(2)})^T$,

---

Algorithm 1. Coordinate Descent Algorithm for Latent Matching Model with Semantic Knowledge

---

1. Input: C, R_x, R_y, α, β, θ, λ_2, ρ_2, T.
2. Initialization: $t \leftarrow 0$, random matrices $L_x^{(0)}$ and $L_y^{(0)}$.

while *not converge and* $t \leqslant T$ **do**
 Compute $A_x = \theta_2 L_x^{(t)}(L_x^{(t)})^T + \lambda_2 I$ and its inverse A_x^{-1} .
 Compute $A_y = \theta_2 L_y^{(t)}(L_y^{(t)})^T + \rho_2 I$ and its inverse A_y^{-1}.
 Compute $B_x = L_x^{(t)}C + \beta L_y^{(t)}R_y$.
 Compute $B_y = C(L_y^{(t)})^T + \alpha R_x(L_x^{(t)})^T$.
 for $u = 1 : d_x$ **do**
  Select u-th row of B_y as b_{yu}^T,
  Compute $l_{xu}^{(t+1)} = A_y^{-1}b_{yu}$.
 end
 for $v = 1 : d_y$ **do**
  Select v-th row of B_x as b_{xv}^T,
  Compute $l_{yv}^{(t+1)} = A_x^{-1}b_{xv}$.
 end
end

$R_y = \frac{1}{m_y}\sum_{i=1}^{m_y} s_{y,i}\, w_{y,i}^{(1)}(w_{y,i}^{(2)})^T$. Algorithm 1 shows the procedure of the coordinate descent algorithm for latent matching model with semantic knowledge. An alternative algorithm using gradient descent can also be obtained[2].

3.3   Acquisition of Semantic Knowledge

Synonyms are obviously useful semantic knowledge for our matching task. A general dictionary of synonyms such as WordNet is usually not suitable for a real-world setting, however. The reason is that synonyms usually heavily depend on domains. Here we adopt an algorithm for mining synonym pairs by exploiting click-through data. Specifically, we first try to find clusters of queries from a click-through bipartite graph (cf., [7]). Queries in one cluster are regarded as synonyms. Next, for each cluster, we extract pairs of terms sharing the same context as candidates of synonym pairs (cf., [8]). Here the context refers to the surrounding text of a term in the query. For example, given a query "download 2048 apk", the context for term "2048" is "download * apk", where '*' is the wildcard character. Then we go through all the clusters and count the numbers of occurrences (called support) for all the candidate pairs. The candidate pairs with support above a certain threshold are chosen as synonym pairs. Algorithm 2 shows the detailed procedure.

We denote the set of mined synonym pairs as $\{(w_{x,i}^{(1)}, w_{x,i}^{(2)}, s_{x,i})\}$, where $w_{x,i}^{(1)}$ and $w_{x,i}^{(2)}$ are the i-th pair of synonyms. $s_{x,i}$ is the corresponding weight for the pair, which is computed as the logistic transformation of the support. The knowledge about the synonym set for the query domain ($\mathcal{X}$) is formalized as $\sum_i s_{x,i}\,(w_{x,i}^{(1)})^T L_x^T L_x w_{x,i}^{(2)}$ in the optimization function.

---

[2] $l'_{xu} = l_{xu} + \gamma(\sum_v c_{u,v}l_{yv} + \alpha\sum_v r_{x,u,v}l_{xv}) - \gamma(\theta_2\sum_v l_{yv}l_{yv}^T l_{xu} + \lambda_2 l_{xu}), l'_{yv} = l_{yv} + \gamma(\sum_u c_{u,v}l_{xu} + \beta\sum_u r_{y,u,v}l_{yu}) - \gamma(\theta_2\sum_u l_{xu}l_{xu}^T l_{yv} + \rho_2 l_{yv})$, where γ is the learning rate.

Algorithm 2. Synonyms Mining Algorithm on Click Bipartite Graph

---

0. Notation: Q: query set, D: document set, C: click set, q: query, d: document, t: term.
1. Input: click bipartite graph $G = (Q, D, C)$.
2. Initialization: dictionary of candidate synonym pairs $S = [\ ]$.
for d *in* D **do**
    Collect $Q_d = \{q|(q, d) \in C\}$.
    Init $T = \{\ \}$.
    for q *in* Q_d **do**
        for t *in* q **do**
            Extract context c_t of t in q
            Add (t, c_t) to T
        end
    end
    Find $P_d = \{(t_i, t_j)|c_{t_i} = c_{t_j}, (t_i, c_{t_i}) \in T, (t_j, c_{t_j}) \in T\}$
    for (t_i, t_j) *in* P_d **do**
        if (t_i, t_j) *not in* S **then**
            Add (t_i, t_j) to S and set $S[(t_i, t_j)] = 1$
        else
            Set $S[(t_i, t_j)] = S[(t_i, t_j)] + 1$
        end
    end
end
3. Sort S by value in descending order.
4. Return top K pairs of S as the synonym pairs.

In addition to synonyms, we also utilize categories or tags in a taxonomy as semantic knowledge for the document domain. For example, in our experiment of the mobile app search, apps are given various tags by users. An app named "tiny racing car" is tagged "action, mario, racing, sports, auto, adventure, racing track". For each tag, we have a list of associated documents. We represent the title of each document as a tf-idf vector and calculate the average vector of the tf-idf vectors for each tag. We select the top k terms in the average vector and view them as the relevant terms to the tag. A set of 'tag-term' pairs is then obtained from all the tags and their relevant terms, and it is denoted as $\{(w_{y,i}, w_{y,ij}, s_{y,ij})\}$, where $w_{y,i}$ is the i-th tag, and $w_{y,ij}$ is the j-th relevant term to the i-th tag, and $s_{y,ij}$ is the corresponding average tf-idf value. We can formalize the knowledge for the document domain ($\mathcal{Y}$) as $\sum_i \sum_j s_{y,ij} (w_{y,i}^{(1)})^T L_y^T L_y w_{y,ij}^{(2)}$ in the objective function of learning of latent matching model.

4 Experiments

4.1 Experimental Setup

We take app search as example and use data from an app search engine. Each app is represented by its title and description and can be viewed as a document. Click-through logs at the search engine are collected and processed. We create two datasets from the click-through logs, one containing one week data and the other containing one month data. Table 1 reports some statistics of the two datasets. Each dataset consists of query-document pairs and their associated clicks, where

Table 1. Statistics of two training datasets

	#clicks	#queries	#apps
one-week click-through data	1,020,854	190,486	110,757
one-month click-through data	3,441,768	534,939	192,026

a query and a document are represented by a term-frequency vector and a tf-idf vector of the title, respectively. The queries and documents are regarded as heterogeneous data in two different spaces, because queries and documents have different characteristics.

In addition, we randomly sample two sets of 200 queries from a time period different from that of training datasets, and take them as two test datasets.

Each test set is composed of 100 head queries and 100 tail queries, according to the frequencies of them. In the following sub-sections, performance on the whole random test set as well as the head and tail subsets will be reported. For each query in the test sets, we collect top 20 apps retrieved by the app search engine and then label the query-app pairs at four levels of matching: Excellent, Good, Fair, and Bad.

As evaluation measures, Normalized Discounted Cumulative Gain (NDCG) at positions 1, 3, 5, 10 are used. We choose the conventional IR model of BM25 (with the parameters tuned for best performance in the training set), and two latent matching models of PLS (Partial Least Square) and RMLS (Regularized Mapping to Latent Structures) as the baseline methods. Our basic model is denoted as LMM (Latent Matching Model) and our augmented models are denoted as LMM-X where X stands for the type of incorporated semantic knowledge.

4.2 Experimental Results

Latent Matching Model: We conduct a series of experiments to test the performances of LMM, LMM-X and the baseline models. For RMLS, LMM, and LMM-X, the results with latent dimensionalities of 100 and 500 are reported. For PLS, only the performance with latent dimensionality of 100 is reported, due to its scalability limitation.

Table 2 report the performances of the models trained using one-week click-through data, evaluated on the test tests: random queries, head queries and tail queries respectively. From the results, we can see that: (1) all the latent matching models significantly outperform the conventional BM25 model in terms of all evaluation measures; (2) among the latent space models with the same dimension, LMM achieves the best performances in many cases. The improvements of LMM over BM25 and RMLS are statistically significant (paired t-test, p-value < 0.05); (3) the improvements of LMM over the other baseline models are larger on tail queries than on head queries, which indicates that LMM can really enhance matching performance for tail queries; (4) for LMM, the performance increases as the dimensionality of latent space increases. Note that PLS requires

Table 2. Matching performance on one week data

Model	NDCG on Random queries				NDCG on Head queries				NDCG on Tail queries			
(Dimension)	@1	@3	@5	@10	@1	@3	@5	@10	@1	@3	@5	@10
BM25	0.687	0.700	0.707	0.741	0.729	0.754	0.758	0.786	0.645	0.645	0.656	0.696
PLS(100)	0.715	0.733	0.738	0.767	0.756	0.780	0.787	0.809	0.675	0.686	0.689	0.726
RMLS(100)	0.697	0.727	0.732	0.765	0.740	0.767	0.772	0.801	0.653	0.686	0.692	0.729
LMM(100)	0.713	0.727	0.741	0.771	0.744	0.771	0.785	0.813	0.681	0.684	**0.697**	0.729
RMLS(500)	0.709	0.720	0.731	0.762	0.742	0.765	0.777	0.805	0.677	0.674	0.686	0.719
LMM(500)	**0.727**	**0.737**	**0.738**	**0.772**	**0.766**	**0.783**	**0.787**	**0.812**	**0.689**	**0.690**	0.688	**0.731**

Table 3. Matching performance on one-month data

Model	NDCG on Random queries				NDCG on Head queries				NDCG on Tail queries			
(Dimension)	@1	@3	@5	@10	@1	@3	@5	@10	@1	@3	@5	@10
BM25	0.644	0.681	0.714	0.740	0.721	0.738	0.756	0.771	0.567	0.624	0.672	0.710
PLS(100)	0.692	0.727	0.749	0.772	0.735	0.757	0.774	0.788	0.649	0.698	0.724	0.756
RMLS(100)	0.668	0.703	0.727	0.752	0.736	0.746	0.762	0.779	0.600	0.660	0.693	0.726
LMM(100)	0.692	**0.733**	**0.751**	0.775	0.744	0.765	**0.779**	0.793	0.640	0.700	0.724	0.758
RMLS(500)	0.687	0.725	0.745	0.774	**0.753**	**0.767**	0.772	0.798	0.620	0.684	0.719	0.751
LMM(500)	**0.704**	0.730	0.749	**0.780**	0.745	0.756	0.770	**0.795**	**0.662**	**0.704**	**0.729**	**0.765**

SVD and thus becomes practically intractable when the dimension is large. In that sense, RMLS and LMM exhibit their advantages over PLS on scalability.

Table 3 show the comparison results of models trained using one-month click-though data, evaluated on the tested random queries, head queries and tail queries respectively, which follows the same trends as that of one-week data, especially on tail queries.

Incorporating Semantic Knowledge: Next, we test the performances of the LMM-X models which incorporate semantic knowledge into the model. The LMM-X models have the ability to leverage multiple sources of semantic knowledge by adding regularization terms to the objective function. We consider two methods of acquiring and utilizing semantic knowledge. In the first method we mine and use synonym pairs from the click-through logs, In the second method we collect and use over 50,000 tags in the app search engine, which are described in 3.3.

We conduct experiments using LMM model and the two types of knowledge. We summarize the results in Table 4 for one-week data and Table 5 for one-month data evaluated on random queries, head queries and tail queries respectively. For each training dataset, we first separately train the LMM model augmented with the synonyms dictionary and the tag-term pairs, denoted as LMM-Synonyms and LMM-Tags, respectively. Then we train the LMM model augmented with both types of knowledge, denoted as LMM-Both. From the results we can see: (1) with knowledge embedded, the performances of the LMM model can be consistently improved; (2) the improvements of LMM-Both over LMM are statistically significant (paired t-test, p-value < 0.05) in terms of most evaluation measures; (3) more significant improvements are made on tail queries than on head

Table 4. Matching performance on one-week data

Model (Dimension)	NDCG on Random queries				NDCG on Head queries				NDCG on Tail queries			
	@1	@3	@5	@10	@1	3	5	10	@1	@3	@5	@10
LMM(100)	0.713	0.727	0.741	0.771	0.744	0.771	0.785	0.813	0.681	0.684	0.697	0.729
LMM-Synonyms(100)	0.730	0.743	0.747	0.772	0.757	0.791	0.794	0.815	0.704	0.695	0.700	0.729
LMM-Tags(100)	0.727	0.746	0.747	0.773	0.757	0.789	0.796	0.817	0.697	0.699	0.699	0.728
LMM-Both(100)	0.735	0.750	0.752	0.772	0.762	0.798	0.799	0.815	**0.709**	0.702	0.705	0.729
LMM(500)	0.727	0.737	0.738	0.772	0.766	0.783	0.787	0.812	0.689	0.690	0.688	0.731
LMM-Synonyms(500)	0.743	0.749	0.758	0.781	0.779	**0.795**	**0.802**	0.819	0.707	0.703	0.714	0.743
LMM-Tags(500)	0.743	0.747	0.759	**0.783**	0.779	0.793	0.801	**0.820**	0.707	0.702	0.716	**0.745**
LMM-Both(500)	**0.743**	**0.750**	**0.759**	0.781	**0.779**	0.793	0.801	0.819	0.707	**0.708**	**0.718**	0.743

Table 5. Matching performance on one-month data

Model (Dimension)	NDCG on Random queries				NDCG on Head queries				NDCG on Tail queries			
	@1	@3	@5	@10	@1	3	5	10	@1	@3	@5	@10
LMM(100)	0.692	0.727	0.749	0.772	0.735	0.757	0.774	0.788	0.649	0.698	0.724	0.756
LMM-Synonyms(100)	0.708	0.738	0.749	0.780	0.741	0.771	0.770	0.795	0.676	0.705	0.729	0.765
LMM-Tags(100)	0.707	0.734	0.750	0.779	0.738	0.760	0.767	0.795	0.676	0.708	0.733	0.763
LMM-Both(100)	0.715	0.739	0.745	0.779	0.738	0.760	0.767	0.795	0.676	0.708	0.733	0.760
LMM(500)	0.704	0.730	0.749	0.780	0.745	0.756	0.770	**0.795**	0.662	0.704	0.729	0.765
LMM-Synonyms(500)	0.719	0.741	**0.762**	**0.783**	**0.752**	0.761	0.775	0.793	0.686	0.723	**0.748**	**0.773**
LMM-Tags(500)	0.719	0.741	0.762	0.781	0.752	0.759	**0.778**	0.794	0.686	0.723	0.746	0.769
LMM-Both(500)	**0.721**	**0.745**	0.761	0.782	0.751	**0.763**	0.777	0.793	**0.691**	**0.728**	0.745	0.771

queries; (4) the improvements of semantic knowledge augmentation are slightly less when the latent dimensionality is high (500) than when it is low (100).

We investigate the latent spaces of LMMs learned with and without incorporating synonym dictionary. The latent representations of some randomly selected words are plotted on a 2-D graph using the multidimensional scaling technique, in Fig. 1. By comparing the distributions of words in Fig. 1(a) and (b), we can clearly see that similar words are clustered closer in LMM-Synonyms than in LMM. This clearly indicates that knowledge about synonyms can be effectively incorporated into LMM-Synonyms and thus the model can further improve matching. For the latent spaces of LMMs learned with and without incorporating category tags, we observe a similar phenomenon.

We make analysis of the ranking results of LMM and LMM-X. In many cases, we find that the semantic relations embedded in LMM-X can indeed improve relevance ranking. For example, the terms "best", "most", "hardest", and "strongest" are mined as synonyms from the log, and these terms are clustered together in the latent space induced by LMM-Synonyms. In search, for the query of "best game in the history", documents about "most popular game", "hardest game" and "strongest game" are promoted to higher positions, which enhances the relevance as well as the richness of search result. However, there are also some bad cases, mainly due to noises in the synonym dictionary. For example, in one experiment our mining algorithm identifies "google" and "baidu" as synonyms. Then for the query of "google map", a document about "baidu map" is ranked higher than a document about "google earth". Therefore, improving

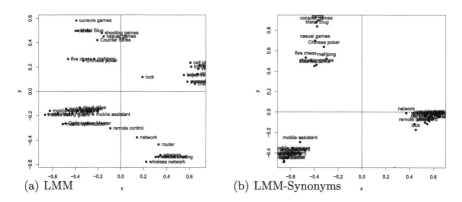

Fig. 1. Representations of query words in latent space.

the quality of the mined semantic knowledge is one issue which we need to address in the future.

5   Conclusion

In this paper, we have studied the problem of latent semantic matching for search. We have proposed a linear latent semantic model that leverages not only click-through data, but also semantic knowledge such as synonym dictionary and category hierarchy. The semantic knowledge is incorporated into the model by imposing regularization based on synonym and/or category information. We employ two methods to acquire semantic knowledge. One is to mine synonyms from the click bipartite graph, and the other is to utilize categories of documents. We have developed a coordinate descent algorithm and a gradient descent algorithm to solve the learning problem. The algorithms can be employed depending on the settings. We have conducted experiments on two large-scale datasets from a mobile app search engine. The experimental results demonstrate that our model can make effective use of semantic knowledge, and significantly outperform existing matching models.

References

1. Bai, B., Weston, J., Grangier, D., Collobert, R., Sadamasa, K., Qi, Y., Chapelle, O., Weinberger, K.: Supervised semantic indexing. In: Proceedings of the 18th ACM Conference On Information and Knowledge Management, pp. 187–196. ACM (2009)
2. Blei, D.M., Ng, A.Y., Jordan, M.I.: Latent Dirichlet allocation. J. Mach. Learn. Res. **3**, 993–1022 (2003)
3. Deerwester, S., Dumais, S.T., Furnas, G.W., Landauer, T.K., Harshman, R.A.: Indexing by latent semantic analysis. J. Am. Soc. Inf. Sci. **41**(6), 391–407 (1990)

4. Hardoon, D., Szedmak, S., Shawe-Taylor, J.: Canonical correlation analysis: an overview with application to learning methods. Neural Comput. **16**(12), 2639–2664 (2004)
5. Hofmann, T.: Probabilistic latent semantic indexing. In: Proceedings of the 22nd Annual International ACM SIGIR Conference on Research and Development in Information Retrieval, pp. 50–57. ACM (1999)
6. Huang, P.-S., He, X., Gao, J., Deng, L., Acero, A., Heck, L.: Learning deep structured semantic models for web search using click through data. In: Proceedings of the 22nd ACM International Conference on Information and Knowledge Management, pp. 2333–2338. ACM (2013)
7. Jiang, D., Pei, J., Li, H.: Mining search and browse logs for web search: a survey. ACM Trans. Intell. Syst. Technol. (TIST) **4**(4), 57 (2013)
8. Lamping, J., Baker, S.: Determining query term synonyms within query context. US Patent 7,636,714, December 2009
9. Li, H., Xu, J.: Semantic matching in search. Found. Trends Inf. Retr. **8**, 89 (2014)
10. Lu, Z., Li, H.: A deep architecture for matching short texts. In: Advances in Neural Information Processing Systems, pp. 1367–1375 (2013)
11. Rosipal, R., Krämer, N.: Overview and recent advances in partial least squares. In: Saunders, C., Grobelnik, M., Gunn, S., Shawe-Taylor, J. (eds.) SLSFS 2005. LNCS, vol. 3940, pp. 34–51. Springer, Heidelberg (2006). doi:10.1007/11752790_2
12. Wang, Q., Xu, J., Li, H., Craswell, N.: Regularized latent semantic indexing: a new approach to large-scale topic modeling. ACM Trans. Inf. Syst. **31**(1), 5 (2013)
13. Wu, W., Lu, Z., Li, H.: Learning bilinear model for matching queries and documents. J. Mach. Learn. Res. **14**(1), 2519–2548 (2013)

Keyqueries for Clustering and Labeling

Tim Gollub, Matthias Busse, Benno Stein, and Matthias Hagen$^{(\boxtimes)}$

Bauhaus-Universität Weimar, Weimar, Germany
{tim.gollub,matthias.busse,benno.stein,matthias.hagen}@uni-weimar.de

Abstract. In this paper we revisit the document clustering problem from an information retrieval perspective. The idea is to use queries as features in the clustering process that finally also serve as descriptive cluster labels "for free." Our novel perspective includes query constraints for clustering and cluster labeling that ensure consistency with a keyword-based reference search engine.

Our approach combines different methods in a three-step pipeline. Overall, a query-constrained variant of k-means using noun phrase queries against an ESA-based search engine performs best. In the evaluation, we introduce a soft clustering measure as well as a freely available extended version of the Ambient dataset. We compare our approach to two often-used baselines, descriptive k-means and k-means plus χ^2. While the derived clusters are of comparable high quality, the evaluation of the corresponding cluster labels reveals a great diversity in the explanatory power. In a user study with 49 participants, the labels generated by our approach are of significantly higher discriminative power, leading to an increased human separability of the computed clusters.

1 Introduction

Document clustering is a popular approach to enable the exploration of large collections such as digital libraries, encyclopedias, or web search results. The objective of clustering is to automatically organize a document collection into a small number of coherent classes or *clusters* such that documents in one cluster are more similar to each other than to those in other clusters. Along with short meaningful labels for the clusters (summarizing the cluster content) a user can get a general overview of a collection, start a systematic exploration, or narrow the focus to just a particular subset of the documents meeting an information need.

The document clustering task falls into two steps: (1) unveil the topical structure in a document collection and (2) provide meaningful descriptions that communicate this structure to a user. For the first step, referred to as *clustering*, many effective algorithms are known. However, clustering algorithms such as the popular k-means are usually not capable of producing meaningful cluster labels. This is usually treated in a subsequent second step—the *cluster labeling*. One major drawback of common keyword-based labeling techniques is their limitation to only selecting "statistical" features from the documents; for example,

© Springer International Publishing AG 2016
S. Ma et al. (Eds.): AIRS 2016, LNCS 9994, pp. 42–55, 2016.
DOI: 10.1007/978-3-319-48051-0_4

by concatenating the most prominent keywords occurring in a cluster. However, a list of keywords tends to represent different and unrelated aspects of the documents and will often fail to provide a readable label.

To account for the crucial aspect of meaningful labels for document clustering, we take an information retrieval perspective. Note that a user's perceived suitability of a label for a document set can be seen as similar to a search engine's decision of whether a document matches a query. Thus, we view queries as good candidates for cluster labels—and as good features for the clustering itself. This way, we establish an explicit connection between clustering and search technology. Furthermore, the interplay between information retrieval systems and cluster analysis brings forth an intuitive approach to hierarchical search result clustering: Once the relevant aspects in a document collection are unveiled in form of search queries, each of the corresponding result sets can then serve as input for another iteration of the clustering process, which in turn leads to a new set of now more detailed aspects, i.e. search queries.

Our main contributions are threefold: (1) a flexible three-step processing pipeline for document clustering using search queries as features and labels, (2) an extended and freely available version of the Ambient data set with 4680 manually annotated documents, and (3) a user study with 49 participants comparing the explanatory power of the cluster labels generated by our approach and two often-used baselines.

2   Related Work

One of the first applications of document clustering was to improve retrieval based on the cluster hypothesis stating that "closely associated documents tend to be relevant to the same requests" [16]. Later, clusters were also used as a browsing interface to explore and organize document collections like in the famous Scatter/Gather algorithm [4]. The numerous document clustering algorithms can be classified into three classes [3]: data-centric, description-aware, and description-centric algorithms.

Data-centric algorithms typically are not limited to the text domain. The to-be-clustered data is represented in models that allow to compute similarities between the data objects; one of the most popular such algorithms being *k*-means (cf. Sect. 3.3). A popular data representation in the text domain is the Vector Space Model with $tf \cdot idf$ weights and cosine similarity [18]. However, the generation of a label that can be presented to a user is not part of data-centric algorithms but tackled as an independent, subsequent step. Examples are labels formed from keywords frequently occurring in a cluster's documents [21] or applying Pearson's χ^2 test taking into account other clusters [12] that forms our first baseline. Still, such labels often are a sequence of rather unrelated keywords rendering even the best clustering less useful to users that rely on the labels as readable descriptions—an issue that inspired a second class of cluster algorithms.

Description-aware algorithms try to circumvent the labeling issue of data-centric approaches by ensuring that the construction of cluster labels produces results that are comprehensive and meaningful to a user. One way to achieve this goal is to use algorithms that assign documents to clusters based on a single feature—so-called monothetic clustering—and to then use this feature as a label. One example is the Suffix Tree Clustering [25] that exploits frequently recurring phrases as similarity features. First, base clusters are discovered by shared single frequent phrases utilizing suffix trees. Second, the base clusters are merged by their phrase overlap. However, since the merging step is based on the single-linkage criterion, the combined phrases of merged clusters forming the cluster labels often still tend to be unrelated and therefore misleading for a user. SnakeT [5] tries to enrich the similarly obtained labels by using phrases from a predefined ontology but still, the cluster analysis precedes and dominates the labeling task—a problem the next class of algorithms tries to circumvent.

Description-centric algorithms consider the cluster labels as the crucial elements of a clustering. They assume that if a cluster cannot be described by a meaningful label, it is probably of no value to a user. The description precedes the assignment of a document to a cluster. Description-centric algorithms mainly tackle the use case of clustering web search results, with Lingo being one of the pioneering examples [14]—now part of Carrot2, an open source framework for search result clustering.[1] A singular value decomposition of a frequent term-search result snippet matrix is used to extract orthogonal vectors assumed to represent distinct topics in the snippets. The documents are then assigned to the extracted topic clusters using the Vector Space Model.

With a similar goal, Weiss revisits the data-centric k-means algorithm and adjusts it to a description-centric version: descriptive k-means [24]—our second baseline. First, k-means with $tf \cdot idf$ features is run. Then frequent phrases are extracted from the cluster's centroids as potential cluster labels. As for document assignment, the algorithm searches for cluster documents that are relevant to a phrase utilizing the Vector Space Model.

Description-centric algorithms focus on label quality but still do not use the full potential. Documents not containing a topic label but being just as relevant from an information retrieval perspective are not considered to belong to a topic's cluster. We believe that queries against suited search engines are able to overcome this drawback, exploiting the extensive information retrieval research of the last decades. Some of the respective ideas that inspired our approach are discussed in the following section.

3   Our Approach

Our approach leverages queries in the clustering process and as labels. This way, we exploit the fact that search queries linking keywords to document sets are a concept well-known to users from their daily web search experience. Both

---

[1] http://project.carrot2.org.

Lingo [14] and descriptive k-means [24] can be interpreted to utilize search queries in their algorithms. However, queries are only used for validating a clustering. Instead, our approach considers search queries as the driving force while deriving the clustering; inspired by Fuhr et al.'s more theoretical optimum clustering framework (OCF) that suggests search relevance scores or retrieval ranks as clustering features [6]. Still, the OCF does not address the problem of labeling the resulting clusters.

Our new approach combines the general idea of OCF with Gollub et al.'s concept of keyqueries as document descriptors [8,9] that recently has been used for recommending related documents [10]. We will use keyqueries as clustering features in an OCF-style but then will as well suggest suited keyqueries as labels. Following Stein and Meyer zu Eißen [21], meaningful cluster labels should be comprehensive (appropriate syntax), descriptive (reflect each document in a cluster), and discriminative (minimal semantic overlap between two cluster's labels). Most existing cluster labeling techniques do not sufficiently address the descriptiveness aspect but queries do as our experiments will show.

3.1 Queries as Label Candidates

To model the descriptiveness of cluster labels, we view a user's perception of a label as follows: The presentation of a cluster label activates a concept in the user's mind, and each document that is relevant to this concept should be classified under that label. This process is conceptually very closely related to the standard task of information retrieval—query-based search. This analogy leads us to propose the use of search queries as cluster labels that have to retrieve the documents of the associated cluster. The task of document clustering then can be formulated as the reverse of query-based search as follows: Given a set of documents, find a set of diverse search queries that together retrieve the document set when submitted to a reference search engine. Along with their retrieved documents as cluster contents, the queries then form a labeled clustering. This implies that the potential clusters of a document are given by the queries for which it is retrieved and leads to a first new constraint within the constrained clustering terminology [2]: the *common-query constraint CQ* stating that two documents cannot be in the same cluster if they are not retrievable by a common query.

In order to find the labeling queries, the possible vocabulary has to be defined. The vocabulary generation is an important step in our pipeline since the choice of vocabulary terms determines the comprehensive power of the cluster labels. In case the terms are ambiguous, not comprehensive, or too specific, the cluster labels will inevitably also exhibit such problems and will fail to reflect the content of a cluster. Also the size of the vocabulary has an impact on the overall performance. With respect to the syntax of cluster labels, category names in classification systems or Wikipedia are considered to be ideal [21,22,24]. Category names typically are noun phrases or conjunctions of these; therefore, we consider noun phrases as suitable to serve as cluster labels. For readability reasons, we suggest to restrict the number of conjunctions to one, like in "Digital Libraries and Archives." This forms our second constraint, the *query-syntax constraint QS* stating that a cluster label consists of noun phrases or a conjunction of these.

But not all noun phrases form good candidates for cluster labels. Even though determiners are often viewed as part of a noun phrase, they are not necessary in our scenario. The same holds for post-modifiers, etc. We consider noun phrases to be a concatenation of pre-modifiers and a head noun. Still, pre-modifiers are not yet restricted in length such that arbitrarily long cluster labels could be generated. Following the distribution in the Wikipedia where a category name on average consists of 3.87 terms, we formulate our third constraint, the *query-length constraint QL* stating that a cluster label consists of maximum four terms per at most two noun phrases (i.e., maximum length is eight plus the conjunction). To find suitable phrases, we use Barker and Cornacchia's head noun extractor [1] that provides a phrase ranking from which we choose the top-6 per document (determined in pilot studies) that are then lemmatized using the Apache OpenNLP library to avoid different flections. Other keyphrase extractors can of course also be integrated.

To avoid meaningless phrases like "big issue" or "common example," we also consider a second form of vocabulary generation allowing only noun phrases from a predefined vocabulary. As the source of a controlled and predefined vocabulary consisting of well-formed and suitable phrases we choose the titles of Wikipedia articles following Mihalcea and Csomai's suggestion [13]. Applying the three constraints from above, we select only those titles with a maximum length of four terms. In addition, we discard Wikipedia article titles that solely consist of stopwords, dates, and special or non-latin characters, because they usually do not serve as meaningful cluster labels. Our resulting vocabulary consists of 2,869,974 titles that are also lemmatized. As for ranking possible Wikipedia phrase candidates, we use the keyphraseness score [13] as the ratio of the number of articles that contain the phrase as a link and the total number of articles that contain the phrase.

3.2 Examined Search Engines/Retrieval Models

In the document indexing step of our clustering pipeline, we exploit the research effort on retrieval models of the last decades by using queries as a good means to derive clusters and labels. Of course, different retrieval models may yield different clusterings and labels. In our pipeline, we experiment with the classic Boolean model (queries based on Boolean logic but no ranking possible), the Vector Space Model with $tf \cdot idf$ weighting [18] (documents and queries modeled as vectors), BM25 [17] ("$tf \cdot idf$ + document length"), and ESA [7] with Wikipedia articles as the background collection (topic modeling approach taking semantic similarities into account). Our evaluation will show that the ESA retrieval model is best suited for our task.

For the retrieval models that rank the results, we include two further relevance constraints for setting a cut-off such that lower ranked documents are not considered part of the result set for the purpose of clustering. These relevance constraints reflect the keyquery idea of Gollub et al. [8]: a keyquery for a document is a query that returns the document in its top ranks. Our *top-k constraint* states that only the k topmost results of a query count as the result set—we set $k = 10$ following the original keyquery idea. Since a document at rank $k+1$ could be as relevant as the one at rank k, such a static cut-off might be problematic and also limits the size of the possible clusters in our scenario—difficult if the size of the clusters is not known in advance. Hence, we propose an alternative *score constraint* stating that to be part of the result set, a document must have a retrieval score above some relevance threshold t. In our pilot experiments with different techniques of "averaging" retrieval scores, $t = \sum s_i^2 / \sum s_i$, where s_i denotes the retrieval score of a document, turned out to be a good choice. Compared to the standard mean $t = \sum s_i / N$, the formula emphasizes the highest scores and reduces the influence of low scores.

3.3   Query-constrained Clustering Algorithms

For every document in the to-be-clustered collection, we store all the queries for which the document is retrieved according to our above relevance constraints in a reverted index [15]. The postlists of the documents in the reverted index contain the respective keyqueries and serve as the document features for the clustering. In the following, we describe three different cluster algorithms that satisfy the common-query constraint.

Set Cover Clustering. The first algorithm tackles clustering as a set covering problem (SCP) on the result lists of the query vocabulary. In our scenario, we apply a variant of the greedy set cover algorithm [23]. For up to k iterations, the query q is selected whose result set size is within a certain range, covers the maximum number of documents not yet covered by previous queries, and where the not-yet-covered documents in the result set have a high *positive rate* in a graph that connects documents by an edge when they share a keyquery (i.e., multiple edges between two documents are possible). The positive rate of a new result set is the ratio of actual edges between not-yet-covered documents in the result set and the minimum number of edges if each of these documents would be retrieved by only this one query. Note that this way, documents in the clustering may be part of several result sets.

Agglomerative Clustering. Our second algorithm variant follows the agglomerative strategy of hierarchical clustering. It starts with each document in its own cluster, and then merges pairs of clusters in a "bottom-up" approach. As for the merging, measures for the distance between pairs of documents and a linkage criterion specifying the distance of clusters are needed. We choose the number of shared keyqueries for both distances. As for cluster similarity, we follow a complete-linkage clustering approach (taking into account all document pair similarities between two clusters) since this avoids the chaining phenomenon of single-linkage clustering, where clusters might be merged due to a single pair of documents being close to each other, even though all other documents are very dissimilar. Our algorithm merges those two clusters, whose document pairs share the most keyqueries. In case that the maximum number is shared by more than two clusters, the algorithm decides upon the ratio of shared to non-shared queries of the document pairs. Since the documents of the two merged clusters are not necessarily the only clusters that are retrieved by the shared keyqueries, we additionally include all other remaining clusters that the shared keyqueries retrieve.

When the merging finally leads to the desired number of clusters, the algorithm stops. But simply concatenating the set of queries as the corresponding cluster label would in many cases violate our query-length constraint (e.g., when more than two queries are left in a node). We therefore strive for the query or pair of queries that "best" cover the cluster documents. Since all queries find at least the cluster documents, we choose the query (pair) that retrieves the fewest additional documents from other clusters.

Constrained k-means Clustering. The query-constrained clustering algorithm in this section adopts the popular data-centric k-means algorithm with keyquery features. Given a collection of data points, k-means operates in three steps. (1) In the initialization, k random values within the data domain are chosen as initial cluster representatives (the centroids). In our scenario, each document is represented by a vector with a 1 at position i if the document is retrieved by that query in the reverted index or 0 otherwise. For the initialization, we randomly generate k such vectors. (2) In the assignment phase, each data point is assigned to its nearest centroid and therefore, clusters of data points are formed. In our scenario, the algorithm calculates for each document vector the dot-product to all centroid vectors and assigns the document to the centroid with the highest value. (3) In the update phase, the k centroids of the new clusters are computed and input to the assignment phase until convergence or some threshold of iterations is reached. In our scenario, for each cluster the query is selected whose result set best covers the assigned documents in terms of the F-Measure. The new centroid is computed as the mean vector of the result documents of that best query.

4   Evaluation

We compare the different variants of our three-step query-based clustering pipeline on an extended version of the Ambient dataset against two often-used approaches; among others, we conduct a user study with 49 participants on the explanatory power of the cluster labels.

4.1   AMBIENT++ Dataset

The original Ambient dataset was published by Carpineto and Romano in 2008,[2] and has become popular for document clustering and labeling evaluation [19,20].

---

[2] Claudio Carpineto, Giovanni Romano: Ambient Data set (2008), http://credo.fub.it/ambient/.

It comprises 44 ambiguous topics with about 18 subtopics each, obtained from Wikipedia disambiguation pages. Some of the subtopics are associated with a set of documents (URL, title, snippet) that were collected by submitting every topic as a query to a web search engine, and by manually assigning each URL of the top 100 results to a subtopic. However, the documents were not stored and the subtopics are very uneven in size. Hence, we reconstruct the Ambient dataset as our extended corpus AMBIENT++ as follows.

The documents of the original Ambient URLs form the basis of our corpus extension and are crawled in a first step. The authors of the original data set assigned a total of 2257 URLs to some subtopic; in fact, most of the subtopics did not get any document assigned while others got up to 76 URLs. In early 2016, only 1697 documents of the original dataset could be crawled. After a manual inspection, 611 documents had to be discarded since they did not discuss the originally assigned subtopic anymore—only 1086 documents remain. We thus enrich the data to have at least ten documents in each of the original subtopics. To this end, the descriptions from the Wikipedia disambiguation pages for the subtopics that do not have ten documents were submitted to a web search engine and the result lists manually assessed until ten documents for the subtopic are available (excluding pages that only contain an image, video, table, etc.). In some cases, the subtopic descriptions are no successful queries (e.g., too long and specific). In such cases, our annotators manually formulated a better suited query. But a few topics still did not get ten "valid" documents although we assigned 4506 additional documents to subtopics—a total amount of 5592 documents.

Since not every subtopic could be sufficiently enriched and some subtopics have way more than ten documents, we balance the dataset to subtopics with exactly ten documents. We discard the subtopics with less than ten documents and from the ones with too many documents we keep the ten best-ranked query results only—resulting in 481 subtopics with ten results compared to only 25 subtopics in the original Ambient dataset. During the manual filtering, we also identified a few subtopics with identical meaning (e.g., subtopic 12.11 (globe, a Japanese trance/pop-rock group) and subtopic 12.17 (globe (band), a Japanese pop music group) that are too difficult to separate in a clustering such that we only keep one of these—13 subtopics were removed. In our enriched dataset, each of the 44 topics has at least three subtopics (468 in total) each having ten documents. As for extracting the main content of the 4680 corpus documents, we use the Default Extractor from the Boilerpipe library [11] which performed best in our pilot experiments.

4.2   Soft F-Measure as a New Evaluation Measure

In our experimental framework, we consider each topic of the AMBIENT++ dataset as one to-be-clustered collection where the "optimal" clustering would form clusters identical to the respective subtopics. However, our query-based clusterings can result in clusters that are difficult to evaluate with the traditional F-Measure against the ground truth. For instance, a query animal for the topic "Camel" could retrieve documents about the humped ungulates but also about arachnids (the camel spider, both subtopics of the topic camel) such that the resulting cluster cannot really be evaluated against just one of the two ground truth subtopics/clusters. As for comparing the quality of clusterings with such ambiguous or overlapping clusters, we propose the Soft F-Measure (name inspired by soft clustering algorithms, where a document may be contained in several clusters). The measure computes true/false positives/negatives on the level of document pairs and not document-cluster pairs like the conventional F-Measure does. For each document pair in the clustering, we calculate the association strength s by the ratio of shared clusters to all clusters they are assigned to (maximum association strength is 1). If the two documents are in the same subtopic/cluster in the ground truth, s is added to the true positive score and $1 - s$ to the false negative score; if not, s is added to the false positive score and $1 - s$ to the true negative score. The scores are finally used in the "traditional" F-Measure formula. Note that the Soft F-Measure is not "symmetric" (e.g., only retrieving six of ten documents in one cluster is worse than retrieving all ten documents and four additional false positives).

4.3   Setting up Our Pipeline

For each of the three pipeline steps (vocabulary generation, document indexing, constraint clustering), we compare the performance of the different variants on a training set of ten topics to a "best" clustering possible and an index clustering. The best clustering is obtained by a brute-force analysis that finds the queries from the index that best identify the subtopics with respect to the traditional F-Measure against the ground truth. The index clustering uses every postlist in the index as a cluster. Rationale for this approach is the assumption that the entries of the inverted index can be seen as support for the query-based clustering: the more queries retrieve similar result sets, the more likely these documents are grouped together.

In our pilot experiments, noun phrase vocabulary achieves slightly better best clustering performance than the Wikipedia vocabulary (F-Measure of 0.93 vs. 0.91) and also a slightly higher Soft F-Measure for index clustering (0.26 vs. 0.25) such that we choose noun phrases as the vocabulary. To decide how many phrases to extract per document, we test 1 to 20 extracted phrases per document. Interestingly, the F-Measure of the best clustering saturates at six extracted noun phrases. Hence, we decide to extract six phrases from each document.

To overcome the influence of possibly insufficient phrases for comparing the different retrieval models (Boolean, $tf \cdot idf$, BM25, ESA) and the relevance constraint parameter settings (rank or score), we manually generated appropriate queries for each of the subtopics in our training set and compare the F-measure of the result lists with respect to the subtopic the query belongs to. Not too surprising, in our AMBIENT++ scenario, a fixed cut-off constraint at rank 10 performs much better than a score constraint that would yield clusters with 70+ documents (remember that each subtopic has ten documents). Except a few outliers, all three ranking-based retrieval models outperform the Boolean model

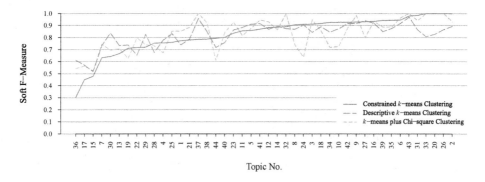

Fig. 1. Comparison of our constrained k-means clustering with the baselines.

while the ESA model outperforms the other models on 8 of 10 topics. As for ESA on our training set, the full Wikipedia articles as the background concept collection perform better than just the first paragraphs of each article.

From the three clustering methods in our pipeline (set cover, agglomerative, constrained k-means) the constrained k-means achieves the highest Soft F-Measure scores with ESA on our training set (0.83 vs. 0.77 for the other two) but is still way below the best clustering with an average Soft F-Measure of 0.94.

Our best pipeline set-up (constrained k-means clustering with six extracted noun phrases per document and top-10 results of ESA with the complete Wikipedia articles as background collection) is now compared to two often-used clustering+labeling approaches: k-means plus χ^2 baseline [12] representing the data-centric algorithms and descriptive k-means [24] representing the description-centric algorithms.

4.4 Clustering Quality

The clustering quality evaluation is performed on all topics of our new Ambient++ dataset employing the Soft F-Measure for the clusters that an algorithm derived for a topic. With their average Soft F-Measure of 0.83 our new constrained k-means and k-means plus χ^2 are slightly better than the 0.82 of descriptive k-means (k always set to the true number of subtopics for every algorithm)—hence, query integration does not harm clustering quality. Figure 1 shows the distribution on a topic level indicating quite different performance for specific topics but similar general trends (topics ordered by our algorithm's performance) as well as some rather difficult to-be-clustered topics (also our set cover and agglomerative clustering methods have similar problems on these).

4.5 Cluster Label Quality

Since our new approach is comparable to two often-used approaches from a cluster quality perspective, we also compare the label quality. As the appropriateness

Judgment	CKM	DKM	χ^2
✓	**213**	180	152
–	**15**	25	39
×	**21**	44	58
F_1	**0.92**	0.84	0.76
p	–	0.005	0.000

Fig. 2. (Left) screenshot of the first user study experiment, (right) judgment distribution (CKM = constrained k-means, DKM = descriptive k-means, χ^2 = k-means + χ^2) indicating that our approach's labels are significantly more discriminative than the baselines' labels.

of a cluster label for a cluster is challenging to evaluate, we conduct a user study with 49 participants (23 female, 26 male, mean age 18.3, $SD = 6.8$) on the AMBIENT++ dataset with two experiments that evaluate (1) the discriminative power and (2) the descriptive power of the cluster labels.

Experiment 1: Discriminative Power. In the first experiment, we examine to what extent the cluster labels can discriminate documents from one cluster to other clusters. We conduct an empirical browser-based study in a within-subjects design meaning that each participant is asked about labels of every approach. For a given subtopic, a participant is given a manually prepared short description of up to five words and a selected identifying image (instead of the often lengthy original disambiguation text) and cluster labels of one algorithm derived for the subtopic's topic. The participant then has to choose the label that best fits the given subtopic (forced-choice). For time constraints, we only consider a subset of 22 random topics from each of which we choose at most four subtopics with the highest average clustering Soft F-Measure over all three approaches (always higher than 0.8 but some topics only have three subtopics (average at 3.77)). At most eight labels are presented to the user (some topics have fewer subtopics, from the others 7 additional random ones are chosen). Each subtopic-algorithm combination in our study was judged by three participants resulting in 747 judgments $((22 \cdot 3.77 \cdot 3) \cdot 3)$; on average around 15 judgments per participant ensuring that no participant judged for the same subtopic twice (not even for another algorithm).

Figure 2 shows a screenshot and the result of the first experiment. In the screenshot, the name of the topic (Jaguar) is shown at the top, the to-be-judged subtopic is presented by an image and a short description at the right-hand side, and a randomly shuffled list of cluster labels for clusters in the topic at the left-hand side. If none of the labels is satisfying, the participant should click the lowermost cross-button.

In the result table, the first row denotes the number of judgments where the selected label is the label generated by the approach (i.e., true positive), the second row lists the number of judgments where the participant selected a different label than the one generated by the approach (i.e., false positive), and the third row gives the judgments where the participant selected neither of the presented labels (i.e., false negatives). A common single measure is the reported F_1-score and to statistically estimate the per-individual effect, we compare the ratio of correct label assignments (true positives) among all assignments given for a subtopic (true positives, false positives, false negatives). Each subtopic is judged by three participants, and the assigned labels split into correct (true positives) and incorrect (false positives and false negatives). In case that all three participants select the correct label, the ratio equals $\frac{3}{3} = 1$. If only one participant decided for the correct label, the ratio is $\frac{1}{3}$. According to a Shapiro-Wilk test, the individual participants' ratios are not normally distributed for either approach such that we choose the non-parametric Wilcoxon signed rank test known as a suitable significance test in our within-subjects design with ratio data and three to-be-compared approaches. For the 49 participants' ratios we get a p-value of 0.005 when comparing the distribution of our approach to descriptive k-means and a p-value below 0.001 compared to k-means plus χ^2 indicating that our approach significantly increases the discriminative power of the cluster labels over the baselines.

Experiment 2: Descriptive Power. In the second experiment, we examine the descriptive power of the cluster labels. A participant is shown the different cluster labels that are generated by the approaches for one subtopic, and has to select that label which best describes the given subtopic. We ensure that the clusters of the approaches cover the same subtopic by calculating their F-measures to the subtopic. Only if the cluster of each approach exceeds the threshold of 0.8 with regard to the subtopic documents, we include that subtopic to the data set of this experiment ensuring that all three approaches derived good clusterings. We obtain judgments by three participants for 226 of the 468 subtopics similar to the setting in Experiment 1; again not showing the same subtopic to the same user twice.

Voting	CKM	DKM	χ^2
✓✓✓	48	45	19
✓✓	52	45	36
✓	51	49	55
–	75	87	116
total votes	**299**	274	184
p-value	–	0.3525	0.0000

Fig. 3. (Left) screenshot of the second user study experiment, (right) judgment distribution (CKM = constrained k-means, DKM = descriptive k-means, $\chi^2 = k$-means $+ \chi^2$) indicating that our approach's labels are more descriptive than the baselines' labels.

The first four rows in the table in Fig. 3 denote the number of judgments where either all three, two, one or no participant(s) voted for the corresponding approach. For all three approaches, the numbers accumulate to the 226 judged subtopics. Our approach is better than descriptive k-means (although not significant on the per-topic vote distribution) and both outperform k-means plus χ^2.

5   Conclusion and Outlook

We have presented a novel query-based clustering pipeline that uses keyqueries as features for the clustering process and as labels for the resulting clusters. The comparison to two often-used baselines shows that our constrained k-means approach with the ESA retrieval model is competitive from a clustering quality perspective and significantly improves the label quality. Thus, our idea of revisiting the clustering problem from an information retrieval perspective combining ideas from the optimal clustering framework and keyquery research is a promising direction for supporting users engaged in exploratory search tasks that need guidance in form of document clusterings with good labels. As part of our evaluation, we have also introduced an enriched AMBIENT++ dataset including 4680 manually annotated documents that will be made publicly available and a Soft F-Measure cluster quality evaluation measure.

Interesting directions for future research could be the inclusion of terms from predefined taxonomies from which we only evaluated Wikipedia titles as a first step. Still, we predict much potential to be explored in that direction as well as in the evaluation on other datasets and with further different retrieval models since the performance of all models still was way below an oracle best query clustering.

References

1. Barker, K., Cornacchia, N.: Using noun phrase heads to extract document keyphrases. In: AI, pp. 40–52 (2000)
2. Basu, S., Davidson, I., Wagstaff, K.: Constrained Clustering: Advances in Algorithms, Theory, and Applications. Chapman & Hall/CRC, Boca Raton (2008)
3. Carpineto, C., Osiński, S., Romano, G., Weiss, D.: A survey of web clustering engines. ACM Comput. Surv. **41**(3), 17:1–17:38 (2009)
4. Cutting, D.R., Karger, D.R., Pedersen, J.O., Tukey, J.W., Scatter, G.: A cluster-based approach to browsing large document collections. In: SIGIR, pp. 318–329 (1992)
5. Ferragina, P., Gullì, A.: The anatomy of SnakeT: a hierarchical clustering engine for web-page snippets. In: Boulicaut, J.-F., Esposito, F., Giannotti, F., Pedreschi, D. (eds.) PKDD 2004. LNCS (LNAI), vol. 3202, pp. 506–508. Springer, Heidelberg (2004)
6. Fuhr, N., Lechtenfeld, M., Stein, B., Gollub, T.: The optimum clustering framework: implementing the cluster hypothesis. Inf. Retr. **15**(2), 93–115 (2012)
7. Gabrilovich, E., Markovitch, S.: Computing semantic relatedness using Wikipedia-based explicit semantic analysis. In: IJCAI, pp. 1606–1611 (2007)

8. Gollub, T., Hagen, M., Michel, M., Stein, B.: From keywords to keyqueries: content descriptors for the web. In: SIGIR, pp. 981–984 (2013)

9. Gollub, T., Völske, M., Hagen, M., Stein, B.: Dynamic taxonomy composition via keyqueries. In: JCDL, pp. 39–48 (2014)

10. Hagen, M., Beyer, A., Gollub, T., Komlossy, K., Stein, B.: Supporting scholarly search with keyqueries. In: Ferro, N., et al. (eds.) ECIR 2016. LNCS, vol. 9626, pp. 507–520. Springer, Heidelberg (2016). doi:10.1007/978-3-319-30671-1_37

11. Kohlschütter, C., Fankhauser, P., Nejdl, W.: Boilerplate detection using shallow text features. In: WSDM, pp. 441–450. ACM (2010)

12. Manning, C.D., Raghavan, P., Schütze, H.: Introduction to Information Retrieval. Cambridge University Press, Cambridge (2008)

13. Mihalcea, R., Csomai, A. Wikify!: Linking documents to encyclopedic knowledge. In: CIKM, pp. 233–242 (2007)

14. Osiński, S., Stefanowski, J., Weiss, D.: Lingo: search results clustering algorithm based on singular value decomposition. In: IIPWM, pp. 359–368 (2004)

15. Pickens, J., Cooper, M., Golovchinsky, G.: Reverted indexing for feedback and expansion. In: CIKM, pp. 1049–1058 (2010)

16. van Rijsbergen, C.J.: Information Retrieval. Butterworth-Heinemann, London (1979)

17. Robertson, S., Zaragoza, H.: The probabilistic relevance framework: BM25 and beyond. FnTIR **3**(4), 333–389 (2009)

18. Salton, G., Wong, A., Yang, C.-S.: A vector space model for automatic indexing. CACM **18**(11), 613–620 (1975)

19. Scaiella, U., Ferragina, P., Marino, A., Ciaramita, M.: Topical clustering of search results. In: WSDM, pp. 223–232 (2012)

20. Stein, B., Gollub, T., Hoppe, D.: Beyond precision@10: clustering the long tail of web search results. In: CIKM, pp. 2141–2144 (2011)

21. Stein, B., Meyer zu Eißen, S.: Topic identification: framework and application. In: I-KNOW, pp. 522–531 (2004)

22. Treeratpituk, P., Callan, J.: An experimental study on automatically labeling hierarchical clusters using statistical features. In: SIGIR, pp. 707–708 (2006)

23. Vazirani, V.V.: Approximation Algorithms. Springer, Heidelberg (2001)

24. Weiss, D.: Descriptive clustering as a method for exploring text collections. PhD thesis, University of Poznan (2006)

25. Zamir, O., Etzioni, O.: Web document clustering: a feasibility demonstration. In: SIGIR, pp. 46–54 (1998)

A Comparative Study of Answer-Contained Snippets and Traditional Snippets

Xian-Ling Mao[1]($^{\boxtimes}$), Dan Wang[1], Yi-Jing Hao[1], Wenqing Yuan[2], and Heyan Huang[1]

[1] Beijing Institute of Technology, Beijing, China
{maoxl,2220150504,hhy63}@bit.edu.cn, wangdan12856@sina.com
[2] Beijing Guzhang Mobile Technology Co., Beijing, China
ywq8876@163.com

Abstract. Almost every text search engine uses snippets to help users quickly assess the relevance of retrieved items in the ranked list. Although answer-contained snippets can help to improve the effectiveness of search intuitively, quantitative study of such intuition remains untouched. In this paper, we first propose a simple answer-contained snippet method for community-based Question and Answer (cQA) search, and then compare our method with the state-of-the-art traditional snippet algorithms. The experimental results show that the answer-contained snippet method significantly outperforms the state-of-the-art traditional methods, considering relevance judgements and information satisfaction evaluations.

Keywords: Answer-contained snippet · Quantitative study · cQA · Information retrieval

1 Introduction

Information retrieval is the process that provides users with the most relevant documents from an existing collection [1]. The semantics of queries submitted by users to search engine, even though specified as questions, are inherently ambiguous. For example, given the query: "what is big bang?", it may be about the band named "big bang", and also may be about "big bang theory", even about the TV play named "big bang". Various retrieval models [2–4] have been proposed. They return a relevant list of query results. However, due to the ambiguity of query semantics, it is impossible to design a fixed ranking scheme which could always perfectly measure the relevance of query results pertaining to users' intentions. To compensate for the inaccuracy of ranking functions, document snippet has become almost standard components for search engines to augment query results. A snippet was used to reflect the relevance between a document and a query [5,6], which we name it as "traditional snippets". If the snippets are high-quality, users could implicitly deduce relevance of the search results. And it could help the search engine better align the user's perceived relevance of documents with the true relevance. Therefore, it is critical for any search engine to

S. Ma et al. (Eds.): AIRS 2016, LNCS 9994, pp. 56–67, 2016.
DOI: 10.1007/978-3-319-48051-0_5

produce high quality snippets to avoid biasing the perceived relevance of documents [7]. If the snippet for a highly relevant document is poorly generated, the user may perceive the document as non-relevant and does not click on the document. What's more, snippet quality is also an important criteria for evaluating search engines [8]. Traditionally, a snippet was used to reflect the relevance between a document and a query [5,6]. For traditional snippets, the basic and most important ability is relevance judgement. The ability of snippets is called *relevance*.

Intuitively, if a snippet contains words which satisfy user's need, users will avoid clicking on the original document and the browsing efficientness would be higher. The property of snippets is called *satisfaction*, and we name this kind of snippet as *answer-contained snippet*. The main goal of satisfaction is to improve the user-friendliness and effectiveness, similar to the goal of the One Click Access Task of NTCIR[1], aiming to satisfy information need with the first system output that's displayed, without clicking any more. However, it's not clear whether answer-contained snippets are better than traditional snippets. To the best of our knowledge, there is no comparative and quantitative study between two kinds of snippets. In this paper, we will investigate the ability of answer-contained snippets, comparing with traditional snippets. We will evaluate their relevance and satisfaction ability on a large-scale archive, by *browsing speed, reference to the full text of the documents* and *feedback from users*.

The contributions of our work include:

– We compare answer-contained snippets and traditional snippets.
– We design a simple answer-contained snippet generating algorithm that can help users judge relevance quickly and satisfy information need fast.
– By our experimental studies, we verify the utility of answer-contained snippets, compared with traditional snippets. The results show that it's promising research direction to contain answer in a snippet in future.

2   Related Work

Early works generated static and query-independent snippets, consisting of the first several sentences of the returned document. Such an approach is efficient but often ineffective [11,12]. Selecting sentences for inclusion in the snippet based on the degree to which they match the keywords has become the state-of-the-art of query-biased snippet generation for text documents [5,13,14]. Generally, there are two kinds of sentence selection methods in query-based snippet generation. One is heuristic method [5,14–17], the importance of sentences are expressed by heuristic rules. For example, if a sentence is the first sentence of paragraph, it may be more important than others. The other is machine learning method [6,18]. Metrics or features are commonly used in those methods. The features include whether the sentence is a heading or the first line of the document, the number of keywords and distinct keywords that appear in the sentence and the

---

[1] http://www.thuir.org/1click/ntcir9/.

number of title words appearing in the sentence, etc. However, snippet generation techniques designed for plain text documents are unable to process structural or semi-structural data well, which can't leverage the structural information of data and therefore do not perform well.

Recently, structural snippets have been explored for XML documents [7]. Huang et al. [7] propose a set of requirements for snippets and design a novel algorithm to efficiently generate informative yet small snippets. Ellkvist et al. [19] have explored how to generate snippet for workflow data by considering fine-grained structural information. However, to the best of our knowledge, there is still no snippet generation methods for cQA data, which still adopted query-biased snippet generation generally and do not perform well, as observed in the user studies.

3 Experimental Setting

In cQA, everybody could ask and answer questions on any topic, and people seeking information are connected to the ones knowing answers. As answers are usually explicitly provided by human, they can be helpful in answering real world questions [9]. Although it is difficult to achieve *satisfaction* for general search, it is relatively easier for cQA search, since a question in a cQA document has corresponding answers. The sentences in these answers can be utilized to satisfy user information needs. Meanwhile, the goal of this paper is to verify the utility of answer-contained snippets, not to extract answers from documents. Thus in this paper, we will choose the cQA archive as our experimental dataset, focusing on our research goal.

3.1 DataSet

To construct a comprehensive dataset for our experiments, we crawled nearly all the question and answer pairs (QA pairs) of two top categories of the most popular(Computers & Internet, Health) with fourty-three sub-categories from 2005.11 to 2008.11 in Yahoo!Answer[2], which produces an archive of 6,345,786 QA pairs. A cQA document includes *question title, question body, best answer, other answers and metadata.*

3.2 Answer-Contained Snippets

To design a reasonable answer-contained snippet method for cQA search, we have to consider two questions: (1) how to rank answers? (2) what if the size of one answer is large?

- **The Answer Importance:** We consider answer quality from 3 angles. Firstly, people could take Best Answer chosed by asker or by cQA system in a cQA document as the highest quality answer. Secondly, using vote number to reflect

---

[2] http://answers.yahoo.com/.

the answers' popularity, the answer with the largest vote number could be thought as the highest quality answer among answers. At last, the answer submitted by the highest authority user could be taken as the highest quality answer.

- **Best Answer Importance:**

$$BestAnsImp(ans) = \mathbf{1}_{IsBestAns}(Type(ans)) \times$$
$$(\alpha \times \mathbf{1}_{ChosedBySys}(BestAnsType(ans)) +$$
$$(1 - \alpha) \times \mathbf{1}_{ChosedByAsker}(BestAnsType(ans))) \qquad (1)$$

where (1) is indicator function; function Type(ans) returns the type of answer ans, i.e. IsBestAns or NotBestAns; function BestAnsType(ans) returns the type of Best Answer ans, i.e. ChosedBySys and Chosed-ByAsker; α denotes the confidence weight for Best Answer chosed by system.

- **Vote Number Importance:**

$$VoteImp(ans_i, doc) = \frac{VoteNum(ans_i)}{\sum_{ans_j \in doc} VoteNum(ans_j)} \qquad (2)$$

where the importance score of the ith answer ans_i in cQA document doc measured by the proportion of the vote number of ans_i.

- **The Authority Importance of the Answerer:** Simply, we compute the authority importance score of the ith user $user_i$ by formula as follows:

$$AuthImp(user_i) = \frac{BestAnsNum(user_i)}{AllAnsNum(user_i)} \qquad (3)$$

where BestAnsNum($user_i$) denote the number of Best Answer submitted by $user_i$ and AllAnsNum($user_i$) the total number of answers submitted by $user_i$.

In our implementation, all these factors are considered to measure the quality of answers. Simply, we combined all these 3 factors together using linear combination as follows:

$$AnswerImp(ans_i, doc) = \alpha BestAnsImp(ans_i) + \beta VoteImp(ans_i, doc) + \gamma AuthImp(user_i) \qquad (4)$$

where α, β, γ denote the weight of each factor; $\alpha + \beta + \gamma = 1$. The answer with highest score computed by Formula (4) would be took as the highest quality answer. In all our experiments, α was set to 0.6, β to 0.3 and γ to 0.1.

- **Size Consideration:** We first obtained the distributions of word number in each part(question body, best answer and answers) of cQA documents in our dataset. We found the distributions follow power laws.

The power law relations show that the size of most of component is small, only a small number of component have large size. Here, component denotes one of question body, answers and Best Answer. The results are quantitatively

Table 1. Statistics of the length (in words) of the question bodies (QBody), the Best Answers (BAns) and All Answers (AAns).

Words	<30	<50	<100	<150	<250	<350
QBody	0.5011	0.6819	0.8935	0.9584	0.9964	0.9986
BAns	0.3653	0.5476	0.7908	0.8883	0.9565	0.9776
AAns	0.5191	0.6965	0.8841	0.9445	0.9804	0.9902

display in Table 1. Thus we can choose a threshold T to filter the answers, if the number of words in an answer is more than T, we simply use the heading T words to represent the answers; if not, we take this answer as the highest quality answer.

Thus, we have designed a answer-contained snippet framework for cQA search consisting of three parts, i.e. title part, question body part and high quality answer part. In summary, the proposed algorithm first parses the cQA document to obtain all parts, including question subject, question body and answers. Then the algorithm ranks all answers to get the highest quality answer by Formula (4). If the word number of the highest quality answer is more than threshold T, the algorithm only trunks the heading T words as the highest quality answer. Thirdly, the algorithm deletes all redundant and less substantial information in question body, and obtain a clean question body. Finally, the algorithm returns the cleaned question body and the highest quality answer as the cQA snippet of the cQA document.

3.3 Baseline Snippet Algorithm

The state-of-the-art snippet generation method proposed by Metzler et al. [18] was chosen as our baseline algorithms, which use gradient boosted decision tree (GBDTs) learning approach for the snippet generation task. The features adopted are *exact match of query, overlap proportion, overlap-syn proportion, sentence language model, sentence length* and *sentence location*. 10 queries sampled from questions in our Dataset and their corresponding top 20 retrieved documents were used as training data. Human evaluator was asked to summarize all the 200 pages by extracting sentences according to the corresponding queries. The number of sentences for a summary is recommended to be five. However, if there are more or less appropriate sentences, they could be selected in spite of the recommendation. For GBDTs, we use the GBM package for R[3].

3.4 Retrieval Model

In order to perform retrieval, we use a ranking function similar to the one proposed by Xue et al. [4], which builds upon previous work on translation-based retrieval models and tries to overcome some of their flaws, formulated as following:

---

[3] http://cran.r-project.org/web/packages/gbm/.

$$P(\mathbf{q}|(q,a)) = \prod_{w \in \mathbf{q}} P(w|(q,a)) \tag{5}$$

$$P(w|(q,a)) = (1-\lambda)P_{mx}(w|(q,a)) + \lambda P_{ml}(w|C) \tag{6}$$

$$P_{mx}(w|(q,a)) = \alpha P_{ml}(w|q) + \beta \sum_{t \in q} P(w|t)P_{ml}(t|q) + \gamma P_{ml}(w|a) \tag{7}$$

where $\mathbf{q}$ is the user question[4], C denotes the whole archive, $C = \{(q,a)_1, (q,a)_2, ..., (q,a)_L\}$. λ is the smoothing parameter for C. And $P_{ml}(w|C) = \frac{\#(w,c)}{|C|}$ is the maximum likelihood estimator while $|C|$ is the length of C. $P(w|t)$ is the probability of translating a question term t to the query term w. We control the relative importance by α, $beta$ and γ. $\alpha + \beta + \gamma = 1$.

One of differences to the original model by Xue et al. [4] is that we use Jelinek-Mercer smoothing for Eq. 6 instead of Dirichlet Smoothing, as it has been done by Delphine et al. [10]. The other is that we take use of the statistical word translation model trained by Delphine et al. [10], which perform better than the one of original model by Xue et al. In all our experiments, α was set to 0.5, β to 0.3 and λ to 0.5.

3.5 Evaluation Criteria

We have two experimental procedures: *experimental procedure for relevance judgement* and *experimental procedure for information satisfaction*. Because of the similar task, the criteria proposed by Tombros et al. [5] would be adopted in this section. Such criteria for the experiment adopted were:

(a) The recall, precision and F_1 of the relevance judgements.
(b) The speed of judgements performing.
(c) The need of the evaluators to seek assistance from the full text of the retrieved documents.
(d) The subjective opinion of the users about the assistance provided by the snippet of each retrieved document.

Recall, Precision and F_1. They are often used to evaluate the effectiveness of the relevance judgements, and are calculated as follows:

$$P = \frac{N_{cr}}{N_{tir}}; R = \frac{N_{cr}}{N_{tr}}; F_1 = \frac{2 \times P \times R}{P + R} \tag{8}$$

where N_{cr} denotes number of relevant documents correctly identified by a evaluator for a query; N_{tir} denotes the total number of relevant documents, within the examined ones, for that query; N_{tr} total number of indicated relevant documents for a query.

The 50 queries (questions) used in two experimental procedures were sampled from the questions in our dataset; meanwhile, top 30 retrieved documents of each query were manually judged to be relevant or not, as our groudtruth.

---

[4] In this paper, the user question has the same meaning as the user query.

4   Experimental Analysis

4.1   Experiments of Relevance

In this section, we examine users' performance in the process of judging relevance between documents retrieved and specific queries (i.e. questions). It includes two tasks: to judge the relevance of the documents in a ranked list, with either baseline or answer-contained snippets. To achieve this, two groups consisting of 10 evaluators each were invited. Evaluators were randomly assigned to a group, and each group was assigned to one task only [5]. For each query, evaluators were presented with the query and a retrieved document list with snippets, and told that the list was the returned retrieval results of a particular query. The only actions evaluators could perform were to move through the list or to click the full text of the cQA document. Thus, their goal was to identify, in 2 min, as many relevant documents as possible.

The results obtained through the experimental procedure are presented and analysed as following section.

Recall, Precision and F_1. As we can see from Table 2[5]. The precision, recall and F_1 values for the group of evaluators using the proposed snippet are all considerably larger than that of the group using the baseline snippet: the performance difference is 20.25 %, 6.3 % and 11.36 %. We conclude that evaluators using proposed snippet in a retrieved cQA document list, performed their relevance judgements significantly better than those using the traditional state-of-the-art snippet. So it shows that proposed snippet algorithm allows users to identify more relevant cQA documents, and identify them more accurately.

Table 2. The P, R, F_1 Value of the Two Groups

	Precision	Recall	F_1
Baseline	0.5944	0.4676	0.5234
Proposed	**0.7969**	**0.5306**	**0.6370**

Speed. The speed result has been shown in Fig. 1. We examined the average numbers of documents using baseline snippet and proposed snippet.

The figure shows that evaluators using the proposed snippet returned on average 15 documents per query, while the other examined on average 12.32 documents. It amounts to a 21.75 % increase in the average number of documents examined. Therefore, there is a definite tendency for users presented with the proposed snippet to perform relevance judgements quicker than users presented with a baseline snippet.

---

[5] The data presented in Table 2 were acquired by averaging the results for each query over the total number of queries, thus producing the average recall, precision and F_1 values per query.

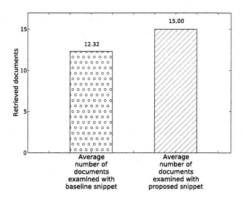

Fig. 1. Speed results

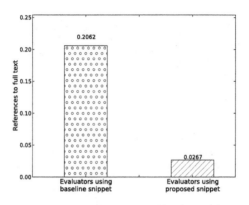

Fig. 2. Average number of references to the full text of the documents (per query)

Reference to the Full Text of the Documents. The data collected on the users' reference to the full text of documents showed that evaluators using the baseline snippet had to refer to 2.54 full texts per query, whereas evaluators from the other experimental group had to refer to 0.4 on average. If we normalise these values to the average number of documents that each experimental group examined for each query, we obtain the results shown in the Fig. 2. The full text of 20.62 % of the documents for each query had to be refered by each evaluator using the baseline snippets, while 2.67 % using the answer-contained snippets.

This difference can be clearly attributed to the snippet information that was companied for each retrieved document. We can find that the proposed method performs better. Users need less clues to establish the relevance of documents, and especially they need clues about the context by which the question-type query are generated. What's more, Our results shows the proposed snippet provided the evaluators with enough evidence to support their relevance judgements.

Opinions of Users. As a form of confirmation of the results obtained in the previous categories, the subjective opinions of the users, gathered from the questionnaire they were asked to till in after their session, rated the utility of the proposed snippet higher than that of the baseline snippet. This result is depicted in Fig. 3 where the scale ranges from 1 (least helpful) to 5 (most helpful). The data indicates that evaluators using baseline snippet rated on average the utility of the accompanying information at 3.25, while evaluators assigned in the other task indicated a rating of 3.65. It means that users require more clues about the relevance of the retrieved documents and the answer-contained snippets have focused on capturing that requirement.

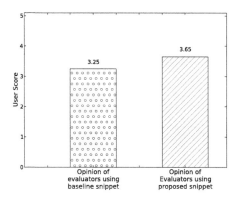

Fig. 3. Subjective opinion of the evaluators

4.2 Experiments of Satisfaction

Here, we would examine two tasks: to get the words that could satisfy information need from a ranked list, with either baseline or answer-contained snippets. To achieve this, two groups consisting of 10 evaluators each were invited. Evaluators were randomly assigned to a group, and each group was assigned to one task only. For each query, evaluators were presented with the query and a retrieved document list with snippets, and told that this list was the result of a retrieval based on a particular query. The only actions evaluators could perform were to move through the list or to click the full text of the cQA document. Thus, their goal was to obtain information that satisfy the query, and stop until the information has been obtained.

The results obtained through the experimental procedure are presented in Fig. 4.

Speed. The Fig. 4(a) shows that evaluators using the proposed snippet examined on average 3.68 documents per query to satisfy the information need, while evaluators using the baseline snippet examined on average 4.76 documents to satisfy the information need. Although this difference is small, it amounts to a

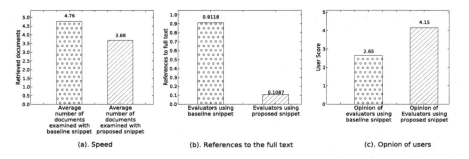

Fig. 4. The speed, reference to the full text, opinion for satisfaction function

22.69 % decrease in the average number of documents examined. Thus we could conclude that there is a definite tendency for users presented with the answer-contained snippets to satisfy information need quicker than users presented with the baseline snippets.

Reference to the Full Text of the Documents. The evaluators using the baseline snippet had to refer to 4.34 full texts per query to satisfy the information need, whereas evaluators from the other experimental group had to refer to 0.4 on average. If we normalise these values to the average number of documents that each experimental group examined for each query, we obtain the results shown in the Fig. 4(b). The full text of 91.18 % of the documents for each query had to be refered by each evaluator using the baseline snippets, while 10.87 % using the answer-contained snippets. This difference can be clearly attributed to the snippet information that was companied for each retrieved document. The result verifies the initial assumption that proposed snippet method could perform better for helping user satisfy information need. If user's information cann't be satisfied from snippet, users refer to the full text of the documents. Our results shows the proposed snippet provided the evaluators with enough evidence to satisfy information need.

Opinions of Users. In Fig. 4(c), the scale ranges from 1 (least helpful) to 5 (most helpful). The data shown in this figure indicates that evaluators using baseline snippet rated on average the utility of the accompanying information at 2.65, while evaluators assigned in the other task indicated a rating of 4.15.

During the post-experimental discussions, users presented with a baseline snippet expressed their dissatisfaction regarding the information they were presented with. More specifically, they emphasised on the fact that they had to refer to the full text for almost every document they were examining to satisfy information need. Hence, the outcome of the post-experimental discussions is yet another indication in favour of the assumption made, that users require the words that could satisfy information need contained in snippet. The answer-contained snippets that include high quality answer have focused on capturing that requirement.

5   Conclusions and Future Work

To the best of our knowledge, this is the first work that addresses the problem of generating answer-contained snippets for cQA search; Meanwhile, our quantitative study shows that the answer-contained snippet method significantly outperforms the state-of-the-art traditional methods in terms of relevance judgements and information satisfaction evaluations, which shows that it's promising research direction to contain answer in a snippet in future.

Acknowledgments. This work was supported by 863 Program (2015AA015404), China National Science Foundation (61402036, 60973083, 61273363), Beijing Technology Project (Z151100001615029), Science and Technology Planning Project of Guangdong Province (2014A010103009, 2015A020217002), Guangzhou Science and Technology Planning Project (201604020179).

References

1. Campos, R., Dias, G., Jorge, A.M., Jatowt, A.: Survey of temporal information retrieval and related applications. ACM Comput. Surv. **47**(2), 1–41 (2015)
2. Jeon, J., Croft, W.B., Lee, J.H.: Finding similar questions in large question and answer archives. In: ACM International Conference on Information and Knowledge Management, pp. 84–90. ACM (2005)
3. Lee, J.T., Kim, S.B., Song, Y.I., Rim, H.C.: Bridging lexical gaps between queries and questions on large online qa collections with compact translation models. In: Proceedings of the Conference on Empirical Methods in Natural Language Processing, EMNLP 2008, 25–27 October 2008, Honolulu, A Meeting of Sigdat, A Special Interest Group of the ACL, pp. 410–418 (2008)
4. Xue, X., Jeon, J., Croft, W.B.: Retrieval models for question and answer archives. In: International ACM SIGIR Conference on Research and Development in Information Retrieval, SIGIR, Singapore, pp. 475–482, July 2008
5. Tombros, A., Sanderson, M.: Advantages of query biased summaries in information retrieval. In: Proceedings of ACM SIGIR, pp. 2–10 (1998)
6. Wang, C., Jing, F., Zhang, L., Zhang, H.J.: Learning query-biased web page summarization. In: Sixteenth ACM Conference on Information and Knowledge Management, CIKM, Lisbon, pp. 555–562, November 2007
7. Huang, Y., Liu, Z., Chen, Y.: Query biased snippet generation in XML search. In: ACM SIGMOD International Conference on Management of Data, pp. 315–326. ACM (2008)
8. He, J., Shu, B., Li, X., Yan, H.: Effective time ratio: a measure for web search engines with document snippets. In: Cheng, P.-J., Kan, M.-Y., Lam, W., Nakov, P. (eds.) AIRS 2010. LNCS, vol. 6458, pp. 73–84. Springer, Heidelberg (2010)
9. Zhou, G., Zhou, Y., He, T., Wu, W.: Learning semantic representation with neural networks for community question answering retrieval. Knowl. Based Syst. **93**, 75–83 (2015)
10. Bernhard, D., Gurevych, I.: Combining lexical semantic resources with question and answer archives for translation-based answer finding. In: ACL 2009, Proceedings of the, Meeting of the Association for Computational Linguistics and the, International Joint Conference on Natural Language Processing of the AFNLP, 2–7 August 2009, Singapore, pp. 728–736 (2009)

11. Edmundson, H.P.: New methods in automatic extracting. J. ACM **16**(2), 264–285 (1969)
12. Gomez-Nieto, E., San, R.F., Pagliosa, P., Casaca, W., Helou, E.S., Oliveira, M.C., et al.: Similarity preserving snippet-based visualization of web search results. IEEE Trans. Vis. Comput. Graph. **20**(3), 457–470 (2014)
13. Silber, H.G., Mccoy, K.F.: Efficiently computed lexical chains as an intermediate representation for automatic text summarization. Comput. Linguist. **28**(4), 487–496 (2002)
14. Turpin, A., Tsegay, Y., Hawking, D., Williams, H.E.: Fast generation of result snippets in web search. In: SIGIR 2007: Proceedings of the International ACM SIGIR Conference on Research and Development in Information Retrieval, Amsterdam, pp. 127–134, July 2007
15. Goldstein, J., Kantrowitz, M., Mittal, V., Carbonell, J.: Summarizing text documents: sentence selection and evaluation metrics. In: Research and Development in Information Retrieval, pp. 121–128 (1999)
16. Joho, H., Hannah, D., Jose, J.M.: Emulating query-biased summaries using document titles. In: International ACM SIGIR Conference on Research and Development in Information Retrieval, pp. 709–710. ACM (2008)
17. Ichikawa, K., Morishita, S.: A simple but powerful heuristic method for accelerating k-means clustering of large-scale data in life science. IEEE/ACM Trans. Comput. Biol. Bioinf. (TCBB) **11**(4), 681–692 (2014)
18. Metzler, D.: Machine learned sentence selection strategies for query-biased summarization. In: SIGIR Learning to Rank Workshop (2008)
19. Ellkvist, T., Strmbck, L., Lins, L.D., Freire, J.: A first study on strategies for generating workflow snippets. In: International Workshop on Keyword Search on Structured Data, pp. 15–20(2009)

Local Community Detection
via Edge Weighting

Weiji Zhao[1,2], Fengbin Zhang[1(✉)], and Jinglian Liu[2]

[1] School of Computer Science and Technology, Harbin University of Science
and Technology, Harbin, People's Republic of China
sdzhaoweiji@163.com, zhangfb@hrbust.edu.cn
[2] School of Information Engineering,
Suihua University, Suihua, People's Republic of China
{sdzhaoweiji,datamining}@163.com

Abstract. Local community detection aims at discovering a community from a seed node by maximizing a given goodness metric. This problem has attracted a lot of attention, and various goodness metrics have been proposed in recent years. However, most existing approaches are based on the assumption that either nodes or edges in network have equal weight. In fact, the usage of weights of both nodes and edges in network can somewhat enhance the algorithmic accuracy. In this paper, we propose a novel approach for local community detection via edge weighting. In detail, we first design a new node similarity measure with full consideration of adjacent nodes' weights. We next develop an edge weighting method based on this similarity measure. Then, we define a new goodness metric to quantify the quality of local community by integrating the edge weights. In our algorithm, we discover local community by giving priority to shell node which has maximal similarity with the current local community. We evaluate the proposed algorithm on both synthetic and real-world networks. The results of our experiment demonstrate that our algorithm is highly effective at local community detection compared to related algorithms.

Keywords: Local community detection · Community structure · Edge weighting · Node similarity

1 Introduction

Network is a data structure composed of a series of nodes interconnected by edges, and widely used to model many complex systems, such as social networks [6, 8, 20], collaboration networks [13], the Internet [4], and E-mail networks [21]. A common property of these networks is community structure. Community structure refers to the division of network nodes into groups within which the edges are dense but between which they are sparse [5, 6, 17, 18]. Community detection has many applications in the field of analyzing online social networks, collaborative tagging systems, biological networks [23].

Traditional community detection methods aim at discovering all communities in network based on the global network structure [3, 6, 14, 16, 19, 21]. For some huge

© Springer International Publishing AG 2016
S. Ma et al. (Eds.): AIRS 2016, LNCS 9994, pp. 68–80, 2016.
DOI: 10.1007/978-3-319-48051-0_6

networks, such as social network and Web network, they are too huge to get the entire network structure nowadays [7]. The global based methods do not work on these huge networks. For solving this problem, local community detection was proposed.

Local community detection aims at discovering a community from a seed node requiring only the information of local network structure, and several algorithms have been proposed in recent years [1, 2, 7, 11, 12, 23]. These algorithms explore local community by maximizing a certain goodness metric. However, most existing goodness metrics are based on the assumption that either nodes or edges in network have equal weight. To ignore the weight of both nodes and edges in network is to throw out a lot of data that could help us to detect local community more accurately.

In this paper, we first design a new node similarity measure with full consideration of adjacent nodes' weights. We next develop an edge weighting method based on this similarity measure. Via edge weighting, every edge in network is assigned with a weight which represents the similarity between two nodes associated with this edge. Furthermore, we propose a new *Closeness-Isolation* metric to quantify the quality of a local community by integrating the edge weights. Finally, we propose our local community detection algorithm. We evaluate the proposed algorithm on both synthetic and real-world networks with ground-truth community structure. The results of our experiment demonstrate that our algorithm is highly effective at local community detection compared to related algorithms.

The rest of the paper is organized as follows. Section 2 introduces the problem definition of local community detection and reviews the existing methods. Section 3 introduces the edge weighting method and a novel local community quality metric *Closeness-Isolation*. We describe our algorithm in Sect. 4 and report experimental results in Sect. 5, followed by conclusions in Sect. 6.

2   Related Work

During the past decades, several local community detection algorithms have been proposed, such as [1, 2, 7, 11, 12, 23]. Most existing algorithms discover local community from a seed node by maximizing a goodness metric. How to design the goodness metric becomes a core problem in local community detection algorithms. In this section, we first introduce the problem definition of local community detection in network, and then review some representative goodness metrics.

2.1   Definition of Local Community in Network

In this subsection, we first give the definition of network, and then present the problem of local community detection in network.

Definition 1 (Network). Let $G = (V, E)$ be an undirected graph, V is the set of nodes and E is the set of edges in G. $n = |V|$ is the number of nodes in G. For two nodes, x, $y \in V$, $(x, y) \in E$ indicates there is an edge between nodes x and y. $m = |E|$ is the number of edges in G. The set of nodes adjacent to node x is denoted by $\Gamma(x)$, $\Gamma(x) = \{y \mid y \in V, (x, y) \in E\}$. The degree of node x is the number of nodes in $\Gamma(x)$, denoted by k_x.

The problem of local community detection can be presented as: For a network $G = (V, E)$, given a goodness metric for local community quality, local community detection starts from a seed node s ($s \in V$), the work is to discover the community D that s belongs to. As shown in Fig. 1, we can dynamically divide the entire network into three parts: local community D, D's shell node set N and the unknown node set U, $U = V - D - N$. Each node in N has at least one adjacent node in D. D has two subsets: the core node set C and the boundary node set B. The nodes in C are only connected by nodes in D, but any node in B has at least one neighbor node in N.

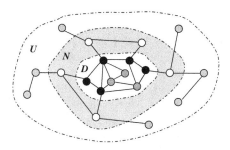

Fig. 1. An illustration of division of a network into local community D (Core Node Set C (green nodes) and Boundary Node Set B (black nodes)), D's Shell Node Set N (white nodes) and Unknown Node Set U (grey nodes) (Color figure online)

During the process of detecting local community, we have perfect knowledge of the connectivity of nodes in $D \cup N$, but have no knowledge of the connectivity of nodes in U. When local community detection algorithm starts, $D = \{s\}$, $N = \Gamma(s)$. In general, local community detection algorithm continuously starts from D and expand outward by absorbing external nodes from N into D until the given goodness metric stops improving [22]. Finally, D is the local community that node s belongs to. Similar definitions of local community detection can be found in [2, 10, 22].

2.2   Goodness Metrics that Assume All Edges are Equal

This kind of goodness metrics assume that all edges in network have equal weight. The representative goodness metrics are R and M.

Clauset [2] defined a local community quality metric called R by only considering the linkage information of boundary nodes in B.

$$R = \frac{B_{in}}{B_{in} + B_{out}} \tag{1}$$

where B_{in} is the number of inward edges that connect boundary nodes in B to other nodes in D, while B_{out} is the number of edges that connect boundary nodes in B to nodes in N. R measures the fraction of inward edges in all edges with one or more nodes in B.

Luo et al. [11] defined a local community quality metric called M, which focuses on the ratio of the number of internal edges and external edges.

$$M = \frac{E_{in}}{E_{out}} \tag{2}$$

where E_{in} is the number of edges with two nodes in local community D, while E_{out} is the number of edges with one node in D and the other in N. M measures the fraction of edges with both nodes in D in edges with one node in D and the other in N.

Both M and R assume that the edges in network have equal weight. In fact, the edge weights are different due to the similarities between each pairs of connected nodes are different. To ignore the edge weights is to throw out a lot of data that could help us to discover local community structure better.

2.3 Goodness Metrics that Assume All Nodes are Equal

This kind of goodness metrics focus on the internal similarity and external similarity of local community. For a local community D, the internal similarity of D is the sum of similarities between any two adjacent nodes both in D, while the external similarity of D is the sum of similarities between any two adjacent nodes with one node in D and the other in N. The representative metrics are *tightness* and *Compactness-Isolation*.

Huang et al. [7] adopted the node similarity measure, as shown in Formula (3), to evaluate the similarity between nodes x and y. Based on this measure, they introduced a metric for local community quality called *tightness*, and present a local community detection algorithm LTE via local optimization of the *tightness* measure.

$$s_{xy} = \frac{|\Gamma(x) \cap \Gamma(y)|}{\sqrt{|\Gamma(x)| \cdot |\Gamma(y)|}} \tag{3}$$

Ma et al. [12] introduced a d-neighbors based similarity measure called s_{xy}^d which takes into account non-adjacent nodes within a distance away. s_{xy}^d is defined in Formula (4). Based on this measure, they introduced a metric for local community quality called *Compactness-Isolation*, and proposed a local community detection algorithm called GMAC by maximizing *Compactness-Isolation*.

$$s_{xy}^d = \frac{\left|\Gamma(x)^d \cap \Gamma(y)^d\right|}{\left|\Gamma(x)^d \cup \Gamma(y)^d\right|} \tag{4}$$

$\Gamma(x)^d$ is a set of nodes whose shortest path length to node x is within d.

There are other node similarity measures, such as Common Neighbors and Jaccard Index [25]. The measure of Common Neighbors is directly counting the number of common neighbors two nodes have, while Jaccard Index is the ratio of the number of their common neighbors to the number of their union. All these methods assume that all adjacent nodes have equal weight. In fact, for any node, it has different similarities with its adjacent nodes.

3 Preliminaries

There are two subproblems in local community detection algorithm: how to design a proper goodness metric for local community quality and how to choose node in N as a member of D. The third problem hidden in these two subproblems is that how to evaluate the edge weights more accurately. In this section, we focus on these three subproblems, and give our solutions.

3.1 Edge Weighting

The weight of edge depends on the similarity between two nodes associated with this edge. In this subsection, we first give a new node similarity measure based on weighted neighbor nodes, and then introduce our edge weighting method.

Definition 2 (Node Similarity Based on Weighted Neighbor Nodes). Let $G = (V, E)$ be a network, for a node x, $o \in \Gamma(x)$, we define the weight of o as s_{xo}. s_{xo} can be calculated by methods in Subsect. 2.3. For a pair of nodes, x, $y \in V$, we define the similarity between x and y based on weighted neighbor nodes as ws_{xy}. ws_{xy} is defined as follows.

$$ws_{xy} = \frac{\sum\limits_{z \in \Gamma(x) \cap \Gamma(y)} (s_{xz} + s_{yz})}{\sum\limits_{u \in \Gamma(x)} s_{xu} + \sum\limits_{v \in \Gamma(y)} s_{yv}} \tag{5}$$

where the numerator is the sum of their common neighbors' weights, and the denominator is the sum of their neighbors' weights. The range of ws_{xy} is $[0, 1]$. When nodes x and y have no common neighbors, $ws_{xy} = 0$, and while they share the same neighbor nodes, $ws_{xy} = 1$.

Based on the above node similarity measure, we introduce our edge weighting method. For a pair of nodes, x and y, the similarity measure ws_{xy} considers the weights of their adjacent nodes, but neglects the fact that whether nodes x and y are directly connected or not. Two nodes tend to have higher similarity if they are directly connected. So our edge weighting method is given as follows.

Definition 3 (Edge Weighting). Let $G = (V, E)$ be a network, for any edge $(x, y) \in E$, let $w_{(x, y)}$ denote the weight of edge (x, y). $w_{(x, y)}$ is defined as follows.

$$w_{(x,y)} = ws_{xy} + \frac{k_x \times k_y}{2m} \tag{6}$$

For the weight of edge (x, y), on the basis of ws_{xy}, we plus the probability of these two nodes being connected to each other to $w_{(x, y)}$. Via edge weighting, we assign every edge in network with a weight.

3.2   Our Local Community Quality Metric *Closeness-Isolation*

Inspired by [7, 11, 12], we propose a new local community quality metric *Closeness-Isolation* (*CI* for short) based on the edge weights.

Definition 4 (*Closeness-Isolation* Metric). Let $G = (V, E)$ be a network, the weight of edge (x, y) is $w_{(x,y)}$. For a local community D with shell node set N, the *Closeness-Isolation* Metric of D, denoted by $CI(D)$, is defined as

$$CI(D) = \frac{\sum\limits_{x,y \in D, (x,y) \in E} w_{(x,y)}}{1 + \sum\limits_{u \in D, v \in N, (u,v) \in E} w_{(u,v)}} \tag{7}$$

where the numerator is the sum of weights of all edges in D, and the denominator is one plus the sum of weights of all edges with one node in D and the other in N.

Instead of assuming the edges with equal weights, *CI* takes into account the edge weights, and is more reasonable than the other metrics. We use *CI* to measure local community quality in our algorithm.

3.3   Similarity Between Shell Node and Local Community

We define the similarity between shell node and local community in weighted network as follows.

Definition 5 (Similarity Between Shell Node and Local Community). Let $G = (V, E)$ be a network, the weight of edge (x, y) is $w_{(x,y)}$. For a local community D with shell node set N, for a node $z \in N$, we denote the similarity between z and local community D by $sim(z, D)$. $sim(z, D)$ can be calculated as follows.

$$sim(z, D) = \sum\limits_{v \in \Gamma(z) \cap D} w_{(z,v)} \tag{8}$$

$sim(z, D)$ is the sum of weights of all edges connecting z and nodes in D. Inspired by the fact that nodes in the same community are more likely to have higher similarities with each other, we choose the node in N which has highest similarity with local community D as candidate node.

4 Discover Local Community via Edge Weighting

With the edge weighting based local community quality metric CI, we propose our local community detection algorithm.

4.1 Our Local Community Detection Algorithm

Our local community detection algorithm starts from a given node s without any manual parameters. The pseudo code is described in Algorithm 1. Firstly, initialize $D = \{s\}$ and $N = \Gamma(s)$ (line (1)). In the while-loop (lines (2)–(16)), our algorithm keeps choosing the node $a \in N$ which has maximal similarity with local community D as candidate node (lines (3)–(10)), the similarity between node in N and local community D is calculated by Formula (8). If agglomerating the candidate node into D will cause an increase in CI, then add it to D and update N, otherwise, remove it from N (lines (11)–(15)), repeat until N is empty. Finally, return D as the local community of s (line (17)).

Algorithm 1: Local Community Detection

Input: a given node s, a network $G = (V, E)$;
Output: local community D;
Describe:
  1) initialize $D=\{s\}$, $N=\Gamma(s)$;
  2) while N is not empty do
  3)   create a new dictionary variable dic_sim to store the similarities of nodes belonging to N with D;
  4)   for each $i \in N$ do
       //calculate similarity between node i and local community D
  5) $dic\_sim[i] = 0$;
  6)   for each $j \in \Gamma(i) \cap D$ do
  7) $dic\_sim[i] += w_{(x,y)}$; // see Formula (6)
  8)   end for
  9)   end for
 10)   find a such that $dic\_sim[a]$ is maximum;
 11)   if $CI(D \cup a) > CI(D)$ then
 12)   add a to D and update N;
 13)   else
 14)   remove a from N;
 15)   end if
 16) end while
 17) return D;

4.2   Time Complexity Analysis

In our algorithm, each node in network is denoted by a unique identifier. A network is stored by a hash table of nodes in the graph, and each node associates with a vector of its adjacent nodes. The values in vectors are sorted for faster access.

The running time of our algorithm depends on the size of the union of local community and its shell node set rather than that of the entire graph. Let t denote the size of $D \cup N$, E_{in} denote the number of edges with two nodes in D, E_{out} denote the number of edges with one node in D and the other in N, k denote the mean node degree of nodes in $D \cup N$. The computational cost of our algorithm mainly consists of two parts: calculating the weight of edges with one or more node in D and choosing a node in N as candidate node. For calculating the weight of edges, we need to compute t nodes' neighbor nodes, the weight of neighbor nodes of t nodes, and then compute $(E_{in} + E_{out})$ edges' weights. Their time complexity is $O(k \cdot t)$, $O(k^2 \cdot \log k \cdot t)$ and $O(k \cdot \log k \cdot (E_{in} + E_{out}))$ respectively. Adding these together, the time complexity is $O((k + k^2 \cdot \log k) \cdot t + k \cdot \log k \cdot (E_{in} + E_{out}))$. The most computational expensive steps is in lines (4)–(9), which is the time to find $a \in N$ having the maximal similarity with the current local community D. In each while-loop, the time complexity is $O(|N| \cdot |D| \cdot \log k)$.

5   Experiments

In this section, we evaluate the effectiveness of our algorithm on synthetic as well as real-world networks.

5.1   Related Methods and Evaluation Criteria

We compare our algorithm with three representative local community detection algorithms: (1) Clauset's algorithm [2] is a classic algorithm by maximizing metric R. Note that the same as [12, 22], we improve its stopping criteria by detecting changes in R. (2) Luo et al.'s algorithm [11] (LWP for short) is a famous algorithm to find the sub-graph with maximum metric M. (3) GMAC algorithm [12] is the most popular algorithm which uses d-neighbors to represent node. We fix $d = 3$ as suggested by authors. Our algorithm uses Jaccard Index to calculate neighbor nodes' weights.

We use three evaluation measures to compare algorithmic performance: *precision*, *recall* and *F-score*, which are widely adopted by other community detection methods [10, 12]. The *precision* and *recall* are calculated as follows.

$$Precision = \frac{|C_F \cap C_R|}{|C_F|} \tag{9}$$

$$recall = \frac{|C_F \cap C_R|}{|C_R|} \tag{10}$$

where C_R is the set of nodes in real local community which contains the given node and C_F is the set of nodes discovered by local community detection algorithm which starts from the given node.

F-score is the harmonic mean of *precision* and *recall*. Its formula is as follows.

$$F - score = 2 \times \frac{precision \times recall}{precision + recall} \tag{11}$$

5.2   Evaluation on Synthetic Networks

For comparing the performance of various local community detection algorithms, we first generate 10 synthetic networks with ground-truth community structure. There are 5000 nodes in every network.

LFR benchmark networks, introduced by Lancichinetti et al. [9], are widely used to test community detection methods [7, 12]. The important properties of this network generating model are defined as follows: the number of nodes is denoted by n, the average degree of nodes is denoted by k, the maximum degree is denoted by k_{max}, mixing parameter is denoted by μ, minus exponent for the degree sequence is denoted by $t1$, minus exponent for the community size distribution is denoted by $t2$, number of overlapping nodes is denoted by on, number of memberships of the overlapping nodes is denoted by om, minimum for the community sizes is denoted by $minc$, maximum for the community sizes is denoted by $maxc$. The parameters are set as follows: $n = 5000$, $k = 10$, $k_{max} = 50$, others except μ use default values. Mixing parameter u is the fraction of edges of each node outside its community, which is used to control the difficulty of community detection [19]. So we generate 10 networks with different mixing parameter μ ranging from 0.05 to 0.5 with a span of 0.05. These LFR benchmark networks are generated together with ground-truth community structure.

For every network in our experiments, we use each node in this network as a seed node once, and repeat the local community detection experiments for 5000 times which start from different node every time, then report algorithmic average *precision*, *recall* and *F-Score* on this network. Figure 2 shows the comparison results of *precision*, *recall*, *F-score* for four algorithms on these networks, respectively.

We discuss the experiments result in detail. Firstly, along with μ becomes larger, all the four algorithms suffer varying degrees of performance degradation and become ineffective to detect community structure. This is because the higher the mixing parameter u of a network, the weaker community structure it has.

Secondly, along with μ becomes larger, the performance of both LWP and Clauset drops rapidly, meanwhile GMAC and our algorithm drop slowly. This is because both LWP and Clauset simply depend on the number of edges incident to the node, neglect the fact that the weight of external edges are smaller than the internal edges.

Thirdly, our algorithm takes node weights in account, so it outperforms GMAC algorithm which neglects the node weights.

The *precision*, *recall*, and *F-score* of the LWP algorithm is zero or nearly zero when $\mu \geq 0.35$. This is because all the local communities discovered by LWP

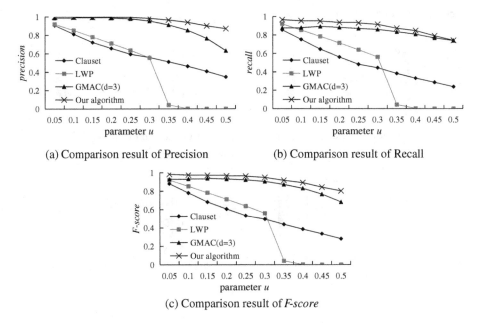

(a) Comparison result of Precision   (b) Comparison result of Recall

(c) Comparison result of *F-score*

Fig. 2. Comparison results on LFR benchmark networks

algorithm satisfy $M > 1$, which means the number of internal edges should be more than the number of external edges. However, almost no local community can satisfy $M > 1$ when $\mu \geq 0.35$, so LWP algorithm performs badly in this case. This conclusion is in accordance with the results reported in Ref. [7].

In general, because our algorithm takes into account the weight of nodes and edges, it performs best on LFR benchmark networks.

5.3   Evaluation on Real-World Networks

So far, we have presented the experimental results of the proposed algorithm on synthetic networks. In this subsection, we use additional three real-world networks to evaluate the performance of our algorithm. (1) The first network is Zachary Karate Club Network (Karate for short) [24], in which $n = 34$ and $m = 78$. It describes the friendships among 34 members of a karate club at a US university. (2) The second is NCAA football network (NCAA for short) [6], in which $n = 115$ and $m = 613$. It describes American football games between Division IA colleges during regular season Fall 2000. (3) The third is Books about US politics (Polbooks for short) [15], in which $n = 105$ and $m = 441$. It is a network of books about US politics published around the time of the 2004 presidential election and sold by Amazon.com. All of them are available at http://www-personal.umich.edu/ ∼ mejn/netdata/.

In our experiment, we use every node in these network as a seed node once, and repeat the local community detection experiments for n times which start from different node every time, where n is the number of nodes in this network, and report algorithmic

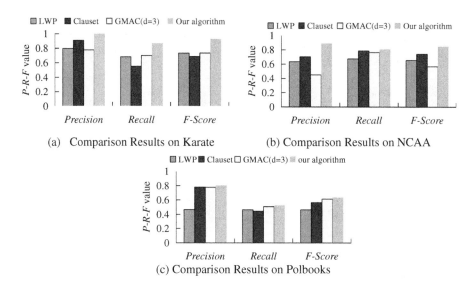

Fig. 3. Comparison results on real-world networks

average *precision*, *recall* and *F-Score* on this network. The comparison result on real-world networks is reported in Fig. 3. Compared with the other three algorithms, our algorithm has highest *precision*, *recall*, and *F-score* at the same time on these real-world networks. Our algorithm makes use of the weight of both nodes and edges in network, and enhances the algorithmic accuracy. So it outperforms the other algorithms on real-world networks.

6   Conclusion and Future Work

Discovering local community is an important work in network analysis and many algorithms have been proposed to identify local community from a given node. Different from the existing local community detection methods that neglect the weight of both nodes and edges, we take into account the information to enhance the algorithmic accuracy. In this paper, we first propose an edge weighting method based on a new node similarity measure. Then, we introduce a framework for local community detection based on the edge weights. This framework opens a rich space for research, all algorithms can be embedded into this framework differing only in the similarity measures. Compared with other related algorithms, our algorithm doesn't need any manual parameters, and achieves good performance on both synthetic and real-world networks.

In future, we will apply our algorithm on real-world networks to discover local community and also study the community detection problem in heterogeneous networks.

Acknowledgments. The project is supported by National Natural Science Foundation of China (61172168).

References

1. Bagrow, J., Bolt, E.: A local method for detecting communities. Phys. Rev. E **72**(4), 046108-1–046108-10 (2005)
2. Clauset, A.: Finding local community structure in networks. Phys. Rev. E **72**(2), 026132 (2005)
3. Clauset, A., Newman, M.E., Moore, C.: Finding community structure in very large networks. Phys. Rev. E Stat. Nonlin. Soft Matter Phys. **70**(6), 264–277 (2004)
4. Faloutsos, M., Faloutsos, P., Faloutsos, C.: On power-law relationships of the internet topology. In: SIGCOMM 1999, pp. 251–262 (1999)
5. Fortunato, S.: Community detection in graphs. Phys. Rep. **486**(3/5), 75–174 (2010)
6. Girvan, M., Newman, M.: Community structure in social and biological networks. Proc. Natl. Acad. Sci. USA **99**(12), 7821–7826 (2002)
7. Huang, J., Sun, H., Liu, Y., Song, Q., Weninger, T.: Towards online multiresolution community detection in large-scale networks. PLoS ONE **6**(8), 492 (2011)
8. Jia, G., Cai, Z., Musolesi, M., Wang, Y., Tennant, D.A., Weber, R.J., Heath, J.K., He, S.: Community detection in social and biological networks using differential evolution. In: Hamadi, Y., Schoenauer, M. (eds.) LION 2012. LNCS, vol. 7219, pp. 71–85. Springer, Heidelberg (2012)
9. Lancichinetti, A., Fortunato, S., Radicchi, F.: Benchmark graphs for testing community detection algorithms. Phys. Rev. E **78**(4), 046110-1–046110-5 (2008)
10. Liu, Y., Ji, X., Liu, C., et al.: Detecting local community structures in networks based on boundary identification. In: Mathematical Problems in Engineering, pp. 1–8 (2014). http://dx.doi.org/10.1155/2014/682015
11. Luo, F., Wang, J., Promislow, E.: Exploring local community structures in large networks. Web Intell. Agent Syst. (WIAS) **6**(4), 387–400 (2008)
12. Ma, L., Huang, H., He, Q., Chiew, K., Wu, J., Che, Y.: GMAC: a seed-insensitive approach to local community detection. In: DaWaK, pp. 297–308 (2013)
13. Newman, M.: The structure of scientific collaboration networks. Work. Pap. **98**(2), 404–409 (2000)
14. Newman, M.: Fast algorithm for detecting community structure in networks. Phys. Rev. E Stat. Nonlin. Soft Matter Phys. **69**(6), 066133-1–066133-5 (2004)
15. Newman, M.: Modularity and community structure in networks. Proc. Natl. Acad. Sci. **103**(23), 8577–8582 (2006). http://www-personal.umich.edu/ ~ mejn/netdata/
16. Newman, M., Girvan, M.: Finding and evaluating community structure in networks. Phys. Rev. E Stat. Nonlin. Soft Matter Phys. **69**(2), 026113-1–026113-15 (2004)
17. Radicchi, F., Castellano, C., Cecconi, F., et al.: Defining and identifying communities in networks. Proc. Natl. Acad. Sci. USA **101**(9), 2658–2663 (2004)
18. Schaeffer, S.: Graph clustering. Comput. Sci. Rev. (CSR) **1**(1), 27–64 (2007)
19. Shao, J., Han, Z., Yang, Q., Zhou, T.: Community detection based on distance dynamics. In: Proceedings of the 21st ACM SIGKDD International Conference on Knowledge Discovery and Data Mining, pp. 1075–1084 (2015)
20. Takaffoli, M.: Community evolution in dynamic social networks - challenges and problems. In: ICDM Workshops 2011, pp. 1211–1214 (2011)

21. Tyler, J.R., Wilkinson, D.M., Huberman, B.A.: Email as spectroscopy: automated discovery of community structure within organizations. Inf. Soc. **21**(2), 143–153 (2005)
22. Wu, Y., Huang, H., Hao, Z., Chen, F.: Local community detection using link similarity. J. Comput. Sci. Technol. (JCST) **27**(6), 1261–1268 (2012)
23. Wu, Y., Jin, R., Li, J., Zhang, X.: Robust local community detection: on free rider effect and its elimination. In: VLDB 2015, pp. 798–809 (2015)
24. Zachary, W.: An information flow model for conflict and fission in small groups. J. Anthropol. Res. **33**(4), 452–473 (1977)
25. Zhou, T., Lü, L., Zhang, Y.: Predicting missing links via local information. Eur. Phys. J. B **71**(4), 623–630 (2009)

Machine Learning and Data Mining
for IR

Learning a Semantic Space of Web Search via Session Data

Lidong Bing[1(✉)], Zheng-Yu Niu[2], Wai Lam[3], and Haifeng Wang[2]

[1] Tencent Inc., Shenzhen, China
lyndonbing@tencent.com
[2] Baidu Inc., Beijing, China
{niuzhengyu,wanghaifeng}@baidu.com
[3] Department of Systems Engineering and Engineering Management,
The Chinese University of Hong Kong, Hong Kong, China
wlam@se.cuhk.edu.hk

Abstract. In Web search, a user first comes up with an information need and issues an initial query. Then some retrieved URLs are clicked and other queries are issued if he/she is not satisfied. We advocate that Web search is governed by a hidden semantic space, and each involved element such as query and URL has its projection, i.e., as a vector, in this space. Each of above actions in the search procedure, i.e. issuing queries or clicking URLs, is an interaction result of those elements in the space. In this paper, we aim at uncovering such a semantic space of Web search that uniformly captures the hidden semantics of search queries, URLs and other elements. We propose session2vec and session2vec+ models to learn vectors in the space with search session data, where a search session is regarded as an instantiation of an information need and keeps the interaction information of queries and URLs. Vector learning is done on a large query log from a search engine, and the efficacy of learnt vectors is examined in a few tasks.

1 Introduction

In the study of word embedding, words are mapped into a vector space such that semantically relevant words are placed near each other [1,16,17]. Word vectors are helpful for a wide range of NLP tasks by better capturing syntactic and semantic information than simple lexical features [8,14,23]. In this work, we explore to apply embedding methodology to model the intrinsic hidden semantic space of Web search. Figure 1(a) gives an example to illustrate the intuition. The user has an information need in mind which can be represented as a 4-dimension vector, and each dimension indicates the relevance of his need with a particular semantic topic. Although the user intends to formulate queries conveying his need on the third dimension, the first two queries are not precise enough. Then the user issues the last query that well matches his need, and accordingly, the

This work is substantially supported by a grant from the Research Grant Council of the Hong Kong Special Administrative Region, China (Project Code: CUHK413510).

S. Ma et al. (Eds.): AIRS 2016, LNCS 9994, pp. 83–97, 2016.
DOI: 10.1007/978-3-319-48051-0_7

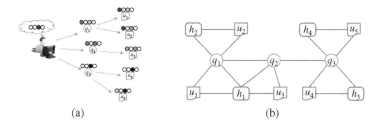

Fig. 1. Search session and session graph.

returned URLs satisfy him. To generalize, websites and query terms could also be involved and projected as vectors in the same space. Obviously, building such a space governing the search procedure could be useful for different tasks, such as query suggestion, result ranking, etc.

We conduct the semantic space learning using search session data since a search session can be regarded as an instantiation of a particular information need. The learning task is cast as a vertex embedding problem on a set of graphs (built from sessions), where the elements in a session are represented as vertices and related vertices are connected by edges. The use of graph seems a suitable choice for session representation because it captures the semantic interactions among elements. Given user's information need, represented as a semantic vector, the probability of obtaining a session is jointly determined by semantic meaning of involved elements, i.e., vertices of the session graph. Then we perform vector learning for vertices via maximizing the log-likelihood of a training data of sessions.

Our main contributions are: (1) a framework is proposed to learn a semantic space of Web search, and different elements (such as queries, URLs, and terms) are projected as vectors in this space. Vectors of different elements are directly comparable for semantic similarity calculation, and our model has good applicability to unseen data; (2) We use graph structure to represent session data and develop an approach for vertex vector learning on session graphs. Our model can capture fine-grained structure information in click-through data. It is flexible to incorporate other types of elements. (3) Our model is trained on a large query log data from a search engine. Extensive experiments are conducted to examine the efficacy of the constructed semantic space, and the results show that the learnt vectors are helpful for different tasks.

2   Related Work

Researchers had observed the potential of generating semantic vectors for search queries and Web pages [7,10,21]. Deep Structured Semantic Model (DSSM) [10] and Convolutional Latent Semantic Model (CLSM) [21] employ deep neural network to map the raw term vector of a query or a document to its latent semantic vector. Both of them use the full text of pages as input, and CLSM also captures the contextual information. The network architecture in our model is different from them and it can be trained more efficiently. Furthermore, our framework also learns vectors of terms and websites and can be easily extended to include other elements, e.g. users. The learnt term vectors enable our model to tackle unseen data. Some other studies attempted to learn binary vectors for queries or URLs and binary values show the relevance to semantic dimensions [18].

Gao et al. [7] proposed Bi-Lingual Topic Model (BLTM) and linear Discriminative Projection Model (DPM) for query-document matching at the semantic level. More specifically, BLTM is a generative model and it requires that a query and its clicked documents share the same distribution over topics and contain similar factions of words assigned to each topic. DPM is learnt using the S2Net algorithm that follows the pairwise learning-to-rank paradigm. Previous works also tried to learn query-document similarity from click-through data with implicit semantic representation, such as bipartite graph or translation models [4,6,24]. Grbovic et al. [8] proposed to learn query and term vectors for query rewriting in sponsored search. Here our framework performs vector learning for a more comprehensive setting, i.e. including URLs, queries, terms, and websites.

Another related area is the study of word embedding. A popular model for estimating neural network language model was proposed in [2]. Word2vec [16] is a development with a simple architecture for efficient training. A development of word2vec maps paragraphs into the same space of words [13], which shares similar architecture with our basic model. In comparison, our work focuses on modeling session graph data, and the session vector is incorporated in the networks to model users' information need. More importantly, our tailor-made enhanced model elegantly projects terms into the same semantic space of search elements. Some other works employed neural networks to learn concept vectors of input text objects for similarity calculation under a supervised setting [25].

3   Problem Formulation

Given a set of search sessions $S = \{s_i\}_{i=1}^n$ as training data, we aim at finding a semantic space to model Web search scenario so that the probability of observing the sessions in S is maximized. Let θ denote the parameters of the space (i.e., the semantic vectors of elements in sessions). The log-likelihood objective is defined as follows:

$$\ell(\theta; S) = \sum_{s_i \in S} \log P(s_i; \theta), \tag{1}$$

where $P(s_i; \theta)$ is the probability of observing s_i in the space.

Let e_j denote an element such as a query or a URL in the session s_i, and $C(e_j)$ denote the context elements of e_j in s_i. Let $\mathbf{v}(e_j)$ denote the vector of e_j, and $\mathbf{v}(s_i)$, having the same dimensionality, denote the information need of the user corresponding to s_i. $\mathbf{v}(s_i)$ is also called session vector. We assume that the probability of e_j only depends on $C(e_j)$ and $\mathbf{v}(s_i)$, and it is denoted as $P(e_j; C(e_j), \mathbf{v}(s_i))$. Therefore, $P(s_i; \theta)$ is calculated as:

$$P(s_i; \theta) = \prod_{e_j \in s_i} P(e_j; C(e_j), \mathbf{v}(s_i)). \tag{2}$$

$P(e_j; C(e_j), \mathbf{v}(s_i))$ is calculated with the element vectors of $C(e_j)$ and $\mathbf{v}(s_i)$ (described later). To summarize, our task is to learn vectors for elements in search sessions so that the objective in Eq. 1 is maximized. To do so, we should transform each session into training instances of the form $(e_j, C(e_j), s_i)$ for calculating $P(e_j; C(e_j), \mathbf{v}(s_i))$, and a major task is to define the context $C(e_j)$ of e_j in s_i. For better capturing the structure information in click-through data, we introduce a graph representation of session data. Then we develop two models for our learning task by extending an algorithm of word2vec [16].[1]

4 Basic Model for Vector Learning

4.1 Session Graph and Training Instances

In a search session, there are several types of elements. A user first issues a query, and some URLs are clicked in the result list. To obtain better results, she may issue more queries. During browsing a clicked page, the user may also browse other pages in the same website. Thus, the website is also involved as an element of the session.

Session Graph. *A search session graph $\mathcal{G} = \{V, E\}$ is defined as an undirected graph. The vertex set V includes search query, clicked URL, and website. The edges are added between (1) two successive queries; (2) a clicked page and the corresponding query; (3) a website and pages from it; (4) a website and a query that results in pages of this website clicked.*

An example is given in Fig. 1(b). With a query q_1, the user clicked two URLs, u_1 and u_2. Thus, the edges (q_1, u_1) and (q_1, u_2) are added. The websites h_1 and h_2 of u_1 and u_2 are involved. Accordingly, we have the edges (u_1, h_1), (q_1, h_1), (u_2, h_2), and (q_1, h_2). After browsing u_1 and u_2, the user issued q_2 and q_3 and clicked more URLs. $C(e_j)$ is defined as adjacent elements of e_j. For example, we have $C(q_1) = \{q_2, u_1, u_2, h_1, h_2\}$. Each training instance $(e_j, C(e_j), s_i)$ means that the target e_j comes from session s_i with context $C(e_j)$.

---

[1] Existing methods for vector representation learning [2,15,16,20] cannot be readily applied here due to: (1) our training data is a set of sessions and each of them is represented as a graph, while the training data of existing methods is a set of word sequences; (2) a vector capturing users' information need is incorporated into our learning procedure. Moreover, we intend to learn a space that uniformly embeds elements of different types such as queries and URLs.

4.2 Basic Learning Model

The objective of our basic model can be written as follows:

$$\ell(\theta; S) = \sum_{s_i \in S} \log P(s_i; \theta) = \sum_{s_i \in S} \sum_{e_j \in s_i} \log P(e_j; C(e_j), \mathbf{v}(s_i)). \tag{3}$$

The network, called **session2vec** (s2v for short), for calculating $P(e_j; C(e_j),$ $\mathbf{v}(s_i))$ is given in Fig. 2(a), which basically introduces an auxiliary vector into CBOW model [16], as previously did in [13,17]. The input layer takes the element vectors of $C(e_j)$ and session vector $\mathbf{v}(s_i)$. In the projection layer, the average of the element vectors[2] is summed with $\mathbf{v}(s_i)$ to get $\mathbf{x}_j$. The output layer contains a Huffman tree with each distinct element in training sessions as a leaf. The more frequent an element is, the shorter its Huffman code is.

Let p denote the path from the root to a leaf e_j and L denote the length of p. Let $v^p_{1:L}$ denote the vertices on p and we have $v^p_L = e_j$. Let $c_{1:L-1}$ be the sequence of binary codewords on p. Let $\boldsymbol{\gamma}_{1:L-1}$ denote the vectors associated with the inner vertices $v^p_{1:L-1}$ on p, each of them has the same dimensionality as $\mathbf{x}_j$. $P(e_j; C(e_j), \mathbf{v}(s_i))$ is calculated as the probability of reaching the leaf e_j along p (going through $L-1$ binary selections). Specifically, at vertex v^p_l, we select the branch having the codeword c_l with probability $P(c_l; \boldsymbol{\gamma}_l, \mathbf{x}_j)$, which is defined with the sigmoid function σ:

$$P(c_l; \boldsymbol{\gamma}_l, \mathbf{x}_j) = \{\sigma(\mathbf{x}_j \boldsymbol{\gamma}_l)\}^{1-c_l} \cdot \{1 - \sigma(\mathbf{x}_j \boldsymbol{\gamma}_l)\}^{c_l}. \tag{4}$$

$P(e_j; C(e_j), \mathbf{v}(s_i))$ is calculated as:

$$P(e_j; C(e_j), \mathbf{v}(s_i)) = \prod_{l=1}^{L-1} P(c_l; \boldsymbol{\gamma}_l, \mathbf{x}_j). \tag{5}$$

Thus, combining Eqs. 3 and 5, the objective can be calculated with the network in Fig. 2(a).

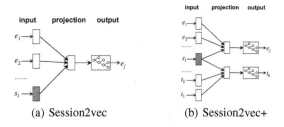

(a) Session2vec   (b) Session2vec+

Fig. 2. Our models.

---

[2] The number of contextual elements varies, so we calculate the average of contextual vectors.

We use the SGD algorithm to learn the vectors of elements, inner nodes of Huffman tree, and sessions. During learning, each instance generated in Sect. 4.1 is fed into the network and its related parameters are updated. The learning procedure is performed by scanning all training instances one or a few times depending on efficiency requirement.

5 Enhanced Model for Vector Learning

With session2vec, each element (i.e., query, URL, and website) in training data is projected as a vector. However, s2v cannot deal with unseen elements in new data. To solve this problem, we propose an enhanced learning model, **session2vec+** (s2v+), as depicted in Fig. 2(b). The upper part of s2v+ is the same as s2v. The lower part, having the same architecture, incorporates the term-based training instances in the form of $(t_k, C(t_k), s_i)$ (t_k is a term in s_i). The session vector is shared by two parts as a bridge to align the dimensions of element vectors and term vectors, learnt by the upper and lower parts, respectively. Thus, terms and elements are embedded in the same space, and term vectors can be utilized to build vectors for new elements such as unseen queries. Another advantage of s2v+ is that these term vectors can help solve the sparsity issue in s2v, since the vectors of sparse elements learnt in s2v might be unreliable.

5.1 Training Instances for Term Vector Learning

We build term-based training instances by post-processing element-based instances. Specifically, if e_j of $(e_j, C(e_j), s_i)$ is a query or a URL, it is transformed into a set of term-based instances. Let t_k denote a term in query e_j or the URL title of e_j. Each t_k corresponds to one term-based instance $(t_k, C(t_k), s_i)$ where $C(t_k)$ is the context of t_k containing all terms of queries or URL titles in $C(e_j)$. Noun phrase chunking is done and a single term here may refer to a phrase, e.g. "New York Times". Because t_k could also come from URL titles, our model is augmented to handle unseen query terms with title terms.

5.2 Enhanced Learning Model

For s2v+, we define a new objective function as follows:

$$\ell'(\theta; S) = \sum_{s_i \in S} \log P(s_i; \theta) + \sum_{s_i \in S} \log P'(s_i; \theta), \qquad (6)$$

where $P'(s_i; \theta)$ is the probability of s_i calculated with the term-based instances:

$$P'(s_i; \theta) = \prod_{t_k \in s_i} P(t_k; C(t_k), \mathbf{v}(s_i)), \qquad (7)$$

where $P(t_k; C(t_k), \mathbf{v}(s_i))$ is the probability of t_k in s_i. Then, $\ell'(\theta; S)$ is written as:

$$\ell'(\theta; S) = \sum_{s_i \in S} \{ \sum_{e_j \in s_i} \sum_{l=1}^{L^e - 1} \log P(c_l^e; \boldsymbol{\gamma}_l^e, \mathbf{x}_j^e) + \sum_{t_k \in s_i} \sum_{l=1}^{L^t - 1} \log P(c_l^t; \boldsymbol{\gamma}_l^t, \mathbf{x}_k^t) \},$$

where superscripts e and t indicate the calculations with element instances and term instances respectively.[3]

Now we derive the gradient of parameters for a single training instance. Two types of training instances from one session are processed separately in each iteration. We first proceed with $(e_j, C(e_j), s_i)$ and let $\ell(j, l) = \log P(c_l^e; \boldsymbol{\gamma}_l^e, \mathbf{x}_j^e)$. After combined with Eq. 4, $\ell(j, l)$ is written as:

$$\ell(j, l) = (1 - c_l^e) \log \{\sigma(\mathbf{x}_j^e \boldsymbol{\gamma}_l^e)\} + c_l^e \log \{1 - \sigma(\mathbf{x}_j^e \boldsymbol{\gamma}_l^e)\}. \tag{8}$$

With some derivations, the partial derivatives with respect to $\mathbf{x}_j^e$ and $\boldsymbol{\gamma}_l^e$ are as follows:

$$\frac{\partial \ell(j, l)}{\partial \mathbf{x}_j^e} = \{1 - c_l^e - \sigma(\mathbf{x}_j^e \boldsymbol{\gamma}_l^e)\} \boldsymbol{\gamma}_l^e, \tag{9}$$

$$\frac{\partial \ell(j, l)}{\partial \boldsymbol{\gamma}_l^e} = \{1 - c_l^e - \sigma(\mathbf{x}_j^e \boldsymbol{\gamma}_l^e)\} \mathbf{x}_j^e. \tag{10}$$

Therefore, the update formula of $\boldsymbol{\gamma}_l^e$ is:

$$\boldsymbol{\gamma}_l^e \leftarrow \boldsymbol{\gamma}_l^e + \eta \{1 - c_l^e - \sigma(\mathbf{x}_j^e \boldsymbol{\gamma}_l^e)\} \mathbf{x}_j^e, \tag{11}$$

where η is the learning rate. $\mathbf{x}_j^e$ is an intermediate vector Our aim is to learn $\mathbf{v}(e')$ for $e' \in C(e_j)$, to do so, $\mathbf{v}(e')$ is updated with the partial derivative of $\mathbf{x}_j^e$:

$$\mathbf{v}(e') \leftarrow \mathbf{v}(e') + \eta \sum_{l=1}^{L^e - 1} \{1 - c_l^e - \sigma(\mathbf{x}_j^e \boldsymbol{\gamma}_l^e)\} \boldsymbol{\gamma}_l^e. \tag{12}$$

Similarly, for a term-based instance $(t_k, C(t_k), s_i)$, let $\ell(k, l) = \log P(c_l^t; \boldsymbol{\gamma}_l^t, \mathbf{x}_k^e)$, and update formulae are:

$$\boldsymbol{\gamma}_l^t \leftarrow \boldsymbol{\gamma}_l^t + \eta \{1 - c_l^t - \sigma(\mathbf{x}_k^t \boldsymbol{\gamma}_l^t)\} \mathbf{x}_k^t, \tag{13}$$

$$\mathbf{v}(t') \leftarrow \mathbf{v}(t') + \eta \sum_{l=1}^{L^t - 1} \{1 - c_l^t - \sigma(\mathbf{x}_k^t \boldsymbol{\gamma}_l^t)\} \boldsymbol{\gamma}_l^t, \tag{14}$$

where $t' \in C(t_k)$. When updating the session vector $\mathbf{v}(s_i)$, both types of instances are considered:

$$\mathbf{v}(s_i) \leftarrow \mathbf{v}(s_i) + \eta \sum_{l=1}^{L^e - 1} \frac{\partial \ell(j, l)}{\partial \mathbf{x}_j^e} + \eta \sum_{l=1}^{L^t - 1} \frac{\partial \ell(k, l)}{\partial \mathbf{x}_k^t}. \tag{15}$$

The learning procedure for s2v can be easily derived by simplifying that of s2v+.

---

[3] One may notice that both $P(s_i; \theta)$ and $P'(s_i; \theta)$ are defined a s probability of s_i and they may be unequal. In fact, refer to Eqs. 2, 4, and 5, the probability of a session is calculated from element vectors and parameter vectors associated with the Huffman tree. Therefore, it is possible that different types of input vectors, term-based or element-based, output different values. We would not restrict $P(s_i; \theta) = P'(s_i; \theta)$ since such constraint will make the model less flexible in learning vectors for different elements. On the other hand, the session vector $\mathbf{v}(s_i)$, as an intermediary, softly aligns the dimensions of element vectors and term vectors.

6   Training Data and Case Study

6.1   Training Data

We employ a query log data set from Baidu search engine, including 10,413,491 unique queries, 13,126,252 URLs, 1,006,352 websites, and 3,965,539 terms (coming from queries and URL titles). Session boundaries are detected with a hybrid method of time-gap-based detection and task-based detection [3,12]: the interval of two consecutive queries is no more than 15 min; and the similarity between two consecutive queries is no less than a threshold. To calculate this similarity, we employ term vectors trained in a baseline system (CBOW of word2vec, described later) to represent query terms and the sum of them is used as query vector. The cosine similarity threshold is 0.5. In total, we collected 23,676,669 sessions, each session contains 2.1 queries and 2.3 clicks on average.

6.2   Case Study

Semantic Dimensions. We show salient elements and terms of three dimensions (manually entitled "Star", "Movie resource" and "Software resource"), generated by s2v+, in Table 1. These terms and elements have the largest values in these dimensions, meanwhile the frequency is ≥ 100. For "Star", five singers/actors from mainland China and Hong Kong are output as salient terms. The queries mainly search for the personal information of stars. For websites, the entertainment homepages of five top websites are listed. In "Movie resource", popular movie titles are output as salient terms and the queries are about movies' showtime and scheme song. Interestingly, although "Star" and "Movie resource" are related, our model generates different salient term sets and query sets for them, focusing on different aspects. Presumably, it is because searching stars and searching movies are two different information needs. The element-based and term-based training instances are generated from individual sessions, thus the two information needs are well identified in learning. The websites involved in these two needs are also different and can help differentiate them to some extent.

Learnt Vectors. The term vector "北京大学 (Peking University)" and the query vector "Peking University" learnt by s2v+ are given in Fig. 3. The two vectors are generally correlated well (cosine similarity is 0.591). Thus, we can reasonably derive the vector of an unseen query with term vectors. The two vectors also show some differences. The reason is that "Peking University" appears in queries or URL titles with diverse meanings, such as "EMBA program in Peking University" and "Peking University Health Science Center". For query "Peking University", the dominant information need is to find the university's homepage or encyclopedia page (Fig. 3).

Table 1. Salient elements and terms in three dimensions.

	Terms	Queries	Websites
Star	刘德华 (Andy Lau)	刘德华女儿 (daughter of Andy Lau)	ent.sina.com.cn
	周星驰 (Stephen Chow)	李连杰的师父 (master of Jet Li)	ent.qq.com
	黄渤 (Bo Huang)	周润发老婆 (wife of Yun-Fat Chow)	ent.ifeng.com
	李连杰 (Jet Li)	梁朝伟个人资料 (profile of Tony Leung)	yule.sohu.com
	周润发 (Yun-Fat Chow)	周杰伦女朋友 (girlfriend of Jay Chou)	ent.163.com
Movie resource	那些年 (You Are the Apple of My Eye)	阿凡达上映时间 (showtimes of Avatar)	www.mtime.com
	手机电影 (movie for smartphone)	无间道在线观看 (Infernal Affairs online watching)	movie.douban.com
	非诚勿扰 (If You Are the One)	非诚勿扰主题曲 (theme song of If You Are the One)	www.verycd.com
	影视在线 (online movie)	叶问上映时间 (showtimes of Ip Man)	www.1905.com
	阿凡达 (Avatar)	唐山大地震票房 (box office of Aftershock)	www.iqiyi.com
Software resource	迅雷看看 (Xunlei player)	酷狗音乐下载 (download Kugou music box)	www.wandoujia.com
	多特软件站 (Duote software)	豌豆荚使用 (how to use SnapPea)	www.onlinedown.net
	华军软件园 (Onlinedown software)	搜狐影音下载 (download Sohu player)	www.pconline.com.cn
	酷我音乐盒 (Kuwo music box)	华军软件园下载中心 (download center of Onlinedown)	www.skycn.com
	豌豆荚 (SnapPea)	天空下载站 (Skycn software)	www.zol.com.cn

Fig. 3. The term vector of "Peking University" and the query vector of "Peking University".

7 Quantitative Experimental Results

7.1 Settings

Variants of Our Framework. S2v+ can generate vectors of elements and terms (from queries and URL titles). According to how to use these vectors, we have three variants: **S2v+.A** directly uses element vectors; **S2v+.B** interpolates an element vector and the term vectors from this element. For instance, for query q, we first calculate the sum of its term vectors, then the sum is summed with $\mathbf{v}(q)$, and the result is used as the final vector of q; **S2v+.C** uses the sum of term vectors of an element as its vector, and it is applicable for both existing elements in the training data and new elements.

Comparison Systems. We employ the CBOW algorithm of word2vec[4] (w2v for short) as a baseline and run it on a corpus containing 1 billion Chinese Web pages (much larger than the training data used in our model), and a vector is generated for each term. PLSA [9] is another baseline, and we run it on a pseudo-document corpus generated from our training data. Each pseudo-document is composed of queries and URL titles of a training session, and topic vectors of terms are learnt.

7.2   Results for Query Similarity Prediction

We analyze our framework with a similar query ranking task to illustrate the behaviors of its variants. NDCG [11] is employed as the metric and 100 dimensions are used for all systems.

Task Description and Evaluation Data. Each testing query has 5 candidate queries, and the task is to rank the candidates according to their similarity with the testing query. Cosine similarity is calculated with the learnt vectors.

We employ an annotated data set containing 500 testing queries, each of which has 5 candidate queries. A Likert scale with three levels is employed to assess the candidates. Specifically, 3 means strongly relevant (e.g. "Bill Gates" and "founder of Microsoft"), 2 means relevant (e.g. "Bill Gates" and "Steve Jobs"), and 1 means irrelevant, (e.g. "Bill Gates" and "Spider-Man"). Each candidate is assessed by 3 assessors and the average score is rounded to the nearest relevance level. On average, each testing query has 1.7 strongly relevant candidates and 1.2 relevant candidates. These 500 testing queries are divided into observed part (**Q_obs**) and unobserved part (**Q_unobs**). Q_obs has 129 testing queries, each testing query and its candidate queries are observed in our training data. Q_unobs has 371 testing queries.

Analysis of S2v Results on Q_obs. For s2v, query vectors are directly learnt, and for the baselines, the vector of a query is obtained by summing its term vectors. The results of different methods are given in the left part of Table 2. S2v can outperform the baselines. Specifically, on NDCG@1, s2v outperforms PLSA and w2v by about 8 % (significant with $P < 0.01$ in paired t-test) and 2 % ($P < 0.05$), respectively. The reasons might be: (1) our training instances are generated from session graphs. In each graph, the elements have similar semantic meanings so that the contextual elements and the target element (i.e., e_j) in a training instance are semantically more cohesive. Such training instances bring in less noise; (2) PLSA and w2v generate query vectors by summing the term vectors, however, their term vectors are learnt without considering query and session semantics and cannot well derive query vector. In contrast, s2v directly generates query vectors; (3) s2v maintains a session vector, and the semantic of a session is normally less ambiguous than a query. Thus, the session vector is

---

[4] https://code.google.com/p/word2vec/.

Table 2. Results of query similarity prediction on Q_obs.

	w2v	PLSA	s2v	s2v+.A	s2v+.B	s2v+.C
NDCG@1	0.769	0.727	0.784	0.792	0.797	0.799
NDCG@3	0.804	0.786	0.824	0.830	0.834	0.835
NDCG@5	0.833	0.810	0.838	0.841	0.849	0.853

helpful to guide vector learning for queries by deriving more precise information need. W2v also performs well, and its large training corpus helps deal with sparse queries more effectively.

Analysis of S2v+ Results. Sparsity will hinder the effectiveness of learnt embeddings by s2v. In addition, if a query was not observed in the training data, s2v cannot learn a vector for it. S2v+ conducts vector learning for terms in a unified model. The learnt term vectors can be used in different variants as described in Sect. 7.1.

Results on Q_obs. To examine the effectiveness of s2v+, we first compare its variants with s2v on Q_obs and the results are given in the right part of Table 2. S2v+.A outperforms s2v by 1 % on NDCG@1 ($P < 0.05$). This shows that the unified learning in s2v+ generates better vector representation for queries. It is because the lower part of the network in Fig. 2(b) for term vector learning can help overcome the sparsity problem to some extent. Specifically, with the term-based learning part, the derived session vectors are more accurate since the sparsity problem of terms is less severe. As a result, accurate session vectors will help learn better query vectors. S2v+.B slightly outperforms s2v+.A, which shows using term vectors to interpolate the query vector can further solve the sparsity problem.

S2v+.C is the most effective one. It shows that the sum of term vectors generated by s2v+ can better derive the query vector. It is probably because the unified learning in s2v+ can better align the semantic meanings of queries and terms with the session vector as bridge. The term vectors from the baselines are not as effective as ours for deriving query vectors. S2v+.C performs better than s2v+.A and s2v+.B. The reason is that s2v+.A and s2v+.B use query vectors, but the sparsity problem affects the reliability of vectors of low-frequency queries. To have a closer look at the sparsity problem, we divide Q_obs into 5 equal buckets, A, B, C, D, and E, according to the descending order of frequency. Similarly, candidates queries are also divided into 5 buckets, A', B', C', D', and E'. We evaluate the variants in different intervals and the results are shown in Table 3. In each cell, the results of s2v+.A, s2v+.B, and s2v+.C are given in the upper, middle, and lower positions. The largest value is underscored, in bold and green, the smallest value is in italic and red. As shown in the upper left of Table 3, S2v+.A and s2v+.B perform better for more frequent queries, When the queries become less frequent, moving toward the lower right corner, the performance

Table 3. Effect of query frequency.

	A' [0–20%]			B' [20%–40%]			C' [40%–60%)			D' [60%–80%)			E' [80%–1]		
A [0–20%]	**0.799**	0.798	*0.791*	0.799	**0.801**	*0.796*	*0.793*	0.794	**0.799**	*0.792*	0.801	**0.803**	*0.790*	0.794	**0.796**
B [20%–40%]	**0.797**	0.796	*0.794*	0.796	**0.797**	*0.793*	**0.790**	0.789	*0.786*	*0.788*	0.796	**0.798**	*0.786*	0.790	**0.797**
C [40%–60%]	*0.791*	*0.791*	**0.792**	*0.793*	0.795	**0.796**	*0.790*	0.796	**0.799**	*0.787*	0.797	**0.802**	*0.784*	0.796	**0.801**
D [60%–80%]	*0.785*	0.791	**0.799**	*0.783*	0.796	**0.797**	0.783	*0.782*	**0.796**	*0.781*	0.785	**0.805**	*0.780*	0.789	**0.800**
E [80%–1]	*0.786*	0.792	**0.807**	*0.779*	0.787	**0.803**	*0.778*	0.790	**0.792**	*0.776*	0.782	**0.790**	*0.771*	0.797	**0.798**

of s2v+.A and s2v+.B declines. Meanwhile, s2v+.C is not affected much and outperforms the other two.

Results on Q_unobs. We also examine the performance of s2v+.C on Q_unobs and compare it with w2v and PLSA baselines. The results are given in Table 4. S2v+.C achieves 8% and 4% improvements ($P < 0.05$) on NDCG@1 compared with PLSA and w2v, respectively. This demonstrates term vectors generated with our model are more effective due to the unified learning and introducing the session vector. Combining the results in Tables 2 and 4, s2v+.C is the most effective system.

7.3 Results for URL Ranking

Setup. Here we examine the performance of our model in the task of URL ranking. The relevance between a query and its candidate URLs is computed as cosine similarity of their vectors. Still, a query vector is obtained by summing the vectors of its terms. For each URL, its vector is obtained by summing the vectors of terms in its title. We introduce another baseline BM25 [19] which is a ranking function commonly used to rank documents according to their relevance to a search query. Specifically, our BM25 baseline is a revision of the original BM25 formula to conduct normalization of term frequency according to [22] and revise inverse document frequency according to [5]. As discussed above, s2v+.C is the most effective variant and it also has better adaptability for unseen data. In addition, URL vectors also face the sparsity problem. Therefore, we conduct the comparison between s2v+.C and baselines.

Evaluation Data. This data set has 1,000 queries of various length and popularity. On average, each query has 19.8 marked URLs retrieved by the query. A Likert scale with five levels is employed to assess the relevance of each URL.

Results. The results are given in Table 5. All vector-based methods can outperform BM25. Our model achieves the best results in all cases. Specifically, it outperforms the baselines by about 4% to 9% on NDCG@1 ($P < 0.01$). Recall that we train s2v+ with term-based training instances (together with element-based) from both URL titles and queries. Presumably, such mixed instances make the learnt term vectors more capable for capturing the similarity between queries and URLs. Another reason might be that s2v+ jointly considers different types of elements (such as queries and URLs) in learning, thus the learnt term vectors can implicitly encode the semantic similarity among these elements to some extent.

Table 4. Results of query similarity on Q_unobs.

	s2v+.C	w2v	PLSA
NDCG@1	0.798	0.766	0.736
NDCG@3	0.836	0.812	0.787
NDCG@5	0.852	0.837	0.815

Table 5. Results for ranking the result URLs.

	s2v+.C	w2v	PLSA	BM25
NDCG@1	0.611	0.587	0.576	0.559
NDCG@3	0.632	0.615	0.607	0.582
NDCG@5	0.640	0.631	0.630	0.616

7.4 Results for Website Similarity Prediction

Setup. In this task, the vectors from different systems are employed to calculate website similarity. For PLSA and w2v, the vector of a website is obtained by summing the terms vectors of its homepage title. Our model has three variants, namely, s2v.S, s2v+.S, and s2v+.T. S2v.S and s2v+.S use the learnt website vectors directly. S2v+.T uses website vectors by summing term vectors, as is done for baselines.

Evaluation Data. This data set contains 500 testing websites with different popularity. Each testing website has 5 candidate websites. A Likert scale with three levels is employed to assess the candidate websites. Specifically, 3 means strongly relevant (e.g. "sports.sina.com.cn" and "sports. qq.com"), 2 means relevant (e.g. "sports.sina.com.cn" and "www.sina.com.cn"), and 1 means irrelevant, (e.g. "sports. sina.com.cn" and "mil.qq.com"). On average, each testing website has 1.6 strongly relevant candidates, and 1.4 relevant candidates. All the testing and candidate websites are covered by our training data set.

Results. The results are given in Table 6. The variants of our model outperform the baselines. Specifically, s2v+.S achieves 3 % to 10 % improvements ($P < 0.05$) on NDCG@1 compared with baselines. Among the variants, s2v+.S and s2v.S perform better than s2v+.T. It shows that the directly learnt website vectors are more effective than summing term vectors of titles for similarity prediction. This observation is different from that of queries. One reason might be that the sparsity problem for websites is not severe in training data. Another possible reason is that homepage titles, such as "The best car website in China", contain irrelevant terms.

Table 6. Results for the prediction of website similarity.

	s2v+.S	s2v+.T	s2v.S	w2v	PLSA
NDCG@1	0.794	0.786	0.791	0.772	0.719
NDCG@3	0.855	0.843	0.849	0.832	0.763
NDCG@5	0.883	0.880	0.881	0.870	0.794

8   Conclusions and Future Work

In this paper, we proposed a framework to uncover a semantic space for Web search. We develop two neural-network-based models, i.e. session2vec and session2vec+, to learn vectors for elements and terms. Compared with previous studies, our framework can perform hidden semantic learning for different types of elements. Moreover, our models enable the learning of vector representation on graph data. Experimental results indicate that the learnt vectors are helpful for a few tasks in Web search. For the future work, one direction is to extend our framework to model the interest profile of users. Another direction is to enhance the session graph with the information of click order and dwell time. A third direction is to derive the real-time information need with the partial information of the current session.

References

1. Baroni, M., Dinu, G., Kruszewski, G.: Don't count, predict! A systematic comparison of context-counting vs. context-predicting semantic vectors. In: ACL, pp. 238–247 (2014)
2. Bengio, Y., Ducharme, R., Vincent, P., Janvin, C.: A neural probabilistic language model. J. Mach. Learn. Res. **3**, 1137–1155 (2003)
3. Bing, L., Lam, W., Wong, T.L., Jameel, S.: Web query reformulation via joint modeling of latent topic dependency and term context. ACM Trans. Inf. Syst. **33**(2), 1–38 (2015)
4. Craswell, N., Szummer, M.: Random walks on the click graph. In: SIGIR, pp. 239–246 (2007)
5. Fang, H., Tao, T., Zhai, C.: A formal study of information retrieval heuristics. In: SIGIR, pp. 49–56 (2004)
6. Gao, J., He, X., Nie, J.Y.: Clickthrough-based translation models for web search: from word models to phrase models. In: CIKM (2010)
7. Gao, J., Toutanova, K., Yih, W.T.: Clickthrough-based latent semantic models for web search. In: SIGIR, pp. 675–684 (2011)
8. Grbovic, M., Djuric, N., Radosavljevic, V., Silvestri, F., Bhamidipati, N.: Context- and content-aware embeddings for query rewriting in sponsored search. In: SIGIR 2015 (2015)
9. Hofmann, T.: Unsupervised learning by probabilistic latent semantic analysis. Mach. Learn. **42**(1–2), 177–196 (2001)
10. Huang, P.S., He, X., Gao, J., Deng, L., Acero, A., Heck, L.: Learning deep structured semantic models for web search using clickthrough data. In: CIKM, pp. 2333–2338 (2013)
11. Järvelin, K., Kekäläinen, J.: Cumulated gain-based evaluation of IR techniques. ACM Trans. Inf. Syst. **20**(4), 422–446 (2002)
12. Jones, R., Klinkner, K.L.: Beyond the session timeout: automatic hierarchical segmentation of search topics in query logs. In: CIKM 2008, pp. 699–708 (2008)
13. Le, Q.V., Mikolov, T.: Distributed representations of sentences and documents. In: ICML, pp. 1188–1196 (2014)
14. Lee, S., Hu, Y.: Joint embedding of query and ad by leveraging implicit feedback. In: EMNLP, pp. 482–491 (2015)

15. Mikolov, T.: Statistical language models based on neural networks. Ph.D. thesis (2012)
16. Mikolov, T., Sutskever, I., Chen, K., Corrado, G.S., Dean, J.: Distributed representations of words and phrases and their compositionality. In: NIPS, pp. 3111–3119 (2013)
17. Pennington, J., Socher, R., Manning, C.D.: Glove: Global vectors for word representation. In: EMNLP, pp. 1532–1543 (2014)
18. Ren, X., Wang, Y., Yu, X., Yan, J., Chen, Z., Han, J.: Heterogeneous graph-based intent learning with queries, web pages and wikipedia concepts. In: WSDM, pp. 23–32 (2014)
19. Robertson, S.E., Walker, S., Jones, S., Hancock-Beaulieu, M., Gatford, M.: Okapi at TREC-3. In: TREC, pp. 109–126 (1994)
20. Schwenk, H.: Continuous space language models. Comput. Speech Lang. **21**(3), 492–518 (2007)
21. Shen, Y., He, X., Gao, J., Deng, L., Mesnil, G.: A latent semantic model with convolutional-pooling structure for information retrieval. In: CIKM, pp. 101–110 (2014)
22. Singhal, A., Buckley, C., Mitra, M.: Pivoted document length normalization. In: SIGIR, pp. 21–29 (1996)
23. Socher, R., Bengio, Y., Manning, C.D.: Deep learning for NLP (without magic). In: Tutorial Abstracts of ACL 2012, p. 5 (2012)
24. Wu, W., Li, H., Xu, J.: Learning query and document similarities from click-through bipartite graph with metadata. In: WSDM, pp. 687–696 (2013)
25. Yih, W.t., Toutanova, K., Platt, J.C., Meek, C.: Learning discriminative projections for text similarity measures. In: CoNLL, pp. 247–256 (2011)

TLINE: Scalable Transductive Network Embedding

Xia Zhang$^{(\boxtimes)}$, Weizheng Chen, and Hongfei Yan

Peking University, Beijing, China
zhangxia9403@gmail.com, cwz.pku@gmail.com, yhf1029@gmail.com

Abstract. Network embedding is a classical task which aims to project a network into a low-dimensional space. Currently, most existing embedding methods are unsupervised algorithms, which ignore the useful label information. In this paper, we propose TLINE, a semi-supervised extension of LINE algorithm. TLINE is a transductive network embedding method, which optimizes the loss function of LINE to preserve both local and global network structure information, and applies SVM to maximize the margin between the labeled nodes of different classes. By applying the edge-sampling and the negative sampling techniques in the optimizing process, the computational complexity of TLINE is reduced. Thus TLINE can handle the large-scale network. To evaluate the performance in node classification task, we test our methods on two real world network datasets, which are Citeseer and DBLP. The experimental result indicates that TLINE outperforms state-of-the-art baselines and is suitable for large-scale networks.

Keywords: Network embedding · Node classification · Transductive learning

1 Introduction

Life is full of information. The links between the information form all sorts of information networks, such as social network formed by people's interactions on social media, citation network generated by the reference relationship between the papers in academic science and the famous WWW (World Wide Web). The basic composition unit of network is a node, which can be a user, a paper, or a webpage. Apperently, the edge has different meaning in different networks.

Network embedding is a very important component of network analysis and study. The large scale and high dimension network can be mapped to a low dimensional space for certain optimization goal. The embedding node vectors preserve the original network's global features and local features, and have a lot more than the network node original representation [3]. After learning the embeddings of nodes, the embedding vectors are applied into various important data mining tasks, like node classification [19], network visualization [14] and link prediction [15].

© Springer International Publishing AG 2016
S. Ma et al. (Eds.): AIRS 2016, LNCS 9994, pp. 98–110, 2016.
DOI: 10.1007/978-3-319-48051-0_8

A widely-studied problem in network analysis area is the node classification task, which can be regarded as learning a mapping function from the nodes to a set of pre-defined and non-overlapping categories. The mapping relationship is the classifier. When the classification task is applied to the network, traditional methods embed the network first, and then use some algorithms like Support Vector Machine (SVM) [13] to do the classification. This is typically a kind of unsupervised learning method. This label attribute should be considered as well for it can distinguish different nodes, which is usually ignored in the previous algorithms.

Network embedding learning is a very challenging research, and its difficult mainly consists of the following two points: on the one hand, for the real network contains a huge amount of data, the learning algorithm should handle the large-scale network. Unfortunately, many existing network embedding algorithms [1, 4, 17] perform well on small networks, but could not deal with large scale networks due to their high computational complexity. On the other hand, adding label information to the embedding learning process may improve the discriminability of node embeddings, but it's a worth thinking problem about where and how to add the label information.

In this paper, inspired by the LINE (Large Scale Information Network Embedding proposed by [16]) method, we propose a new transductive algorithm named TLINE, which uses the SVM (support Vector Machine) as the training classifier. Unlike previous unsupervised network embedding methods, the node embeddings and the SVM classifier are optimized simultaneously in TLINE. By using the edge sampling and the negative sampling techniques in the stochastic gradient descent process, the algorithm complexity of TLINE is greatly reduced. So our model is able to learn embeddings of the large networks at a very small time and memory cost. We test TLINE algorithm on Citeseer and DBLP datasets. The performance of TLINE is compared with three competitive baselines, including two popular unsupervised baselines, Deepwalk and LINE, and a state-of-the-art transductive method, MMDW (Max-Margin DeepWalk) [18]. The experimental results show that the performance of TLINE in node classification task is significantly better than other baselines. The stability of TLINE is also shown in the parameter sensitivity experiments.

The rest of this paper are organized as follows. Section 2 discusses other related work about this problem. Section 3 introduces some notations that will be used in the following paper. And Sect. 4 introduces TLINE model which is inspired by LINE and SVM. Section 4 talks about our experiments of TLINE, and compares the results with other algorithms. Section 5 draws a conclusion about this paper and provides the direction for future work.

2   Related Work

Network embedding aims to create feature representations in low-dimensional space, which preserves the original network structure.

Traditionally, PCA (Principle Component Analysis) and SVD (Singular value decomposition) are the common methods to project data into low-dimensional space. And many other primitive network representation learning methods, such as MDS (multi-dimensional scaling) [4], LLE [12], Laplacian Eigenmap [1] and DGE [3], are also based on spectral factorization. And there is still another kind of method based on the probabilistic graphical models. The key point of this kind of algorithm is modeling generative process of the network and associated texts information by sampling. Some representative algorithms are Link-PLSA-LDA [9], RTM [2] and PLANE [7]. However, the high computational complexity prevents them from being applied to large scale networks.

Recently, inspired by the widely used distributed representation learning techniques in NLP domain, like Skip-Gram [8], researchers propose some novel network embedding methods to learn distributed representations for networks. DeepWalk [10] is proposed by Perozzi and his colleagues, which uses the truncated random walks on the networks to generate node sequences and feeds the sequences to the Skip-Gram model as pseudo sentences to obtain node representations. In order to handle the large-scale networks, Tang et al. propose LINE [16], which optimizes the objective function to preserve both the local and the global network structures. However, both DeepWalk and LINE are unsupervised models, which means that they are not able to utilize the label information or the category information in the network. Actually, label or category information is common in network data, such as the conference or journals a paper publish on in a paper citation network, or the affiliation of an author in a coauthor network. Therefore, the distinguishability of the learnt representations is limited in these unsupervised frameworks.

To take advantage of the label information, semi-supervised learning approaches are employed to learn node representations. LSHM [5] and MMDW [18] are two representative methods. LSHM can be applied in heterogeneous networks, which learns node representations in a common latent space for all the different node types. MMDW utilizes the label information and max-margin principle to learn node representations. However, MMDW is hard to be applied for large networks because it is a unified learning framework based on matrix factorization.

Our motivation is optimizing the loss function of LINE and applying SVM at the same time to make full use of the label information in networks.

3   Problem Formulation

For a smoother and easier read, we first introduce some notations which will be used in this paper. Consider a partly labeled network $G = (V, E)$, V is the set of nodes and E is the set of edges. For each edge $e_{i,j} \in E$, $w_{i,j}$ is the weight of the edge. $\{v_1, v_2, ..., v_L\}$ is used to denote the labeled nodes, while $\{v_{L+1}, ..., v_{L+U}\}$ is the unlabeled nodes. And we also assume there are K label types in the network. If v_i is in class k, we set $y_i^k = 1$, otherwise $y_i^k = -1$.

Traditional way for predicting labels of the unlabeled nodes based on unsupervised learning methods have two steps, which are embedding and classification. Embedding means to learn a vector in low-dimension space $\mathbb{R}^d$, where

$d \ll |V|$. The traditional way is only embedding the node without considering the label information, and in the next step, we use the embeddings of the labeled nodes to train a classifier and make predictions for the unlabeled nodes. But for these transductive learning methods, given the training set $\{(v_i, y_i)\}, i = 1, 2, ..., L$ and testing set $\{v_j\}, j = L + 1, ..., L + U$, the goal of transductive learning is to find the node representation u_v and classification function $f : u_v \rightarrow y$ which can have a good performance on training set. The difference between the traditional unsupervised way and the transductive way is that transductive manner merges the two steps embedding nodes and training the classifier into one step.

4   Model

4.1   Large Scale Information Network Embedding

In the real world network, the direct relations between different nodes observed by us are actually a small part of the network information. If we only take the edges between nodes into account, there will be a considerable proportion of the information loss. From a global view, in the social network, if two people have many mutual friends, even they are not friends, they are likely to get to know each other through one mutual friend and become friends because of the same hobbies or interests. This global network structure is also called as second-order proximity.

To preserve the global structure of information networks, Tang [16] propose an algorithm named LINE (Large Scale Information Network Embedding), which uses $p_u = (w_{u,1}, ..., w_{u,|V|})$ to denote that the connect situation of v_u to all other $|V| - 1$ nodes and uses the similarity of p_u and p_v to measure the proximity of the global network structure.

Each node will be treated as a specal context, and the node with the similar contexts are assumed to have the close embeddings. So every node has two roles, one is the node itself, denoted as a vector u_i, the other one represents the impact on the other nodes as a context, which is denoted as a vector u'_i. For every directed edge $e_{i,j}$, we can define the probability of v_j's context generated by v_i as:

$$p_2(v_j|v_i) = \frac{exp(u'^T_j \cdot u_i)}{\sum_{k=1}^{|V|} \exp(u'^T_k \cdot u_i)}. \tag{1}$$

So we assume that the learned embeddings of two nodes will be similar to each other if they have the similar distribution of contexts. Then the empirical distribution of $p_2(v_j|v_i)$ is defined as:

$$\hat{p_2}(v_j|v_i) = \frac{w_{i,j}}{d_i}$$
$$= \frac{w_{i,j}}{\sum_{k \in N(i)} w_{i,k}} \tag{2}$$

where $w_{i,j}$ presents the weight of the edge $e_{i,j}$, d_i denotes the out-degree of node v_i, and $N(i)$ is the set of the nodes which have the edge where the starting point is v_i. In order to preserve the global structure of the information network, the algorithm should resemble p_2 and $\hat{p}_2$ as closely as possible, which also means that we would minimize the KL distance:

$$O_{line2} = D_{KL}(\hat{p}_2||p_2).\tag{3}$$

After omiting some constants, we have the final objective function:

$$O_{line2} = - \sum_{(i,j)\in E} w_{i,j} \log p_2(v_j|v_i)\tag{4}$$

After using the negative sampling technique to reduce the computational complexity, the Eq. (4) is rewritten as:

$$O_{line2} = - \sum_{(i,j)\in E} \left\{ \log \sigma(\boldsymbol{u'}_j^T \cdot \boldsymbol{u}_i) + \sum_{m=1}^{M} E_{v_n \sim P_n(v)} \left[\log \sigma(-\boldsymbol{u'}_n^T \cdot \boldsymbol{u}_i) \right] \right\},\tag{5}$$

where $P_n(v) \propto (d_v)^{0.75}$ is the noisy node distribution, M is the number of negative edges, and σ presents the sigmoid function. Just like the second-order proximity, the first-order proximity is defined as below (See [16] for details):

$$O_{line1} = - \sum_{(i,j)\in E} w_{i,j} \log \sigma(\boldsymbol{u}_i^T \cdot \boldsymbol{u}_j)\tag{6}$$

4.2 Classification Based on Support Vector Machine

For the binary classification problem of the label k, linear support vector machine is equivalent to the optimization problem as below:

$$\min_{\boldsymbol{\omega}^k} \sum_{i=1}^{L} [1 - y_i^k \boldsymbol{\omega}^{k^T} \boldsymbol{u}_i]_+ + \lambda \| \boldsymbol{\omega}^k \|^2\tag{7}$$

where the first part is the empirical loss, and we use hinge loss function $\mathscr{L}(y_i^k \boldsymbol{\omega}^{k^T} \boldsymbol{u}_i) = [1 - y_i^k \boldsymbol{\omega}^{k^T} \boldsymbol{u}_i)]_+$ here. The second part is the regularization term and is expressed as the L2-norm with coefficient λ of the parameter vector $\boldsymbol{\omega}^k$ for label k.

Binary classification is just a special case of multi-class classification. We can also expand the optimization problem to multi-class classification:

$$O_{svm} = \sum_{i=1}^{L} \sum_{k=1}^{K} \max(0, 1 - y_i^k \boldsymbol{\omega}^{k^T} \boldsymbol{u}_i) + \lambda \| \boldsymbol{\omega}^k \|^2.\tag{8}$$

4.3   Transductive Network Embedding

Given a network where only some of the nodes have labels, the task is to tag the label to the unlabeled nodes. The traditional way have two parts, which can be clearly seen in Fig. 1, the first step is embedding the node in the information network to a low-dimensional vector, and the second step is using the training set to train the classifier and then do the classification.

Fig. 1. Traditional unsupervised embedding and classification

Transductive embedding learning is a semi-supervised learning algorithm. Just as Fig. 2 shown below, the embedding learning and classifier training are proceed simultaneously. The process of embedding learning and classification has influence on each other. As a result, the information of labels will contribute to the quality of node low-dimensional vectors, and the embedding of nodes also have influence on the parameters of the classifier. So the node embedding is more explicit, and the meaning of it is richer.

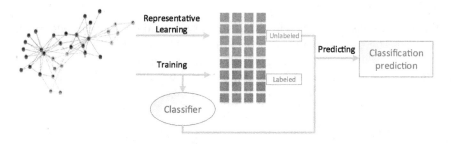

Fig. 2. Transductive embedding learning and classification

In order to have a better expression of the network structure and meanwhile improve the results of classification, we combine LINE and SVM together, which means that

$$O_{TLINE} = O_{line} + \beta O_{svm} \qquad (9)$$

where β is the trade-off parameter used to balance LINE and SVM. For the second-order proximity, we substitute loss function (5) and loss function (8) into the formula (9). Finally we have the objective function of TLINE(2nd) as:

$$
\begin{aligned}
O_{TLINE(2nd)} =& O_{line2} + \beta O_{svm} \\
=& \sum_{(i,j)\in E} w_{i,j} \Bigg\{ -\log \sigma(\boldsymbol{u'}_j^T \cdot \boldsymbol{u}_i) - \sum_{m=1}^{M} E_{v_n \sim P_n(v)} \big[\log \sigma(-\boldsymbol{u'}_n^T \cdot \boldsymbol{u}_i) \big] \\
& + \beta \mathbf{I}(i \leq L) \Big(\sum_{m=1}^{M+1} \sum_{k=1}^{K} \max(0, 1 - y_i{}^k \boldsymbol{\omega}^{k^T} \boldsymbol{u}_i) + \lambda \| \boldsymbol{\omega} \|^2 \Big) \Bigg\}
\end{aligned}
\tag{10}
$$

The same as the first-order proximity:

$$
\begin{aligned}
O_{TLINE(1st)} =& O_{line1} + \beta O_{svm} \\
=& \sum_{(i,j)\in E} w_{i,j} \Bigg\{ -\log \sigma(\boldsymbol{u}_j^T \cdot \boldsymbol{u}_i) \\
& + \beta \mathbf{I}(i \leq L) \sum_{m=1}^{M+1} \Big(\sum_{k=1}^{K} \max(0, 1 - y_i{}^k \boldsymbol{\omega}^{k^T} \boldsymbol{u}_i) + \lambda \| \boldsymbol{\omega} \|^2 \Big) \\
& + \beta \mathbf{I}(j \leq L) \Big(\sum_{k=1}^{K} \max(0, 1 - y_j{}^k \boldsymbol{\omega}^{k^T} \boldsymbol{u}_j) + \lambda \| \boldsymbol{\omega} \|^2 \Big) \\
& - \sum_{m=1}^{M} E_{v_n \sim P_n(v)} \Big[\log \sigma(-\boldsymbol{u}_n^T \cdot \boldsymbol{u}_i) \\
& \qquad - \beta \mathbf{I}(n \leq L) \Big(\sum_{k=1}^{K} \max(0, 1 - y_n{}^k \boldsymbol{\omega}^{k^T} \boldsymbol{u}_n) + \lambda \| \boldsymbol{\omega} \|^2 \Big) \Big] \Bigg\}
\end{aligned}
\tag{11}
$$

We use ASGD [11] (asynchronous stochastic gradient descent algorithm) to optimize the objective function of TLINE(1st) and TLINE(2nd). And the learning rate is dynamically computed by using the method mentioned in [16]. Specifically, the learning rate $\rho_0 = 0.025$ in the beginning, then it changes as $\rho_t = \rho_0(1 - \frac{t}{T})$, where T is the amount of sampling edges.

5   Experiments

5.1   Data Sets

We select the following two typical datasets to evaluate our approaches.

- **Citeseer.** Citeseer is a paper citation network data used in [18]. It contains 3312 nodes, 4732 edges and 6 labels. It is an unweighted network, where the citation relationships between papers form a typical network.
- **DBLP.** DBLP is a coauthor network data used in [16]. It contains 18058 nodes, 103011 edges, and 3 labels. The co-author relationships between authors form the network. Two nodes are connected by an edge if and only if they are

coauthors. Compared with Citeseer data, the DBLP network is a weighted network, and the weight of the edge is encoded by the number of co-authored papers.

5.2   Compared Methods

In the experiments, we compare the following 6 methods to exam the performance of our approaches.

- **DeepWalk** [10]. DeepWalk is an unsupervised method proposed by Perozzi et al. in 2014, which learns latent representations of vertices in a network. We set parameters as follows, the sliding window size $w = 10$, the length of each node sequence t = 40, and the number of node sequences for each node $\gamma =$ 80. We use liblinear to do the classification, while LINE(1st), LINE(2nd) and MMDW also use this lib.
- **LINE(1st)** [16]. We employ the LINE with first-order proximity for comparison. We sample 5 million edges for Citeseer data, and 50 million edges for DBLP data. The edge sampling for LINE(2nd), TLINE(1st) and TLINE(2nd) is the same. We set the dimension $d = 200$ for LINE(1st), LINE(2nd) and TLINE(2nd).
- **LINE(2nd)** [16]. LINE algorithm with second-order proximity, which assumes that nodes with similar neighbors distributions will have similar embedding vectors.
- **MMDW** [18]. MMDW is a semi-supervised transductive network learning method based on matrix decomposition. MMDW employs the labeling information and max-margin principle to learn vertex representations. MMDW also use SVM as its classifier. We use the code provided by [18] and set the dimension $d = 200$.
- **TLINE(1st)**. TLINE with first-order proximity. We set $\beta = 0.5$, $\lambda = 0.02$, the dimension $d = 10$.
- **TLINE(2nd)**. TLINE with second-order proximity. We set $\beta = 0.5$, $\lambda = 0.02$.

5.3   Node Classification

We evaluate the quality of the node embeddings learned by different models when the training ratios vary from 10 % to 90 %. Tables 1 and 2 show the classification accuracies with different training ratios on the two datasets. All results listed are averaged over 20 runs. From the results, we have following observations:

(1) The proposed method TLINE consistently outperforms all the baseline methods on both the two datasets. It is worth noting that the superiority of TLINE tends to increase with more training data.
(2) MMDW fails to generate results on DBLP data when our workstation has only 64 G memory, which means that it is difficult for the most promising baseline method MMDW to handle large scale networks because of memory constraint. In contrast, the TLINE consumes only 0.5 G memory for DBLP network on the same machine. Our method can scale to large networks and performs well.

Table 1. Accuracy (%) of node classification on Citeseer data.

% labeled nodes	10 %	20 %	30 %	40 %	50 %	60 %	70 %	80 %	90 %
DeepWalk	52.62	56.62	56.63	57.42	57.48	57.27	58.47	56.81	55.05
LINE(1st)	45.70	51.22	54.55	56.28	57.02	58.05	58.94	59.77	59.37
LINE(2nd)	46.68	51.23	53.36	55.41	57.55	58.14	58.37	59.00	59.04
MMDW	54.72	59.64	62.60	64.10	65.83	68.96	69.56	69.58	69.16
TLINE(1st)	**49.33**	**55.91**	**60.38**	**63.66**	**65.55**	**67.17**	**67.54**	**67.06**	**63.58**
TLINE(2nd)	**53.72**	**59.22**	**61.95**	**64.97**	**68.03**	**69.37**	**70.78**	**72.50**	**73.80**

Table 2. Accuracy (%) of node classification on DBLP data.

% labeled nodes	10 %	20 %	30 %	40 %	50 %	60 %	70 %	80 %	90 %
DeepWalk	83.18	83.64	84.06	84.10	84.46	84.16	84.51	84.28	84.94
LINE(1st)	77.93	79.77	80.16	80.46	80.61	80.75	80.74	81.08	81.35
LINE(2nd)	79.46	80.29	80.66	81.05	81.22	81.10	81.45	81.14	81.25
MMDW	-	-	-	-	-	-	-	-	-
TLINE(1st)	**76.64**	**81.20**	**83.56**	**85.24**	**86.71**	**87.56**	**88.23**	**88.32**	**88.05**
TLINE(2nd)	**81.08**	**83.57**	**84.89**	**85.72**	**86.31**	**86.84**	**87.23**	**87.43**	**87.64**

(3) Transductive network embedding methods perform better than unsupervised network embedding methods in most cases. For example, compared with LINE(1st), TLINE(1st) obtains around 7 % improvement on the Citeseer data and nearly 10 % improvement on the DBLP data. It suggests that label information is crucial to network representation learning and can improve the classification accuracy.

5.4 Parameter Sensitivity

This section presents a series of experiments about parameter sensitivity in TLINE model. The Figs. 3 and 4 show the sensitivity experiment results of the trade-off parameter β and the regularization coefficient λ for TLINE(1st) and TLINE(2nd). In the experiments of TLINE(1st) and TLINE(2nd) on Citeseer dataset, λ varies from 0.001 to 10 while β varies from 0.001 to 10. With the increasing of λ, the Micro-F1 has an obvious increase from the very beginning but a slight decrease at end. And in DBLP dataset, when $\lambda \in [0.01, 1]$ and $\beta \in [10, 100]$, the Micro-F1 of TLINE(1st) and TLINE(2nd) both have a relatively good performance. Through this experiment, we find that the two parameters have some correlation, and when β gets a better value, λ is less sensitive. In the rest experiments of this paper, we set $\beta = 0.5$ and $\lambda = 0.02$ to get better performances on both datasets.

From Fig. 5 we can see, TLINE(1st) is a little sensitive to the vector space dimension in Citeseer while TLINE(2nd) is insensitivite in both datasets. It

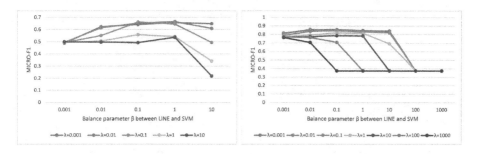

Fig. 3. Parameter β and λ sensitivity study of TLINE(1st) on different datasets (left: Citeseer, right: DBLP).

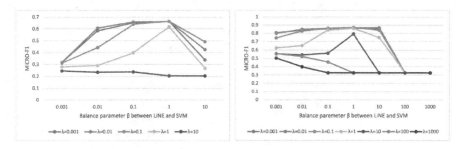

Fig. 4. Parameter β and λ sensitivity study of TLINE(2nd) on different datasets (left: Citeseer, right: DBLP).

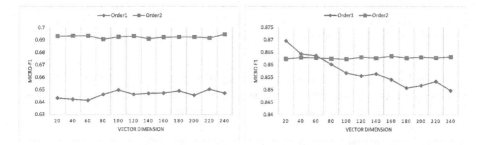

Fig. 5. Parameter d sensitivity study of TLINE(1st) and TLINE(2nd) on different datasets (left: Citeseer, right: DBLP).

means TLINE(2nd) is more universal and robust than TLINE(1st) for vector dimensions. In the other experiments of this paper, we set $d = 10$ for TLINE(1st) and $d = 200$ for TLINE(2nd).

5.5 Network Visualization

Visualization is an intuitive way to verify whether the learnt representations is discriminative. In this section, we use t-SNE [6] to display the 2D representations

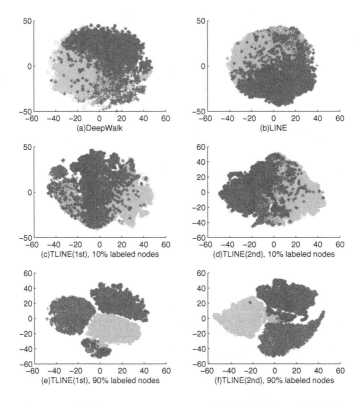

Fig. 6. Visualization of 2D representations on the DBLP data. Red, blue and green represent authors labeled "data mining", "machine learning" and "computer vision" respectively. (Color figure online)

of vertices. Figure 6 shows the results of DeepWalk, LINE and TLINE on DBLP data. In this figure, each dot represents a vertex while colors are encoded into categories. In this case, we choose red, blue and green to indicate authors labeled "data mining", "machine learning" and "computer vision" respectively.

From Fig. 6, we observe that neither DeepWalk nor LINE create clear boundaries among three different communities, and there are plenty of overlaps in Fig. 6(a) and (b). However, Fig. 6(c–f) indicates that the boundaries are becoming clear gradually with the increase of the training ratio. Particularly, we can obtain well-separated clusters when the training ratio equals 0.9, as shown in Fig. 6(e) and (f). The node embeddings learnt by TLINE are much more discriminative, which indicates effectiveness and improvements of our method.

6   Conclusion

This paper proposes a new transductive algorithm named TLINE which is inspired by LINE. TLINE uses the SVM classifier in node embeddings learning process to improve the nodes' distinguish degree. By adopting the stochastic gradient descent algorithm, the edge sampling and the negative sampling techniques to our method, the complexity of TLINE is greatly reduced, so TLINE is able to handle large scale network at a very small time and memory cost.

At the end of the paper, we test TLINE and other state-of-the-art baseline algorithms on Citeseer and DBLP datasets. Compared with the newest semi-supervised learning algorithm MMDW, TLINE achieves significantly higher accuracy in node classification task. And the parameter sensitivity experiments also show the stability of TLINE.

For future work, we may shift our focus to the heterogeneous network. In real world, homogeneous networks are just a small part of various information networks, while heterogeneous networks are more common. And we may also have a try to optimize our classification algorithm. In recent years, the deep learning techniques, such as convolution neural network and recursive neural network, outperform the traditional classification algorithms on various categorization tasks. So the next step of our work is to replace the old classifier SVM with the more complex deep neural networks.

Acknowledgments. This work is supported by 973 with Grant No. 2014CB340400, NSFC with Grant No. U1536201 and NSFC with Grant No. 61472013. And we also thank the three anonymous reviewers for their comments.

References

1. Belkin, M., Niyogi, P.: Laplacian eigenmaps and spectral techniques for embedding and clustering. In: NIPS, vol. 14, pp. 585–591 (2001)
2. Chang, J., Blei, D.M.: Relational topic models for document networks. In: International Conference on Artificial Intelligence and Statistics, pp. 81–88 (2009)
3. Chen, M., Yang, Q., Tang, X.: Directed graph embedding. In: IJCAI, pp. 2707–2712 (2007)
4. Cox, T.F., Cox, M.A.: Multidimensional Scaling. CRC Press, Boca Raton (2000)
5. Jacob, Y., Denoyer, L., Gallinari, P.: Learning latent representations of nodes for classifying in heterogeneous social networks. In: Proceedings of the 7th ACM International Conference on Web Search and Data Mining, pp. 373–382. ACM (2014)
6. Laurens, V.D.M., Hinton, G.: Viualizing data using t-SNE. J. Mach. Learn. Res. **9**(2605), 2579–2605 (2008)
7. Le, T., Lauw, H.W.: Probabilistic latent document network embedding. In: IEEE International Conference on Data Mining (ICDM), pp. 270–279. IEEE (2014)
8. Mikolov, T., Sutskever, I., Chen, K., Corrado, G.S., Dean, J.: Distributed representations of words and phrases and their compositionality. In: Advances in Neural Information Processing Systems, pp. 3111–3119 (2013)
9. Nallapati, R.M., Ahmed, A., Xing, E.P., Cohen, W.W.: Joint latent topic models for text and citations. In: Proceedings of the 14th ACM SIGKDD International Conference on Knowledge Discovery and Data Mining, pp. 542–550. ACM (2008)
10. Perozzi, B., Al-Rfou, R., Skiena, S.: Deepwalk: online learning of social representations. In: Proceedings of the 20th ACM SIGKDD International Conference on Knowledge Discovery and Data Mining, pp. 701–710. ACM (2014)

11. Recht, B., Re, C., Wright, S., Niu, F.: Hogwild: a lock-free approach to parallelizing stochastic gradient descent. In: Advances in Neural Information Processing Systems, pp. 693–701 (2011)
12. Roweis, S.T., Saul, L.K.: Nonlinear dimensionality reduction by locally linear embedding. Science **290**(5500), 2323–2326 (2000)
13. Suykens, J.A., Vandewalle, J.: Least squares support vector machine classifiers. Neural Process. Lett. **9**(3), 293–300 (1999)
14. Tang, J., Liu, J., Zhang, M., Mei, Q.: Visualizing large-scale and high-dimensional data. In: Proceedings of the 25th International Conference on World Wide Web, pp. 287–297. International World Wide Web Conferences Steering Committee (2016)
15. Tang, J., Lou, T., Kleinberg, J., Wu, S.: Transfer link prediction across heterogeneous social networks. ACM TOIS, **9**(4), 1–42, Article 43 (2015). http://citeseerx.ist.psu.edu/viewdoc/download?doi=10.1.1.696.2188&rep=rep1&type=pdf
16. Tang, J., Qu, M., Wang, M., Zhang, M., Yan, J., Mei, Q.: Line: large-scale information network embedding. In: Proceedings of the 24th International Conference on World Wide Web, pp. 1067–1077. International World Wide Web Conferences Steering Committee (2015)
17. Tenenbaum, J.B., De Silva, V., Langford, J.C.: A global geometric framework for nonlinear dimensionality reduction. Science **290**(5500), 2319–2323 (2000)
18. Tu, C., Zhang, W., Liu, Z., Sun, M.: Max-margin DeepWalk: discriminative learning of network representation. In: Proceedings of the Twenty-Fifth International Joint Conference on Artificial Intelligence (IJCAI 2016), pp. 3889–3895 (2016)
19. Yang, Z., Cohen, W., Salakhutdinov, R.: Revisiting semi-supervised learning with graph embeddings. arXiv preprint arXiv:1603.08861 (2016)

Detecting Synonymous Predicates from Online Encyclopedia with Rich Features

Zhe Han$^{(\boxtimes)}$, Yansong Feng, and Dongyan Zhao

Institute of Computer Science and Technology, Peking University, Beijing, China
{hanzhe1992,fengyansong,zhaodongyan}@pku.edu.cn

Abstract. The integration of Linked Open Data faces great challenges on the semantic level, despite unified data models. Inappropriate use of ontology concepts, namely predicates, impedes knowledge discovery. Although predicate unification is one of the most crucial steps when building structured knowledge base, little effort has been put forward. In this paper, we propose a supervised approach to detect synonymous predicates. Our detection focuses on feature selection and their effectiveness analysis. We not only leverage different resources such as Wikipedia, Freebase, but also use different word embeddings to represent predicates. The experimental results indicate that wikitext defined by Wikipedia and predicate surface form are most useful features.

1 Introduction

1.1 Motivation

Recently, much effort has been devoted to automatically building structured KBs from English version of Wikipedia, such as Freebase [1], DBpedia [2] and YAGO [3]. These KBs, unlike Wikipedia itself, extract facts in structured form of subject-predicate-object (s-o-p) triples from infoboxes in Wikipedia. However, Wikipedia pages are maintained by volunteers around the world. People from different backgrounds may use different expressions to convery the same idea, causing different surface forms of predicates which are semantically identical. The incorrect use of predicates causes either low recall of extractor or redundant relations in structured KBs.

There are 14,000 different predicate surface forms in Chinese Wikipedia[1]. Many predicates are in fact synonymous. For example, there are 17 predicates containing 邮政 (*postcode*) in Chinese Wikipedia infoboxes, shown in Table 1. Most of them represent 'postcode'. When an editor is submitting a new attribute of an entity, the system should provide the editor with candidate predicates. Besides, query expansion also requires synonymous predicates recommendation. The tremendous predicate list makes it impossible to get rid of duplicate predicates manually. It's urgent to put forward an auto-detect method to find synonymous predicates in online encyclopedias, which helps improve the quality of structured KB's extractions.

---

[1] Based on Chinese Wikipedia web pages in August 2014.

© Springer International Publishing AG 2016
S. Ma et al. (Eds.): AIRS 2016, LNCS 9994, pp. 111–122, 2016.
DOI: 10.1007/978-3-319-48051-0_9

Table 1. Predicates containing 邮政 (postcode) in Chinese Wikipedia infoboxes

predicate	INSEE/邮政编码	邮政编码	邮政区号	邮政编号	邮政简称	ISO 3166-2 邮政简写	邮政信箱	美国邮政编号	美国邮政编码
frequency	7257	5671	127	14	10	3	2	2	2
predicate	邮政编码首字母	邮政分区	邮政	邮政代码	邮政号码	邮政编码 FSA	邮政缩写	邮政号字母	
frequency	1	1	1	1	1	1	1	1	

1.2 Challenges

Predicate detection in online encyclopedias differs from that on Linked Open Data(LOD), such as DBpedia and Freebase. Objects in online encyclopedias are often non-standard, making it difficult to use. Moreover, predicates are various due to different backgrounds of editors. Usage of global synonym databases is not sufficient as predicates are used in various KBs for various purpose by various editors. There are also different interpretations of predicates. Some predicates are too concrete while others are too general. Besides, many Chinese characters share a similar pronunciation, causing typos (mistaking characters with the same pronunciation) and transliteration differences (different characters chosen to represent the same pronunciation). As for Chinese predicates, there are fewer external resources like WordNet. The long tail [4] causes little information could be extracted from low frequent predicates.

1.3 Contribution

Earlier studies on structured KBs are not appropriate in the case for online encyclopedias. Our method is exactly designed for semi-structured online encyclopedias where objects are seldom linked entities. The contribution of this paper is fourfold. First, we leverage Wikipedia's wikitext for the first time to describe predicates. Besides, we extract many detailed information in Wikipedia and joint dumps and web pages together for the first time. Second, we propose various word-embeddings, varying from predicate types to predicate semantics. Third, we use linking information between Freebase schema and Wikipedia schema and use the better organized Freebase schema to describe predicates. Finally, we understand the predicate from these features.

The rest of this paper is organized as follows: In the next section we present related work with regard to synonymous predicates discovery. Next in Sects. 3 and 4, we introduce the resources and features used in our experiment. We evaluate the features from different perspectives in Sect. 5 and conclude in Sect. 6.

2   Related Work

Although it is a fundamental step in building structured KB, rare work has been done on this intractable problem. Many released KBs avoid predicate unification by using a predefined and limited predicate list, such as YAGO. There are less than 200 predicates in YAGO. You cannot find the screenwriter of any movies in YAGO because this relation has not be defined yet. Freebase indeed detect synonyms based on user domain expertise and co-occurrence of objects and subjects [5]. However, this method calls for user logs and well-structured KB, which can not be utilized by other KBs.

Most techniques for synonym detection derive from schema matching as data mining in the Semantic Web, associated with query expansion and synonym discovery. Others are based on different language processing and information retrieval techniques.

Mature methods in Semantic Web, such as frequent subgraph or subtrees analysis [6], are not suitable because no two different nodes in an RDF graph have the same URI. Instead, we consider the corresponding type of each URI as different URI may belong to the same type. Cafarella [7] presents a approach to detect synonyms among table attributes. However, the authors restrict attributes and ignore instance-based method because they concentrate only on extracted table schemata. So far, Abedjan [8] treats synonymous predicates detection as a association rule mining problem. Note that he works on structured DBpedia using linking information of objects and does not understand the predicates. This method is not appropriate for encyclopedias.

Baroni [9] and Wei [10] propose a common approach using co-occurrence of synonym candidates in web documents, based on the idea of synonymous word co-occur [11]. Naumann [12] proves the effectiveness of aggregate features and Li's work [13] shows that the performance using dictionaries only in real data is poor. In this case, we use multi features to capture predicate semantics.

Since only subject is normalized in encyclopedias, we use subject schema and NLP tools to discover synonyms, leveraging both the benefit of schema matching and semantic understanding.

3 Resources

Various resources have been used as features to present predicates, from inside and outside the KB. We leverage both web pages and dumps of Wikipedia in the experiment. Besides, bilingual dictionary and Freebase schema are used to represent each predicate. Our main dataset is a semi-structured KB (See footnote 1) (only subject is defined as an entity) with 3.5 millions s-p-o triples extracted from 33.8 thousands of infoboxes in Chinese Wikipedia [14], which contains 14,000 different predicates. The KB is open-domain and predicates in the KB that have same surface forms are considered the same in our experiment.

3.1 Wikipedia

In Wikipedia, there are mainly three parts of data to help evaluate the similarity between predicates, including *section names* in Wikipedia web pages and *wikitext-predicates*, *infobox names* in Wikipedia dumps[2]. Figure 1 shows all the information in Wikipedia.

---

[2] Available at https://dumps.wikimedia.org/zhwiki/.

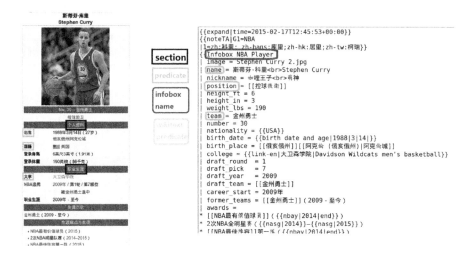

Fig. 1. The web infobox (left) and wikitext infobox (right) of Stephen Curry in Chinese Wikipedia. wikiSection and wikiInfobox refer to 'section' and 'infobox name'. 'wikitext predicate' will be aligned to 'predicate' based on their attribute values. e.g. wikitext predicate 'nationality' is aligned to web predicate '国籍' because 'USA' '美国' and are the same.

wikiText. Wikitext, also known as wikicode, is a lightweight markup language used to write pages in Wikipedia. Infobox [15] is a template used to collect and present a subset of information about its subject. All the wikitexts and infoboxes [16] mentioned in this paper are referred to Wikipedia's wikitexts and infoboxes.

wikiSection. Attributes are often divided into different sections in Wikipedia infoboxes. As shown in Fig. 1, 个人资料 (*personal information*) is the section name of predicate 出生 (*birth*) and 登陆身高 (*listed hight*) while predicate 大学 (*college*) and 选秀 (*NBA draft*) are in section 职业生涯 (*career information*).

wikiInfobox. Actually, each infobox of an entity has a name, which can be extracted from wikitext. For example, predicate 国籍 (*nationality*) usually appears in infoboxes concerning people while predicate 编剧 (*writer*) appears in infoboxes concerning drama.

3.2 Freebase

Different predicates are usually associated with different kinds of entities [17]. Predicate categories can be represented by their corresponding subject categories. On the one hand, Wikipedia's build-in categories are too detailed to use. There are more than 190,000 categories in Chinese Wikipedia. In addition, there exist many confusing but frequent categories, such as 优良条目 (*good articles*) and 含有希伯来语的条目 (*articles containing Hebrew-language text*). On the other hand, Freebase provides find-grained category information for most entities and

Table 2. Categories of actor 张家辉 in Freebase (mid = m.03cp9fl)

people.person	award.award_winner	film.actor	film.editor
tv.tv_actor	award.award_nominee	common.topic	

fortunately many Freebase entities have been linked to Wikipedia entities[3]. For example, the category information of Hongkong film actor 张家辉 [4,5] (*Nick Cheung*) in Freebase is shown in Table 2. We collect all the categories of Freebase entities that correspond to Chinese Wikipedia entities[6].

4   Features

Features are of great importance in our experiment. We not only use the surface form of predicate, but also extract many latent features inside Wikipedia and Freebase. Table 3 presents all the features used in the experiment.

Table 3. All features

surfaceForm	1.unigram$_{(0,1)}$	3.edit_distance$_{(0,1)}$	5.length_ratio
	2.unigram$_{(1,0)}$	4.edit distance$_{(1,0)}$	
Pinyin	6.pinyin_unigram$_{(0,1)}$	8.pinyin_edit_distance$_{(0,1)}$	10.pinyin_length_ratio
	7.pinyin_unigram$_{(1,0)}$	9.pinyin_edit_distance$_{(1,0)}$	
wikiText	11.wikiText-embedding	12 wikiText$_{(0,1)}$	13.wikiText$_{(1,0)}$
wikiSection	14.wikiSection-embedding	15.wikiSection$_{(0,1)}$	16.wikiSection$_{(1,0)}$
wikiInfobox	17.wikiInfobox-embedding	18.wikiInfobox$_{(0,1)}$	19.wikiInfobox$_{(1,0)}$
Freebase category	20.Freebase_SVD-embedding	21. Freebase-embedding	

4.1   SurfaceForm

The most straightforward features would be those extracted from surface forms of predicates. This kind of features express the character level similarity between two predicates. We first consider unigram overlap and explore two metrics, unigram$_{(0,1)}$ (feature 1) and unigram$_{(1,0)}$ (feature 2). Unigram$_{(1,0)}$ and unigram$_{(0,1)}$ scores between a predicate pair $(pred_1, pred_2)$ are defined as follows, while other features containing subscript $_{(1,0)}$ or $_{(0,1)}$ are defined in the same way as Eqs. (1) and (2):

---

[3] The linking property in Freebase rdf dump is *Wikipedia.zh-cn_id* while the Freebase category predicate is *rdf:type*.

[4] https://zh.wikipedia.org/wiki?curid=472824.

[5] http://www.freebase.com/m/03cp9fl.

[6] The version of Freebase used in the experiment is 2013-06-02 (1.37 billion triples). We collected categories of 337042 entities in Freebase.

$$unigram_{(1,0)}(pred_1, pred_2) = \frac{character\_overlap(pred_1, pred_2)}{character\_count(pred_1)} \quad (1)$$

$$unigram_{(0,1)}(pred_1, pred_2) = \frac{character\_overlap(pred_1, pred_2)}{character\_count(pred_2)} \quad (2)$$

$$edit\_distance_{(0,1)}(pred_1, pred_2) = \frac{edit\_distance(pred_1, pred_2)}{character\_count(pred_1)} \quad (3)$$

We also compute edit distances of each pair of predicates (feature 3 and 4) in Eq. (3). Synonymous predicates usually have similar length in characters, which is taken into account by *length(shorter predicate)/length(longer predicate)* as character length ratio (feature 5).

4.2   Pinyin

Pinyin is the official phonetic system (and ISO standard) for transcribing Mandarin pronunciations into the Latin alphabets. There are many words in Chinese with different writing forms, conveying the same meaning. For example, 坐标 (*coordinate*) and 座标 (*coordinate*) are different predicate forms but actually the same. We use the most frequent pinyin string of each Chinese character to construct the pinyin representation for a predicate. Features in *Pinyin* (feature 6–10) are similar to features in *surfaceForm*. Compared to features in *surfaceForm*, characters are replaced with their corresponding pinyin strings while calculating the similarity scores.

4.3   WikiText

Wikipedia uses a large amount of rules to translate particular wikitext templates to the infoboxes we see on web pages. In our case, predicates in wikitext are aligned to predicates in web pages. The alignment is based on manual rules calculating the similarities between objects in web page triples and the attribute values of dumps' wikitexts. Accordingly, given a predicate, we can collect a set of aligned wikitext-predicates, with their alignment frequencies to this predicate. The alignment is not a one-to-one mapping, causing noise in the alignment. For example, the wikitext-predicates and their frequency aligned to predicate 面积 (*area*) are shown in Table 4. We defined it the *wikitext-predicate distribution* of predicate 面积.

Let $freq(p_i, wp_j)$ be the frequency of predicate p_i aligning to wikitext-predicate wp_j. Let $WL(p_i)$ be the wikitext-predicate set that has aligned to p_i. The wikiText$_{(0,1)}$ (feature 12) and wikiText$_{(1,0)}$ (feature 13) further characterize the overlap between the two predicates in an asymmetric way, defined in Eqs. (4) and (5):

$$wikiText_{(0,1)}(p_1, p_2) = \frac{\sum_{wp_j \in (WL(p_1) \cap WL(p_2))} freq(p_1, wp_j)}{\sum_{wp_j \in WL(p_1)} freq(p_1, wp_j)} \quad (4)$$

Table 4. The wikitext-predicate distribution of predicate 面积 [**]

wikitext	面积	area	areatotal	*arearank*	*population total*	tarea	面积排名	area imperial	总面积	...
frequency	2860	1251	272	163	124	93	72	24	19	...

[**]Alignment errors are in red

$$wikiText_{(1,0)}(p_1, p_2) = \frac{\sum_{wp_j \in (WL(p_1) \cap WL(p_2))} freq(p_2, wp_j)}{\sum_{wp_j \in WL(p_2)} freq(p_2, wp_j)} \qquad (5)$$

The wikitext-embedding of each predicate p_i is a unit, sparse vector $v_i = (v_1^i, v_2^i, ..., v_M^i)$. M is equal to the number of different wikitext-predicates. v_j^i is the normalized frequency between predicate p_i and wikitext-predicate wp_j, representing their co-occurence, defined in Eq. (6). Feature 11 of each predicate pair is the cosine similarity of the two wikitext-embedding vectors.

$$v_j^i = freq(p_i, wp_j) / \sqrt{\sum_j freq^2(p_i, wp_j)} \qquad (6)$$

4.4   WikiSection and WikiInfobox

The predicate synonyms should have similar sections and infobox names. We collect all the wikiSections and wikiInfoboxes of predicates and convert them to distribution vectors. For example, The wikiSection and wikiInfobox distribution of predicate 国家 (*country*) is shown in Tables 5 and 6. wikiSection and wikiInfobox features are calculated in a similar way as *wikiText* features.

4.5   Bilingual Dictionary

Some synonymous predicates are in different surface forms mainly because of translation differences. Thus we translate the original Chinese predicates to their corresponding English expressions. This kind of features works well when one or both predicates is low frequent and less information could be extracted by other kind of features.

Table 5. The wikiSection distribution of predicate 国家 (country)

wikiSection	概览	地理位置	基本资料	废除前地理位置	概况	赛事信息	概要	基层政权	位置	...
frequency	42483	2107	721	582	180	113	113	93	86	...

Table 6. The wikiInfobox distribution of predicate 国家 (country)

wikiInfobox	infobox_settlement	infobox_city	东亚男性历史人物	infobox_kommune	geobox	infobox_uk_place	...
frequency	30978	1997	430	293	204	202	...

4.6   Freebase Category

For each predicate, we average the category vectors of entities that appear as subject of this predicate to generate category vectors of predicate. Since Freebase has a large mount of different categories, the raw category information will be very sparse. Therefore, we use two kinds of Freebase category embeddings. One uses the original category distribution vector while the other uses singular value decomposition (SVD) to transform each entity's category information to a 100-dimension unit vector.

Let $S_i = \{e_1^i, e_2^i, ..., e_{N_i}^i\}$ be the set of entities that has predicate p_i, $T(e_j^i)$ be the set of types of entity e_j^i in Freebase. The original category embedding of p_i is $F_i = (f_1^i, f_1^i, ..., f_N^i)$. N_i is size of S_i while N is the total number of categories in Freebase. f_i^j is the normalized frequency between predicate p_i and Freebase category $cate_j$, defined in Eq. (7). The Freebase-embedding (Feature 21) of predicate pair (p_i, p_j) is $F_i * F_j$. So does feature 20.

$$f_j^i = (\sum_{e_k \in S_i} \sum_{c_j \in T(e_k)} 1)/(\sum_j (f_j^i * f_j^i)) \tag{7}$$

5   Experiment

We treat this task as a binary classification problem, that is, given a pair of predicates $pred_1$, $pred_2$, predicting whether these two predicates are synonyms.

The dataset is validated by three experts in computer science major. The first expert randomly selects predicate pairs and tag *0* or *1* to represent whether they are synonyms. Since the class distribution is highly skewed with most predicate pairs being negative, we select a balanced set of 1500 pairs with 1000 positive and 500 negative to avoid failures in training. Then the second expert tags on this balanced pairs and the last person only tags the inconsistent pairs. The result training set contains 1000 pairs (464 pairs are tagged *1*) of predicates while the test set contains 500 pairs (240 pairs are tagged *1*).

To evaluate features' validity, we present three experiments: In Sect. 5.1 we evaluate the classification performance using only one kind of features. In Sect. 5.2 we evaluate the classification performance using all except one kind of features at one time. In Sect. 5.3 we evaluate all the feature combinations and seek the feature combination that outperforms others. We use LibSVM (with kernel type of LINEAR and RBF), decision tree (C4.5), voted perceptron and AdaBoost as classifiers in each experiment. Compared to Abedjan's work [8], we deal with different resources (linked open data and online encyclopedia) using different methods, thus, we use different evaluation methods.

5.1   Single Feature Experiment

First we explore the effectiveness of each kind of features. For each classifier, We report the accuracy using only one kind of features, shown in Table 7.

Table 7. The single feature accuracy

Feature(dimension)	Accuracy				
	AdaBoost	LibSVM_RBF	LibSVM_LINEAR	C4.5	VotedPerceptron
Pinyin (5)	**0.662**	**0.664**	**0.610**	**0.666**	0.618
surfaceForm (5)	0.634	0.584	0.586	0.634	**0.626**
Bilingual dictionary (2)	0.594	0.598	0.598	0.596	0.586
Freebase category (2)	0.568	0.580	0.562	0.568	0.582
wikiText (3)	0.562	0.572	0.586	0.562	0.562
wikiSection (3)	0.518	0.526	0.522	0.518	0.532
wikiInfobox (3)	0.518	0.526	0.522	0.518	0.532

wikiSection and *wikiInfobox* features are indistinctive because much of low frequent predicates do not have enough wikiSections and wikiInfoboxes to represent predicates properly. *Pinyin* feature is of great importance as expected. It takes spell mistakes and different forms of expression into consideration. *SurfaceForm* and *bilingual dictionary* are also reported as good single features.

5.2   Minus One Feature Experiment

In the second experiment we have evaluated the redundancy of each kind of predicate comparing other features. We first remove one kind of features and then evaluate the utility of remaining features. The detached kind of features is more likely to be redundant if the remaining features have higher accuracy. The results are shown in Table 8.

Table 8. The minus one feature accuracy

Reduced feature	Accuracy				
	LibSVM_RBF	AdaBoost	LibSVM_LINEAR	C4.5	VotedPerceptron
-surfaceForm	**0.634**	**0.642**	**0.634**	0.604	**0.624**
-wikiText	0.656	0.666	0.648	0.644	0.666
-wikiInfobox	0.680	0.670	0.676	0.664	0.670
-wikiSection	0.680	0.670	0.676	0.664	0.670
-Pinyin	0.688	0.666	0.688	0.668	0.676
-Freebase category	0.684	0.686	0.696	0.672	0.668
-Bilingual dictionary	0.698	0.666	0.678	0.682	0.692

wikiText feature is only inferior to *surfaceForm* feature while *bilingual dictionary* performs poor. It shows the importance of wikiText. *wikiText* includes the bilingual information for its cross-linguistic property. It also indicates that the wikiText defined by *Wikipedia.org* is valid and irreplaceable in representing predicate. No matter what kind of classifier we use, *surfaceForm* and *wikiText* appear the top 2 features in this experiment. What's more, *bilingual dictionary* is usually the most useless kind of features. It is because *bilingual dictionary* and *Pinyin* features can be seen as a coarse combination of *surfaceForm* and *wikiText* features. Comparing to the first experiment, *wikiInfobox* and *wikiSection* take effect in complex feature combinations.

5.3   Best Feature Combination

In our last experiment, we want to find the best feature combination. We use LibSVM with RBF kernel as classifier example. The other classifiers output similar results. The result is shown in Table 9. Our best accuracy is achieved with features: [*pinyin, surfaceForm, wikiText, wikiSection, wikiInfobox, Freebase_category*]. It corresponds to the previous two experiments: *surfaceForm* and *wikiText* are fundamentally useful while *wikiInfobox* and *wikiSection* show their efficacy in complex feature combinations.

Table 9. The top feature combinations by accuracy (RBF)

Features	Accuracy
pinyin, surfaceForm, wikiText, wikiSection, wikiInfobox, Freebase_category	0.698
pinyin, surfaceForm, wikiText, wikiInfobox, Freebase_category	0.694
pinyin, surfaceForm, wikiText, wikiSection, Freebase_category	0.694
surfaceForm, wikiText, wikiInfobox, Freebase_category	0.688
surfaceForm, wikiText, wikiSection, Freebase_category	0.688
surfaceForm, wikiText, wikiSection, wikiInfobox, bilingual_dict, Freebase_category	0.688

6   Conclusion and Future Work

In this paper, we propose a full-fledged method on detecting predicate synonyms, including features extraction and comparison. It is groundwork for building Chinese structured KB. We exploit a mount of features, including linking information between Freebase and Wikipedia. Thorough study has been done on wikitext. Our experiment shows that the wikitext provides unique information comparing to normal features. Subject category information and section information are also essential features, which can be used by other online encyclopedias.

In online encyclopedias, only few predicates will be inserted or changed by editors to entities pages during a short time. Synonymous predicates can be calculated offline and we can only calculate the similarity between recently modified predicates and other predicates, which reduces computation resources. Another way to speedup our system is using part of distribution data. We find that the top three wikitext-predicates in Sect. 4.3 already account for most correct alignment. Hence, the time complexity of feature calculating can be approximately linear.

Objects, or attribute values in the KB have not be leveraged, except for wikitext-predicate alignment. They depict the predicates directly and may contribute much in predicting the predicate synonyms. Future work will explore the use of objects in predicate comparison. Predicate unification between different Chinese encyclopedias, such as baidu baike and Chinese Wikipedia will also be conducted.

Acknowledgement. This work was supported by National High Technology R&D Program of China (Grant No. 2015AA015403, 2014AA015102), Natural Science Foundation of China (Grant No. 61202233, 61272344, 61370055) and the joint project with IBM Research. Any correspondence please refer to Yansong Feng.

References

1. Bollacker, K., Evans, C., Paritosh, P., Sturge, T., Taylor, J.: Freebase: a collaboratively created graph database for structuring human knowledge. In: Proceedings of the 2008 ACM SIGMOD International Conference on Management of Data, pp. 1247–1250. ACM (2008)
2. Bizer, C., Lehmann, J., Kobilarov, G., Auer, S., Becker, C., Cyganiak, R., Hellmann, S.: Dbpedia-a crystallization point for the web of data. Web Semant. Sci. Serv. Agents World Wide Web **7**, 154–165 (2009)
3. Suchanek, F.M., Kasneci, G., Weikum, G.: Yago: a core of semantic knowledge. In: Proceedings of the 16th International Conference on World Wide Web, pp. 697–706. ACM (2007)
4. Wu, F., Hoffmann, R., Weld, D.S.: Information extraction from wikipedia: moving down the long tail. In: Proceedings of the 14th ACM SIGKDD International Conference on Knowledge Discovery and Data Mining, pp. 731–739. ACM (2008)
5. Tan, C.H., Agichtein, E., Ipeirotis, P., Gabrilovich, E.: Trust, but verify: predicting contribution quality for knowledge base construction and curation. In: Proceedings of the 7th ACM International Conference on Web Search and Data Mining, pp. 553–562. ACM (2014)
6. Kuramochi, M., Karypis, G.: Frequent subgraph discovery. In: Proceedings IEEE International Conference on Data Mining, ICDM 2001, pp. 313–320. IEEE (2001)
7. Cafarella, M.J., Halevy, A., Wang, D.Z., Wu, E., Zhang, Y.: Webtables: exploring the power of tables on the web. Proc. VLDB Endow. **1**, 538–549 (2008)
8. Abedjan, Z., Naumann, F.: Synonym analysis for predicate expansion. In: Cimiano, P., Corcho, O., Presutti, V., Hollink, L., Rudolph, S. (eds.) ESWC 2013. LNCS, vol. 7882, pp. 140–154. Springer, Heidelberg (2013). doi:10.1007/978-3-642-38288-8_10
9. Baroni, M., Bisi, S.: Using cooccurrence statistics and the web to discover synonyms in a technical language. In: LREC (2004)
10. Wei, X., Peng, F., Tseng, H., Lu, Y., Dumoulin, B.: Context sensitive synonym discovery for web search queries. In: Proceedings of the 18th ACM Conference on Information and Knowledge Management, pp. 1585–1588. ACM (2009)

11. Harris, Z.S.: Distributional structure. Word **10**, 146 (1954)
12. Naumann, F., Ho, C.T., Tian, X., Haas, L.M., Megiddo, N.: Attribute classification using feature analysis. In: ICDE, vol. 271 (2002)
13. Li, W.S., Clifton, C.: Semint: a tool for identifying attribute correspondences in heterogeneous databases using neural networks. Data Knowl. Eng. **33**, 49–84 (2000)
14. Denoyer, L., Gallinari, P.: The wikipedia XML corpus. In: Fuhr, N., Lalmas, M., Trotman, A. (eds.) INEX 2006. LNCS, vol. 4518, pp. 12–19. Springer, Heidelberg (2007). doi:10.1007/978-3-540-73888-6_2
15. Wu, F., Weld, D.S.: Autonomously semantifying wikipedia. In: Proceedings of the Sixteenth ACM Conference on Information and Knowledge Management, pp. 41–50. ACM (2007)
16. Wu, F., Weld, D.S.: Automatically refining the wikipedia infobox ontology. In: Proceedings of the 17th International Conference on World Wide Web, pp. 635–644. ACM (2008)
17. Cucerzan, S.: Large-scale named entity disambiguation based on wikipedia data. In: EMNLP-CoNLL, vol. 7, pp. 708–716 (2007)

IR Applications and User Modeling

Patent Retrieval Based on Multiple Information Resources

Kan Xu[1], Hongfei Lin[1(✉)], Yuan Lin[2], Bo Xu[1], Liang Yang[1],
and Shaowu Zhang[1]

[1] School of Computer Science and Technology,
Dalian University of Technology, Dalian, China
{xukan, hflin, zhangsw}@dlut.edu.cn,
{xubo2011, yangliang}@mail.dlut.edu.cn
[2] WISE Lab, Dalian University of Technology, Dalian, China
zhlin@dlut.edu.cn

Abstract. Query expansion methods have been proven to be effective to improve the average performance of patent retrieval, and most of query expansion methods use single source of information for query expansion term selection. In this paper, we propose a method which exploits external resources for improving patent retrieval. Google search engine and Derwent World Patents Index were used as external resources to enhance the performance of query expansion methods. LambdaRank was employed to improve patent retrieval performance by combining different query expansion methods with different text fields weighting strategies of different resources. Experiments on TREC data sets showed that our combination of multiple information sources for query formulation was more effective than using any single source to improve patent retrieval performance.

Keywords: Information Retrieval · Query expansion · Learning to rank · Patent retrieval

1 Introduction

The amount of patent information is growing rapidly with an abundant production of digital collection of documents. It is a real challenge to accessing to useful information among this large size dataset. Although patent search engine like Derwent World Patents Index and Google patent search engine have large databases, the search results are not satisfactory. People got very different results when they use different search engines with the same keywords, and they cannot determine which result is more relevant to the keywords. So it is necessary to integrate multiple patent data sources and search methods to improve the performance of patent retrieval.

Automatic query expansion technologies have been widely used in information retrieval (IR) [1–3] In particular, the pseudo-relevance feedback (PRF) which uses query expansion has been proven to be effective [4, 5]. The process of query expansion modified the original keyword query submitted by the user and it would be better represented the underlying intent of the query. The formulated query is then used as an

© Springer International Publishing AG 2016
S. Ma et al. (Eds.): AIRS 2016, LNCS 9994, pp. 125–137, 2016.
DOI: 10.1007/978-3-319-48051-0_10

input to the search engine's ranking algorithm. Thus, the primary goal of query formulation is to improve the overall quality of the ranking presented to the user in response to the query. However, the general query expansion method cannot be introduced directly to special tasks, such as patent retrieval. The patent documents, which are constructed by several special text fields, are different from Web page documents. These fields describe different aspects of patent and have different importance. The traditional expansion methods select candidate terms from the whole document without considering the information from fields which are not suitable for patent retrieval. The existing work [6–8] did not pay enough attention to it. In previous work [9, 10], we proposed a query expansion method, which used patent text fields as the resource of expansion terms, the performance was improved by introducing the field information to query expansion. However, we only use the pseudo-relevance feedback documents for expansion terms. There are still some external information resources which can be used to improve the retrieval performance. It is highly effective to query expansion by using external information resources [11–13].

Learning to rank [14] has become an important research issue for information retrieval. It is an effective approach to improve the ranking performance. The basic premise for learning to rank method is that there are three types of input spaces, they are pointwise, pairwise, and listwise samples. In this paper, we will apply the learning to rank approach to optimize the combination of information sources to improve the performance of patent retrieval.

The remainder of our paper is organized as follows. Section 2 reviews some related work. Section 2 explores the impact of different information resources for patent retrieval. Section 3 proposes the learning to rank based query expansion approach on Derwent World Patents Index and Google search engine for patent retrieval. In Sect. 4, we report the experimental results. Finally, we conclude the paper and discuss future work in Sect. 5.

2 Related Work

2.1 Patent Retrieval

In recent years, researchers show growing interests in patent retrieval. Their research mainly focused on exploring methods on query formulation for topics. Keywords was extracted to form new queries in the early work [15, 16]. Full patent texts were used as the query to reduce the burden on patent examiners which was advocated by Xue and Croft [17]. Text mining, bibliographic coupling and citation analysis were also used in patent retrieval [18, 19]. Chen and Chiu [20] developed an IPC-based vector space model for patent retrieval and achieved a higher accuracy than normal patent search engine. Rusinol et al. [21] presented a flowchart recognition method for patent image retrieval. Recent work showed that the best retrieval results were obtained when using terms from all the fields of the queried patents [22]. It seems that field information is very effective to improve the patent retrieval. However, there are still few works on exploring the text fields to improve query expansion. This paper will use the patent text field information to select candidate terms and improve the results of patent retrieval.

We also investigate the capability of text field of patent in improving the performance of retrieval as promising information for query expansion.

2.2 Query Expansion and External Sources

Pseudo-relevance feedback (PRF) is an effective automatic query expansion method by reformulating the original query using expansion terms from pseudo-relevant documents. Traditional PRF has been implemented in several retrieval models, such as vector space model [23], probabilistic model [24], relevance model [25], mixture model [26], and so on. Meanwhile, there are many research work which focus on improving traditional PRF in different ways. For example, using passages instead of documents [27], using a local context analysis method [1], using a query-regularized estimation method [4], using latent concepts [3], and using a clustered-based re-sampling method for generating pseudo-relevant documents [5]. These methods follow the basic assumption that the top-ranked documents from an initial search contain useful terms that can help discriminate relevant documents from irrelevant ones.

Two external information sources will be employed in our experiment, Google Search Engines and Derwent World Patents Index. Google is one of best search engines in the world, which can provide the accurate information for the users according to the their queries, so we also want to use Google to provide the relevant web pages to expand the query terms for patents. The Derwent World Patents Index (or DWPI) is a database containing patent applications and grants from 44 of the world's patent issuing authorities. Compiled in English by editorial staff, the database provides a short abstract detailing the nature and use of the invention described in a patent and is indexed into alphanumeric technology categories to allow retrieval of relevant patent documents by users. Each record in the database defines a patent family, the grouping of patent documentation recorded at the various patent offices as protection of an invention is sought around the world. Each patent family is grouped around a Basic patent, which is usually the first published example of the invention. All subsequent filings are referred back to the Basic patent as Equivalent patents. The database has some 20 million "inventions", corresponding to ten millions of patents, with almost a million new inventions added each year. Since Derwent database is so effective to the patent research, we will use it as another external information resource to patent query expansion.

2.3 Learning to Rank

Learning to rank approaches can be divided into three categorizations, the pointwise approach, the pairwise approach, and the listwise approach. Different approaches model the process of learning to rank in different ways. They define different input and output samples, using different hypotheses and employ different loss functions. This paper will focus on the construction of samples of listwise approach for further analysis. The listwise approach addresses the ranking problem in a natural way. It takes ranking lists as samples in both learning and prediction. The structure of ranking is maintained and ranking measures is incorporated directly into the loss functions. More specifically, the listwise approach takes the labeled query-document list as one instance. LambdaMART [28] is the boosted tree version of listwise approach of learning to rank, which is based on RankNet. Boosting and LambdaMART have been shown as the best performing learning methods on public data sets. LambdaMART rankers won Track 1 of the 2010 Yahoo Learning To Rank Challenge. It has been proven to be an effective ranking method for merging the ranking features to improve the performance of retrieval. In this paper, we will use this approach to improve the ranking performance of patent retrieval based on multiple query expansion methods and text fields.

3   Query Expansion Using External Information Resources

3.1   Query Expansion Model

In this section, we introduce our method for patent query expansion. Our query expansion model includes two Rocchio models, one is the original Rocchio model [23], and the other is modified Rocchio model [9].

The original Rocchio model is defined as follows:

$$Q_1 = \lambda * Q_0 + (1 - \lambda) \sum_{r \in R} \frac{r}{|R|} \tag{1}$$

where Q_1 is the expansion query, Q_0 is the original query. R is the pseudo relevance document collection, r is the relevant document. The modified Rocchio model is based on patent fields. In this paper, the model is defined as follows:

$$Q_2 = \lambda * Q_0 + (1 - \lambda) \sum_{r \in R} \frac{\sum_{f \in F} r_f * q_{r_f}}{|R|} \tag{2}$$

where Q_2 is the expansion query, Q_0 is the original query. R is the pseudo relevance document collection, r_f is the field f of the relevant document r. q_{rf} is the weight of r_f. We expand the original queries by this formula.

3.2   Information Resources for Patent Retrieval

The common information resource for pseudo-relevance feedback is the top ranked documents from the corpus with a given query. Relevance feedback takes the results that are initially returned from a given query to perform a new query. The content of the assessed documents is used to adjust the weights of terms in the original query and/or to add words to the query. So the first resource is the TREC data for patent. A patent document is composed of several fields of information, in particular the title, the abstract, the description and the claims. We use these content text fields as research objects to improve the quality of expansion terms. The title field contains the title of patent. The abstract field contains the text of summary or main idea of a patent. The description field consists of the some sentences about different aspects of a patent content. The claims are the boundary associated with a patent, which is assumed to describe its limits. All the information from the fields may be related to the relevance, and the terms appear in the different fields have different degrees of relevance. So we try to apply the fields to weight the terms for query expansions.

A common web data source from Google for query expansion of patent retrieval is very effective. When the query is submitted to the search engine, the answer is returned in the form of title and abstract texts. The texts and real user search queries are very similar because most title and abstract texts are succinct descriptions of the destination page. The relevant documents for the given query are the second resource of query expansion. The fields we use to query expansion from Google are title and abstract.

The third resource is based on Derwent World Patents Index. The initial set of candidates associated with a query is restricted by considering only those anchor texts that point to a short set of top ranked patents from a larger set of top-ranked patents. These patents can provide more effective information for query expansion. The patent is represented by title and abstract texts. The fields we use to query expansion are title and abstract.

3.3 Term Selection for Query Expansion

For query expansion, there are two steps: select the pseudo relevance document collection R and evaluate the weight of q_f.

In this paper, the pseudo relevance documents come from three information resource: TREC patent data set, Google and Derwent World Patents Index. For TREC patent data set, the first step is the pseudo feedback document selection, which applies three ranking methods for top-k documents: TF*IDF, BM25, BM25F.

TF*IDF [29] contains two variables, term frequency and inverse document frequency. There are various ways to determine the exact values of both variables. For term frequency, the simplest choice is to use the raw frequency of a term in a document, i.e. the number of times that term occurs in a document.

$$w_{t,d} = tf_{t,d} * \log(\frac{N}{n_t}) \tag{3}$$

where $tf_{t,d}$ is the number of times that term t occurs in document d. n_t is the number of the documents which contain the term t. N is the number of documents in the collection.

BM25 [24] is based on the probability model. The retrieved documents are ranked in the order of their probabilities of relevance to the query. A query term is assigned a weight based on its within-document term frequency and within-query frequency. The weighting function used in our experiments is BM25, shown as follows:

$$\omega = \frac{(k_1 + 1) * tf}{K + tf} * w^{(1)} * \frac{(k_3 + 1) * qtf}{k_3 + qtf} \oplus k_2 * nq * \frac{(avdl - dl)}{(avdl + dl)} \tag{4}$$

$$w^{(1)} = \log \frac{(r + 0.5)/(R - r + 0.5)}{(n - r + 0.5)/(N - n - R + r + 0.5)} \tag{5}$$

w is the weight of a query term, N is the number of indexed documents in the collection, n is the number of documents containing the term, R is the number of documents known to be relevant to a specific topic, r is the number of relevant documents containing the term, tf is within-document term frequency, q_{tf} is within-query term frequency, dl is the length of the document, $avdl$ is the average document length, n_q is the number of query terms, the k_i s are tuning constants (which depend on the database and possibly on the nature of the queries and are empirically determined), K equals to $k_1 * ((1 - b) + b * dl / avdl)$, and $\oplus$ indicates that its following component is added only once per document, rather than for each term.

BM25F [30] is an extension of the BM25 function to a document description over multiple fields. A key property of this function is that it is nonlinear. Since BM25F reduces to BM25 when calculated over a single field, we will refer to both functions as BM25F, where F is a specification of the fields contained in the document description. In this paper, we use BM25F as the initial retrieval method for feedback documents, which considers multiple fields. BM25F is computed as follows for document d, with a document description over fields F, and query q:

$$S = \sum_{t \in q} TF_t * I_t \tag{6}$$

The sum is over all terms t in query q. It is the Robertson-Sparck-Jones form of inverse document.

We apply the BM25F approach as the initial retrieval method, and select the documents ranking on top-k positions as the candidate collection for the second step. TF*IDF and BM25 are used as baselines for comparison, which rank the documents for top-k pseudo feedback documents without field information, i.e. taking the whole document as a field.

The second step is to decompose every pseudo relevant document generated from the first step into several pieces according to the fields of patent, while each field is regarded as an independent short document. We use the BM25 approach to calculate the relevance between the query and the field document. The relevance score can be seen as the importance of field, which we used to weight the fields of the patent. We also evaluate the importance of each term in the short field document by the query expansion methods, such as TF, TF*IDF, BO1 and BO2 [31]. This analogy suggests us to use the other urn model for IR to obtain alternative methods of expansion for the query, which is the Bose-Einstein statistics. Note that one possible approximation of the Bose-Einstein statistics is given by the geometric distribution G. The probability P generating the geometric distribution has the same parameter $\lambda = N$ as the Poisson process. P defined as follows:

$$P = \frac{1}{1 + \lambda} \tag{7}$$

The urn model based on BE can be thus used for measuring the information content of terms in the query expansion process giving us:

$$Weight(t) = -log_2(\frac{1}{1 + \lambda_{E_q}}) - F_{E_q} * log_2(\frac{\lambda_{E_q}}{1 + \lambda_{E_q}}) \tag{8}$$

where F_{E_q} is the frequency of the term and λ_{E_q} is defined by:

$$\lambda_{E_q} = \begin{cases} \frac{F_{E_q}}{N} & [BO1] \\ TotFr_{E_q} \cdot \frac{F_{E_q}}{TotFr_D} & [BO2] \end{cases} \tag{9}$$

where $TotFr_D$ is the total number of term tokens in the collection D. We use these expansion methods to evaluate the relevant importance of a term in the patent fields, which combine the weights of fields to obtain the final weight of the term in the patent document. The finally expanded queries will be used to improve the ranking accuracy.

In this paper, we also take ranking methods and weight evaluation methods as parameters for the patent retrieval method. If there are M optional parameter settings for a method, N ranking methods and K weight evaluation methods, and L information resources, the number of features is $M*N*K*L$. The experiments focus on the effectiveness of different forms of patent retrieval methods on learning a ranking model.

3.4   LambdaMart

The performance of patent retrieval system is also evaluated by IR measures such as MAP and NDCG. Learning to rank approaches can define the ranking loss function such as cross entropy loss according to the relevance judgments. By minimizing the loss, it can learn a ranking model to improve ranking performance directly. The aim of query expansion is also to improve the performance of ranking. Therefore, learning to rank can be used to learn a model for query expansion approaches.

LambdaMART combines MART and LambdaRank. MART is a boosted tree model, a linear combination of the outputs of a set of regression trees. LambdaMART utilizes gradient boosting to optimize its loss function defined in the same way as LambdaRank. Gradient Boosting produces an ensemble of weak learner to form a strong one. LambdaRank constructs its loss function based on RankNet, whose loss function is a differentiable function of the model parameters based on cross entropy objective function. The λ for a given document in the ranking list gets contributions from all other documents under the same query with different labels. The λ can also be interpreted as a force, which indicates whether the document should move up or move

Table 1. Example of learning features of TREC-CHEM

ID	(N/M)	Ranking methods	Weight evaluation
1	10/50	BM25	BO1
2	20/100	TF	IDF
3	20/150	BM25F	BO2

down in this round of optimization and also the distance it will move. The λ for a document is the sum of λ_{ij} computed by using the formula as below.

$$\lambda_{ij} = \frac{\partial C(s_i - s_j)}{\partial s_i} = \frac{-\sigma}{1 + e^{\sigma(s_i - s_j)}} |\Delta NDCG| \tag{10}$$

Loss function C has the same form as RankNet based on a probability function combining the score of each document. LambdaRank modifies the gradient with the variation of NDCG through swapping the rank positions of the two documents. LambdaMART uses λ as the gradient of loss function and use boosted regression tree as its model to decrease ranking loss in iterations as MART does. In this paper, we mainly utilize the multiple query expansion methods to extract features for ranking model. We expect that it is effective to improve the ranking accuracies of patent retrieval.

Feature space is constructed by different parameter settings, different ranking methods, and different weight evaluation methods. Overall, there are 18 features, which can be directly used in learning algorithms. The ranking methods include TF*IDF, BM25, and BM25F, and the weight evaluation methods include BO1 and BO2. The example of feature set is shown in Table 1. Table 1 gives some details of implementation of these features, and for the parameter settings *N/M* means to extract M expansion terms from *N* documents. *N* can be set to be 10, 20, and *M* can be set to be 50, 100, and 150. Ranking methods include BM25, BM25F and TF*IDF. BO1 and BO2 are used as weight evaluation methods.

4   Experiments and Results

In this section, we show the experimental results of query expansion based on patent fields. The TREC-CHEM collection is the experimental data set. We adopt all the topics from TS (Technology Survey) task from TREC-CHEM2010 and TREC-CHEM2011 as our query set. Our research is based on data set of the subtask technology survey. This set contains TS-topics, which is manually created by human experts. Each topic has a description as a natural language expression of information need based on data described in a patent document. The systems should return a set of documents that answer this information need as good as possible. These topics are created to be interesting, so their main priority will be as similar as possible to a genuine information need of an expert searcher. We only use the patent documents in this collection. A patent document is composed of several fields, including title, abstract, description, and claims. These special text fields are used to improve the quality of expansion terms. For the information resources from Google and Derwent, we select expansion query terms from the title and abstract fields. The 6-fold cross validation is used to obtain the average results. The results are evaluated by mean average precision (MAP) and P@n.

4.1   Effectiveness of Query Expansion Based on Patent Fields

In this section, we conduct the experiment based on TREC data patent fields. We compare the method based on text field for expansion terms (short for TFET) with retrieval methods without query expansion (Original) and the oracle method (use the best feature to rank the documents of test topic of every fold). Table 2 lists the results of these methods.

Table 2. Performance comparison of ranking methods (TFET, Original, and Oracle)

Methods	P@5	P@15	P@20	MAP
Original	0.3333	0.1944	0.1708	0.2173
TFET	0.3833	0.2000	0.1750	0.2342
Oracle	0.3833	0.2278	0.1875	0.2608

From Table 2, we can see that TFET method achieves better performance than original method. Especially for MAP and P@5, the ranking performance of TFET method is much better than Original method, and is similar to the performance of Oracle method in terms of P@5. Results show that query expansion approach based on field information is indeed effective in improving the patent retrieval results. However, TFET is not as good as Oracle method in terms of other evaluation methods. The results of Oracle method come from the best ranking feature of test set of every fold. Therefore, it is feasible to develop a method considering the impact of different ranking features other than using a single ranking feature. Based on these results, the optimization of the query expansion based ranking methods for queries could be expected to further improve the retrieval performance. Now our goal is to develop an effective method to construct a ranking model based on different ranking features.

4.2   Effectiveness of Learning to Rank Model

In order to take full advantage of all the ranking methods, we introduce a learning to rank model: LambdaMART to learn a ranking model from the ranking features. In this section the TFET and Original methods serve as baseline approaches. We will examine the effectiveness of LambdaMart model whose features are extracted from TREC data sets. Table 3 lists the results of the ranking methods.

From Table 3, we can see that the LambdaMart ranking model based on TREC data is superior to TFET method in all of the evaluation methods. Moreover, the relative improvement of LambdaMart is even over that of Oracle method for P@5. And in

Table 3. Performance comparison of ranking methods (TFET, Original, TREC, and Oracle)

Methods	P@5	P@15	P@20	MAP
Original	0.3333	0.1944	0.1708	0.2173
TFET	0.3833	0.2000	0.1750	0.2342
TREC	0.4000	0.2167	0.1875	0.2469
Oracle	0.3833	0.2278	0.1875	0.2608

terms of P@20, it also achieves the same results as the Oracle method. As the information of test set is unknown in the training process and the ranking model is learned from training set as well as the feature selection of TFET, it seems that it is effective to take into account the impact of all the ranking features based on text fields for patent retrieval. It also reveals that the query expansion method based on learning to rank model can improve the ranking performance of patent retrieval.

4.3   Effectiveness of External Information Resources

On above experiments, we only use the TREC data sets for query expansion to extract the features for learning to rank approach. In this section, we also apply the Google and Derwent information resources for query expansion in order to obtain the features for the ranking model. From Table 4, we can see that the LambdaMart ranking model based on TREC data is superior to TFET method in all of terms of evaluation methods. It is also effective to improve the ranking performance by using Google and Derwent information resources. Especially when we use all the features from TREC, Google and Derwent information resources, the ranking model learned from that can achieve the best performance. It seems that it is effective to take the impact of all the information resources based on text fields into account for patent retrieval. It also reveals that the query expansion method based on learning to rank model using multiple information resource can improve the ranking performance of patent retrieval.

Table 4. Performance comparison of ranking methods

Methods	P@5	P@15	P@20	MAP
Original	0.3333	0.1944	0.1708	0.2173
TFET	0.3833	0.2000	0.1750	0.2342
TREC	0.4000	0.2167	0.1875	0.2469
Google	0.4333	0.2722	0.1875	0.2375
Derwent	0.3833	0.2555	0.1875	0.2166
All	**0.4833**	**0.3000**	**0.2541**	**0.2727**

5   Conclusion

In this paper, we explored the multiple information resources for query expansion. For TREC topics, we measure the importance of expansion terms on the retrieval performance. Our experiments show that the query expansion method is an effective approach for patent retrieval. Furthermore, we investigate the effectiveness of learning to rank model based on the query expansion ranking features. The experimental results demonstrate that, the ranking model which is based on multiple information resources, can effectively cope with the patent ranking problem. In future work, for the pseudo relevant selection method, we will try other retrieval methods to obtain more relevant documents. For the term ranking model, we plan to explore more term ranking methods for further accuracy of patent retrieval.

There are several important differences between our work and previous work on improving query expansion: (1) we examine the effectiveness of different information resources for the patent query expansion; (2) we cast the combination of information sources as an optimization problem that can be solved under a learning to rank framework; (3) we take different query expansion approaches by different resources as features for learning; (4) we apply learning to rank approach with the ranking features to improve the performance of patent retrieval.

Acknowledgement. This work is partially supported by grant from the Natural Science Foundation of China (No. 61272370, 61402075, 61572102, 61572098, 61272373), Natural Science Foundation of Liaoning Province, China (No. 201202031, 2014020003), State Education Ministry and The Research Fund for the Doctoral Program of Higher Education (No. 20090041110002), the Fundamental Research Funds for the Central Universities.

References

1. Xu, J., Croft, W.B.: Query expansion using local and global document analysis. In: Proceedings of the 19th Annual International ACM SIGIR Conference on Research and Development in Information Retrieval, pp. 4–11. ACM (1996)
2. Cronen-Townsend, S., Zhou, Y., Croft, W.B.: A framework for selective query expansion. In: Proceedings of the Thirteenth ACM International Conference on Information and Knowledge Management, pp. 236–237. ACM (2004)
3. Metzler, D., Croft, W.B.: Latent concept expansion using Markov random fields. In: Proceedings of the 30th Annual International ACM SIGIR Conference on Research and Development in Information Retrieval, pp. 311–318. ACM (2007)
4. Tao, T., Zhai, C.: Regularized estimation of mixture models for robust pseudo-relevance feedback. In: Proceedings of the 29th Annual International ACM SIGIR Conference on Research and Development in Information Retrieval, pp. 162–169. ACM (2006)
5. Lee, K.S., Croft, W.B., Allan, J.: A cluster-based resampling method for pseudo-relevance feedback. In: Proceedings of the 31st Annual International ACM SIGIR Conference on Research and Development in Information Retrieval, pp. 235–242. ACM (2008)
6. Magdy, W., Jones, G.J.: An efficient method for using machine translation technologies in cross-language patent search. In: Proceedings of the 20th ACM International Conference on Information and Knowledge Management, pp. 1925–1928. ACM (2011)
7. Ganguly, D., Leveling, J., Magdy, W., et al.: Patent query reduction using pseudo relevance feedback. In: Proceedings of the 20th ACM International Conference on Information and Knowledge Management, pp. 1953–1956. ACM (2011)

8. Leveling, J., Magdy, W., Jones, G.J.: An investigation of decompounding for cross-language patent search. In: Proceedings of the 34th International ACM SIGIR Conference on Research and Development in Information Retrieval, pp. 1169–1170. ACM (2011)

9. Xu, K., Liu, W., Lin, H., et al.: Patent query expansion using text fields. J. Comput. Inf. Syst. **8**(13), 5607–5614 (2012)

10. Xu, K., Lin, H., Liu, W., et al.: Learning to rank based query expansion for patent retrieval. J. Comput. Inf. Syst. **9**(13), 5387–5394 (2013)

11. Diaz, F., Metzler, D.: Improving the estimation of relevance models using large external corpora. In: Proceedings of the 29th Annual International ACM SIGIR Conference on Research and Development in Information Retrieval, pp. 154–161. ACM (2006)

12. Lin, Y., Lin, H., Jin, S., et al.: Social annotation in query expansion: a machine learning approach. In: Proceedings of the 34th International ACM SIGIR Conference on Research and Development in Information Retrieval, pp. 405–414. ACM (2011)

13. Xu, Y., Jones, G.J., Wang, B.: Query dependent pseudo-relevance feedback based on wikipedia. In: Proceedings of the 32nd International ACM SIGIR Conference on Research and Development in Information Retrieval, pp. 59–66. ACM (2009)

14. Liu, T.-Y.: Learning to rank for information retrieval. Found. Trends Inf. Retrieval **3**(3), 225–331 (2009)

15. Konishi, K.: Query terms extraction from patent document for invalidity search

16. Itoh, H., Mano, H., Ogawa, Y.: Term distillation in patent retrieval. In: Proceedings of the ACL-2003 Workshop on Patent Corpus Processing, vol. 20, pp. 41–45. Association for Computational Linguistics (2003)

17. Xue, X., Croft, W.B.: Transforming patents into prior-art queries. In: Proceedings of the 32nd International ACM SIGIR Conference on Research and Development in Information Retrieval, pp. 808–809. ACM (2009)

18. Liu, S.-H., Liao, H.-L., Pi, S.-M., et al.: Development of a patent retrieval and analysis platform–a hybrid approach. Expert Syst. Appl. **38**(6), 7864–7868 (2011)

19. Mahdabi, P., Crestani, F.: The effect of citation analysis on query expansion for patent retrieval. Inf. Retrieval **17**(5–6), 412–429 (2014)

20. Chen, Y.-L., Chiu, Y.-T.: An IPC-based vector space model for patent retrieval. Inf. Process. Manag. **47**(3), 309–322 (2011)

21. Rusiñol, M., de las Heras, L.-P., Terrades, O.R.: Flowchart recognition for non-textual information retrieval in patent search. Inf. Retrieval **17**(5–6), 545–562 (2014)

22. Wanagiri, M.Z., Adriani, M.: Prior art retrieval using various patent document fields contents. In: CLEF (Notebook Papers/LABs/Workshops) (2010)

23. Rocchio, J.J.: Relevance feedback in information retrieval. In: Proceedings of the Smart Retrieval System, pp. 313–323 (1971)

24. Robertson, S.E., Walker, S., Beaulieu, M., et al.: Okapi at TREC-4. In: Proceedings of the Fourth Text Retrieval Conference, pp. 73–97. NIST Special Publication (1996)

25. Lavrenko, V., Croft, W.B.: Relevance based language models. In: Proceedings of the 24th Annual International ACM SIGIR Conference on Research and Development in Information Retrieval, pp. 120–127. ACM (2001)

26. Zhai, C., Lafferty, J.: Model-based feedback in the language modeling approach to information retrieval. In: Proceedings of the Tenth International Conference on Information and Knowledge Management, pp. 403–410. ACM (2001)

27. Yeung, D.L., Clarke, C.L., Cormack, G.V., et al.: Task-specific query expansion (MultiText Experiments for TREC 2003). In: TREC, pp. 810–819 (2003)

28. Wu, Q., Burges, C.J., Svore, K.M., et al.: Adapting boosting for information retrieval measures. Inf. Retrieval **13**(3), 254–270 (2010)

29. Salton, G., Wong, A., Yang, C.-S.: A vector space model for automatic indexing. Commun. ACM **18**(11), 613–620 (1975)
30. Robertson, S., Zaragoza, H., Taylor, M.: Simple BM25 extension to multiple weighted fields. In: Proceedings of the Thirteenth ACM International Conference on Information and Knowledge Management, pp. 42–49. ACM (2004)
31. Amati, G.: Probability Models for Information Retrieval Based on Divergence from Randomness. University of Glasgow (2003)

Simulating Ideal and Average Users

Matthias Hagen[(✉)], Maximilian Michel, and Benno Stein

Bauhaus-Universität Weimar, Weimar, Germany
{matthias.hagen,maximilian.michel,benno.stein}@uni-weimar.de

Abstract. We propose a framework for deterministic simulation of user behavior that allows to analyze the cost-gain-based performance on single result lists or whole search sessions. The ideal user representing optimal behavior (i.e., most gain with lowest effort) is contrasted with more "average" users that employ the spreading activation model from cognitive theory. On TREC Session Track data, the ideal user achieves about double the gain of real users at the same costs while the average gain of our different simulated users correlates well with the session-DCG metric—another argument for that metric in session-based evaluation.

1 Introduction

Analyzing search logs is a common way to study users and their information needs and also for evaluating search systems in for instance A/B tests—assuming that users more likely click on relevant documents. However, such evaluations require huge user populations that the commercial web search engines certainly have but that are lacking in many other settings (e.g., enterprise search or academic research). To overcome this problem of scarce user data, simulating user behavior got more prominent over the last years [14,15]. We propose a framework to deterministically simulate user behavior over search sessions in cost-gain-based scenarios. Our focus is on the click and result list switching behavior leaving the integration of simulated query formulation for future work. One contribution is the ideal user with optimal behavior (e.g., clicking on only those results that lead to some gain). In contrast, we also contribute more "average" users who employ a cognitive model to base click decisions on the shown result snippets. Furthermore, given pre-defined queries of a search session, the user models also decide when to switch to the next query. Each session is restricted by a predefined cost budget (e.g., time-based), every action (clicking, querying, reading) comes with some costs. Therefore, the simulated users assess each decision not only by its potential benefits in form of information gain, but also according to the accompanying costs. We compare the simulated users to real users on TREC session track data and show that the average information gain of our models highly correlates with the session-DCG measure often used in evaluation. Interestingly, the ideal user achieves about double the performance of real users at the same costs.

© Springer International Publishing AG 2016
S. Ma et al. (Eds.): AIRS 2016, LNCS 9994, pp. 138–154, 2016.
DOI: 10.1007/978-3-319-48051-0_11

2   Related Work

We briefly review the literature on search evaluation and user modeling; more references follow in the sections detailing our approach.

Search Evaluation. Over time, the measures for evaluating search results have changed from precision and recall to more rank-oriented metrics. One first example is MAP (mean average precision): the precision is measured at the ranks of the relevant results. The underlying assumption of MAP in form of a user model would be that the user clicks on only the relevant results and stops when all relevant documents have been visited—a scheme we will use in our simulated ideal user. Alternatives to MAP are normalized discounted cumulative gain (nDCG) [24] where results have different relevance levels (i.e., information gain) and lower ranked results are less likely to be seen (i.e., discounted gain) or expected reciprocal rank (ERR) [16] following a cascading model where the probability that a user views a result depends on its rank position and a stopping criterion. In order to evaluate whole search sessions, Järvelin et al. also introduced a session-variant of nDCG [25] with the results of later queries having discounted gains. In our simulation framework we employ a cascading scheme with cost-based stopping criteria but instead of discounting gain for lower ranks—except that we assume no gain from showing the same or similar results again—we take the higher costs for viewing lower-ranked results into account.

Over the last years, several user studies found that MAP has a weak correlation with real user performance [41], that the information gain of real users correlates with the precision overall [37], and that the preference for some ranking strongly correlates with its nDCG and ERR score [34]. Although the experimentation setup usually does not resemble the process of a real web search, many studies agree that evaluation metrics like ERR resemble the users' performance in general, but they also claim that Cranfield-style evaluation metrics lack realism and sound user models [36]. As a more realistic metric, simulation-based time-biased gain (TBG) was recently proposed [36]. Each user action (view summary or document, save document) comes with a time-based cost in a semi-Markov model (initialized with data from 48 real users who solved some pre-defined tasks within 10 min). The simulation is then used to estimate the information gain for different time limits and rankings and the performance variance. This idea very much inspired our scheme but instead of non-deterministic users we simulate more "general" deterministic user types reflecting the ideas of existing standard evaluation metrics. Our framework allows to compare an optimal or average deterministic user (i.e., perfect or average decisions) to a real user and to measure the spread of the gain differences of optimal and average behavior.

User Modeling. User modeling deals with predicting and explaining user behavior and intentions. For instance, O'Brien and Keane [31] compare clicks predicted by the SNIF-ACT spreading activation model of information scent [21] to real users. They show that a cascading threshold strategy (top-down assessment of search results, clicking if result is above some threshold) is more common among users and that it is favorable to a comparative strategy (first assessing all snippets, then clicking on the most relevant). We will employ both, thresholding and spreading activation, in two of our user models. But in addition to O'Brien and Keane's model we also take switching to another query into account. User click models describe the click behavior while interacting with a search engine. Such models can be used to infer document preferences from the click-through rates in query logs [17]. In contrast, Zhang et al. claim that user behavior is related to the information task as a whole and therefore, the click behavior depends on previous queries and clicks for the same information task [42]. Consequently, task-centric click models use the complete search session in order to infer the relevance of results (e.g., duplicate results are less likely to be clicked again)—an idea we adopt for our simulation. Still, probabilistic click models are not really applicable in our scarce-user scenario since they typically rely on the availability of huge search logs and we aim for deterministic models instead.

3   Our General User Model

An information-seeking user approaches a search engine to satisfy an information need. For non-trivial tasks, the user typically submits several queries, scans their results and clicks on the ones whose snippets appear to be relevant—forming a search session. In this section, we propose a general user model that represents the space of all interaction sequences (we call them *paths*) a user might follow in a search session. Typically, search sessions are characterized by the respective query reformulations [22]. Note however, that we will concentrate on how users navigate through the result lists of a search session and we will not simulate query (re-)formulation.

3.1   The Framework

Basic assumption of our general user model is that a user wants to gain information in order to satisfy an information need against a retrieval system. The respecitve interactions come with certain costs (usually time but it could also be monetary charges for API querying etc.). The user has to find a trade-off between costs and benefits since the total

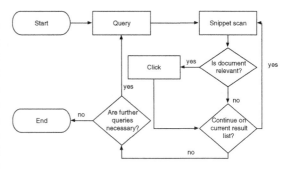

Fig. 1. Flowchart of our general user model.

"budget" for a search session typically is limited; leading to cost-driven behavior [3,4,7,8]. Our set of possible actions is similar to the elementary action types of Baskaya et al. [10]. Each session S consists of at least an initial query q_1, and a potentially empty list of subsequent queries q_2 to q_n. Each query q has an associated cost $cost_q(|q|)$ that depends on the length of the query (assumption: longer queries require more "effort"). After a query is submitted, the retrieval system returns a ranked result list with short snippets. The user starts scanning those snippets from top to bottom. Each scan of a snippet s has an associated cost $cost_{sc}$ that we assume to be a constant (assuming snippets of about equal length but non-constant length-dependence is also possible). In our model, at least one snippet is scanned following a query before another action can be performed. From scanning a snippet, the user estimates the result's relevance. If the result appears to be relevant, the user clicks on it. Each click c has some cost $cost_{cl}$ that we also assume to be constant (variable cost again is not difficult). A click leads to an information gain corresponding to the result's relevance level rel (i.e., the total gain is achieved with just one click assuming the whole document to be "read" at once) with one exception: no gain from a second click on the same or a similar result (cosine similarity). Consequently, relevance and thus click decisions not only depend on snippet relevance assessment but also on the previous clicks. After each snippet scan and after each click, the user decides if they proceed with scanning the next snippet or if they submit a new query. A search session ends when there are no further queries necessary or a given cost budget is reached—of course, the budget should suffice for at least submitting all pre-defined queries. Following others [11,30,38], Fig. 1 depicts an abstract flow-chart of our general user model including three kinds of decisions: (1) whether to click on a result, (2) whether to submit a new query, and (3) whether to end the session. Our simulated users instantiate schemes for those three decisions.

3.2 Restrictions of the General Model

Our general user model forms an abstraction of complex cognitive processes that might differ from user to user; consequently, not all possible search behavior can be expressed within our general model. For instance, the only way to cumulate information gain in our model is to click on a new result after a snippet scan. However, the user may already find the desired information in the snippet—a case we do not include in the current abstraction. We also assume a top-down processing of the result lists, starting with the first item in the result list. Through eye-tracking studies, Klöckner et al. found that this depth-first strategy is used by a majority of users [28]. Still, our user model does not represent the other around 15 % of users. Furthermore, in our general model the users assess a document's relevance right after scanning its snippet and click on it if the relevance exceeds some threshold. This is in line with studies of O'Brien et al. who show that thresholding is the most common user strategy [31] but the information foraging theory, for instance, states that users might also first assess all results and then decide to click on the one with the most gain [32]—a strategy that we do not model. Finally, we assume a cascading scheme where the user does not go back to a previous result list. The only way to see the results again would be to submit the same query (at the same costs). Such back-and-forth switching at lower costs is an interesting future simulation direction—also for query suggestion evaluation.

4   The Ideal User

First, we propose to simulate an ideal user: accumulating the most information gain for a certain cost "budget." Given the interaction costs and a search session with result lists and relevance judgments, the task is to find an optimal sequence of interactions within our general model. We call an interaction sequence a *path* through the state space formed by a session. A state is characterized by the lowest click or snippet scan in the different result lists and by the result currently in focus. Possible interactions form the edges connecting such states. The path of the ideal user shows how deep in the individual result lists a perfect behavior would scan snippets and which results should be clicked.

According to our general user model, three kinds of decisions have to be instantiated: clicking, switching to the next query, and ending the search session. Remember that we do not model query formulation but require pre-defined queries. The knowledge of the query sequence is used for the stopping criterion. We assume that each query of the sequence is submitted such that the user can only finish a session on that last query. Since the ideal user only clicks on results that lead to some gain, the crucial point of modeling the ideal user is the decision of when to change to a new query result list—very recently, independent of our investigations, optimal switching has also been investigated by Smucker and Clarke in a slightly different context [35].

Let l denote the rank in the result list R at which the ideal user stops scanning and switches to the next query (e.g., $l = 10$ means scanning the first 10 snippets). Whenever the ideal user encounters a result $r \in R$ not similar to a previously clicked document with a relevance level $rel(r)$ above a relevance threshold τ_{rel}, a click on the result is performed at the click cost $cost_{cl}$. The document is then added to the list *Clicked* of clicked documents. The accumulated cost $Cost(l, q, R_q)$ and gain $Gain(l, q, R_q)$ for a query q and its result list R_q with limit l is

$$Cost(l, q, R_q) = cost_q(|q|) + \sum_{i=1}^{l} cost(r_i), \text{ where}$$

$$cost(r_i) = \begin{cases} cost_{sc} + cost_{cl}, & \text{if } rel(r_i) \geq \tau_{rel} \text{ and } r_i \text{ not similar to sth. in } Clicked, \\ cost_{sc}, & \text{otherwise.} \end{cases}$$

$$Gain(l, q, R_q) = \sum_{i=1}^{l} gain(r_i), \text{ where}$$

$$gain(r_i) = \begin{cases} rel(r_i), & \text{if } rel(r_i) \geq \tau_{rel} \text{ and } r_i \text{ not similar to sth. in } Clicked, \\ 0 & \text{otherwise.} \end{cases}$$

Determining the ideal search behavior forms a multiple-choice knapsack problem. For each result list R_q of each query q in the session S, we have to choose a limit l_q such that the total cumulated information gain is maximized and a given cost budget $cost_{max}$ is not exceeded.

$$\text{maximize} \sum_{q,R_q \in S} Gain(l,q,R_q) \quad \text{while} \quad \sum_{q,R_q \in S} Cost(l,q,R_q) \leq cost_{max}$$

Multiple-choice knapsack is NP-hard [27]. In order to prune the problem space, we omit *dominated* states that can never be part of an optimal solution: For result list R_q of query q, a limit l is dominated by a limit $l' \neq l$ iff either $Cost(l,q,R_q) > Cost(l',q,R_q)$ and $Gain(l,q,R_q) \leq Gain(l',q,R_q)$ or $Cost(l,q,R_q) \geq Cost(l',q,R_q)$ and $Gain(l,q,R_q) < Gain(l',q,R_q)$. For a sample result list with the relevant entries at ranks 1, 3, and 6, these ranks form the dominating limits. A limit at rank 2 is dominated by the limit at rank 1 since both lead to the same information gain but the limit at rank 2 has higher costs. Limits at ranks 4 or 5 are dominated by the limit at rank 3, etc. For determining the click behavior of the ideal user, each relevant result not similar to something clicked before represents a dominating limit.

In order to derive an optimal interaction sequence (i.e., ideal behavior), we have to choose from each result list in the session the limit that leads to an optimal gain for the whole session (i.e., the highest information gain possible for a given cost budget). There are several algorithmic solutions for such a multiple-choice knapsack problem like a dynamic programming approach [33] or a branch-and-bound strategy [20]. However, we cannot apply these approaches since we do not allow for clicking a relevant document if something similar has been clicked before. Hence, each click has a potential influence on the information gain of later results. If the user clicks on a relevant result in the current list, similar entries are no longer relevant in the next lists. In other words, we cannot treat the result lists independently but every combination of dominating states has to be checked for finding an optimal sequence. Let a path $P = < l_1, \ldots l_n >$ through a search session S be a list of limits for every result list. We call P a d-path, if only dominating limits are included. Let $\mathcal{P}$ be the family of all possible d-paths. In order to find a d-path that represents ideal user behavior, we derive the total cost $Cost(P,S)$ and gain $Gain(P,S)$ for every d-path $P \in \mathcal{P}$ as

$$Cost(P,S) = \sum_{\substack{l_q \in P, \\ q,R_q \in S}} Cost(l_q,q,R_q) \quad \text{and} \quad Gain(P,S) = \sum_{\substack{l_q \in P, \\ q,R_q \in S}} Gain(l_q,q,R_q).$$

From the d-path family we algorithmically choose a d-path that does not exceed the cost limit and that has the highest gain as follows. The dominating limits in every result list in the session are set to the ranks of the relevant results. All the combinations of all dominating limits of every result list then form the family $\mathcal{P}$ of possible d-paths. From this family, an ideal d-path P_{ideal} for a cost budget $cost_{max}$ is derived by first removing from $\mathcal{P}$ all d-paths that exceed the cost limit and then choosing one d-path with the highest gain. Note that clicks on similar results will not be part of such a path as long as the budget is not too high (since they do not yield any gain in our scenario) and that he resulting path is an optimal sequence of interactions given the cost budget—the ideal user behavior.

5   Spreading Activation Users

To simulate ideal click behavior, relevance judgments have to be "known" to the user. When no relevance information is available, we need another strategy for deterministic click decisions. We propose a cognitive approach employing the task description and shown snippets to this end.

5.1   Cognitive Modeling and Spreading Activation

Cognitive models explain basic cognitive processes (e.g., learning and decision making) and their interactions in more complex processes. Their big advantage over statistical models is that instead of inferring a posterior description from generated data, explanations for cognitive processes can be found in an inductive way [13]. One example of cognitive modeling is Pirolli and Card's information foraging theory [32] stating that users searching for information are faced with traces of navigational cues (e.g., links) emitting *information scent* and that the cue with the most information scent will be followed. This rational behavior aims for an effective trade-off between cost and benefit and matches our general user model. However, we will not employ the costly comparison strategy of the original model but only use the cognitive SNIF-ACT architecture [21]; calculating information scent with the help of the spreading activation model.

Fu and Pirolli use the spreading activation model to estimate the utility of navigational choices [21]. The neuronal structure of the brain is modeled as an associative network consisting of interconnected concepts with different association strengths as in Anderson et al.'s cognitive architecture ACT-R [1]. When the user reads a document or a snippet, some of the concepts in the associative network are *activated*. This activation then spreads through the network and may activate other concepts depending on the associative strength. In our context, two regions in the associative network are important for the snippet relevance assessment: the region that is activated by reading the snippet (the perception), and the region that represents the user's focus and intention (i.e., the topic description in TREC scenarios). While scanning a snippet, the user model encounters the concepts in the snippet and these network nodes are activated and spread through the network to eventually activate topic description concepts. The relevance is then assessed according to the total activation level of the description concepts; if the activation is above a certain threshold, the document is perceived as relevant and a click is performed.

Concept Extraction. The head-noun phrase extractor [9] is used to identify concepts in task descriptions and snippets. On average, document snippets contain fewer terms than a TREC task description (34 vs. 42) but both have similar number of concepts (8 vs. 10). We also removed some more "instructional" concepts like `find information` contained in many TREC descriptions.

Spreading Activation Calculation. The concepts extracted from the topic descriptions and the document snippets form the nodes of a network. As for the edges (i.e., the activation strength), we simplify the relevance assessment situation to a bipartite directed graph. The concepts extracted from a scanned snippet form one node subset (the perception) and the concepts from the task description form the other (the focus). We assume that all snippet concepts are connected to all description concepts and omit any activations that may spread between concepts of one side. Based on this simplified network, we compute the total activation level A of the task concepts C_T that spread from the snippet concepts C_S. The activation level of a snippet is modeled as the sum of the attentional weighted association strength of every concept in C_T and every concept in C_S as $A(C_S, C_T) = \sum_{i \in C_T} \sum_{j \in C_S} association(i,j) \cdot attention(j)$ [21]. The formula includes a length normalization preventing unbounded activations and includes a temporal decay of activation following the assumption that a user spends more attention on the first concepts of a snippet. We follow Fu and Pirolli [21] using the exponential decay function $attention(j) = a \cdot e^{b \cdot j}$ and setting the scaling parameter $a = 1$ and the decay parameter $b = -0.1$. As for the association strength $association(i,j)$ between two concepts, we use the pointwise mutual information (PMI) [18] of $\log \frac{p(i,j)}{p(i)p(j)}$ approximating the probabilities $p(i,j)$, $p(i)$ and $p(j)$ with the normalized document frequencies df/N from the English Wikipedia, where N is the total number of Wikipedia articles. In a study comparing PMI to (generalized) latent semantic analysis as measures for association strength, Budiu et al. found that PMI is the most efficient method for identifying semantic similarities [12]. Following their suggestion, we use a window of 16 terms to derive the document frequencies $df(i,j)$.

Relevance Thresholding. The total activation level A indicates how relevant a result appears to the user after the snippet scan. To distinguish between relevant results that should be clicked and non-relevant results that should not be clicked, an activation threshold τ_{act} is part of the spreading activation model. We set the binary relevance of a snippet S and a task description T to

$$rel(C_S, C_T) = \begin{cases} 1 & \text{if } A(C_S, C_T) \geq \tau_{act}, \\ 0 & \text{otherwise}, \end{cases}$$

and propose two ways for setting the threshold τ_{act}: a static constant extracted from user interaction logs and a dynamic variant adapted to the rank bias favoring clicks on the first ranks.

Static Threshold. To determine a static threshold, we use the TREC 2012 Session track logs. We compute the activation level of every result snippet and let the relevance judgments ≥ 2 form the relevant class. The mean activations of relevant and non-relevant results then are significantly different (22.8 vs. 12.8, $p \ll 0.01$ for a t-test). To choose a thresholding strategy, we compare the F-scores of a maximum a posteriori estimation (MAP) threshold, a likelihood comparison variant of MAP ignoring the prior probabilities, and an oracle threshold chosen to yield the best possible F-score. Due to the big difference of the prior probabilities (only 20 % of the results are relevant), the conservative MAP estimation had a lot of false negatives (F-score of 0.19) such that we choose the likelihood estimation as our static threshold that comes pretty close to the artificial best possible F-score method in our pilots (F-score of 0.47 vs. 0.48).

Dynamic Threshold. The underlying assumption of our dynamic threshold is a rank bias on the user side meaning that the users get more and more "skeptical" at lower ranks requiring a higher activation for a click. We model this assumption as follows. The user starts with a fixed activation threshold for the first rank that may very well represent rank bias by setting the initial $\tau_{act} = 0$ resulting in a blindfold click on the first rank. Every further result on a lower rank must have a higher activation level than the last clicked result; hence, the activation τ_{act} is monotonically growing. This dynamic thresholding is inspired by findings of Kean and O'Brien on users' rank bias [26] but in our cost-based model also resembles the fact that a mediocre result accessible at low costs may still be more appealing than a result with high relevance at a low rank. Hence, dynamic thresholding also models the *satisficing* behavior, meaning that the user prefers a fast and sufficient decision over evaluating all possible actions in order to find the optimum [29].

6   Our Analyzed User Models

Our general user model requires two components: (1) the *click behavior* of when to click on a result, and (2) the *stopping strategy* of when to switch to the next query and when to end the session.

6.1   Click Behavior

We propose three kinds of click behavior. First, the *optimal* click behavior of users who only click on relevant results—like the ideal user introduced in Sect. 4. Second, *activation-based* click behavior inspired by the spreading activation model introduced in Sect. 5—potentially leading to non-optimal clicks on non-relevant results. Third, a simple *click all* approach whose click decisions are independent of the relevance of a result: every result that is scanned is also clicked. This click behavior probably is the least cost efficient one, since clicking every result means also clicking every non-relevant result among the scanned results.

6.2   Stopping Strategies

We propose four simple stopping strategies following previous research. Zhang et al. observed that users tend to click more at the end of a session [42]. Their explanation is that with every query reformulation the user improves the quality of the query and eventually ends up with a "best" query. The user probably scans some of the results in earlier queries but invests most of their budget for the last results. Our respective *prefer-last* stopping strategy is formally defined as follows. Let a path P consist of a list of limits $l_1 \ldots l_n$ that represent the lowest rank the user views in each result list. A path P is a prefer-last path iff $l_i \leq l_{i+1}, i = 1, \ldots, n-1$. In contrast to the findings of Zhang et al., the user model of the session-nDCG metric is based on the assumption that results of reformulated queries are less valuable since the user has to invest more effort [25]. According to this model, the user would prefer results of the first queries—yielding a *prefer-first* stopping strategy. A path P is a prefer-first path iff $l_i \geq l_{i+1}, i = 1, \ldots, n-1$. To model the stopping strategy of the ideal user, we propose the highest-gain strategy. A user following this strategy views as many documents that appear to be relevant as possible for a given cost budget. A user model with optimal clicking behavior and highest gain strategy represents the ideal user. Let $\mathcal{P}$ be the family of all possible paths for a given cost limit and search session and let $gain(P)$ the accumulated information gain of a path P. A path P is a highest-gain path iff $gain(P) = max\{gain(P') : P' \in \mathcal{P}\}$. Similarly, to model more "average" users, we also propose a *median-gain* strategy where the user accumulates an information gain that represents the median of all information gains of all possible paths through a search session for a given cost limit. A path P is a median-gain path iff $gain(P) = median\{gain(P') : P' \in \mathcal{P}\}$.

6.3   Combining Clicking and Stopping

In order to simulate a certain click behavior and stopping strategy for a given search session, we identify paths through the search session that do not exceed the cost budget and that represent the stopping strategy. Finding such a path involves three steps. (1) For each result, determine whether it is clicked based on the click behavior. (2) Determine the family of all paths that do not exceed the cost budget. (3) From the path family, choose a path that matches the stopping strategy and has the highest information gain. From the 16 possible combinations, we further investigate all combinations with the highest-gain strategy (the ideal user with optimal clicks, the dynamic/static activation clicks, and the click-all user), the median user with optimal click behavior and median-gain strategy, and the prefer-first/-last users with clicking-all behavior and prefer-first/-last strategies. While the prefer-first/-last users represent assumptions from the literature, the median user somewhat represents an "average" user and the ideal user represents experts with perfect judgments from reading a snippet. The activation user models represent users without perfect click decisions and they can even be simulated in scenarios without relevance judgments. Although the click-all user seems very trivial, we include it in our considerations since it somewhat represents the envisioned user of a perfect retrieval systems. If the click-all user achieves the same information gain at the same costs as the ideal user, the ranking of the result list is perfect.

6.4   The TREC Session Track User

In the course of the TREC Session Track, logged interactions of real users were provided for several topics. We compare our simulated models to these users by modeling the *TREC user* whose behavior follows the originally logged data. In general, we expect the TREC user's performance to differ a lot from the ideal user in terms of information gain since a human user will not be able to optimally assess relevance from snippets, will have a rank bias, and will not make perfect stopping decisions. We instantiate the TREC user model for each search session in the TREC Session Track data as if they were produced following our general user model framework. This assumes top-down scanning, at least one snippet scan per result list, scans of all snippets of ranks above clicked results, etc.

7   Evaluation

We conduct experiments on data from the TREC Session Track 2011–2013 comparing our models with respect to information gain and cost usage, and analyzing the relation to traditional effectiveness metrics.

7.1   Accumulated Information Gain

The budget for a session is set to the time the original user's interactions would need in our general model setup. Following observations of Tran and Fuhr [39] we assume 2 s for a snippet scan and 15 s for a click, and following observations of Arif and Stuerzlinger [2] a query costs 1 s per term. The original TREC data consists of 288 search sessions for 160 topics. However, for 188 sessions none of the models (including the original TREC user) can achieve any information gain given the budget (i.e., no relevant results at all or too low in the lists). For our evaluation, we use the remaining 110 sessions and assume the gain per clicked result to correspond to the relevance score in the TREC Session Track judgments.

Table 1 shows the characteristics of the accumulated information gain distribution. The ideal user performs best, followed by the median user. The click-all user, the static activation user and the TREC user have about the same average performance. Interestingly, the ideal user almost doubles the performance of the original TREC user at the same cost. The prefer-first user is significantly better than the prefer-last user: the TREC search sessions seem to have more relevant documents in the first result lists.

To identify correlating user models, we compute the Spearman's rank correlation coefficient for each of the 136 possible pairs among the TREC user and the

Table 1. Average accumulated gain on the TREC Session Track 2011–2013 data.

	Ideal	Median	Act. St.	Click all	TREC	Act. Dyn.	Pref. First	Pref. Last
mean	2.7	1.9	1.5	1.5	1.4	1.2	1.2	1.0
med	2.0	1.0	1.0	1.0	1.0	1.0	1.0	1.0
std	2.5	1.4	1.5	1.9	1.8	1.3	1.2	1.2
max	18.0	9.0	9.0	16.0	15.0	5.0	4.0	6.0

16 different user models possible from our four click behaviors and four stopping strategies. The user models with the same click behavior correlate more than user models with the same stopping strategy and the choice of the click behavior has a higher impact on the user model's performance than the choice of the stopping strategy. The user models with the highest correlation to the TREC user are the model with dynamic activation clicks and prefer-first stopping strategy (Spearman's rank correlation test $\rho = 0.65$, $p < 0.01$) and the dynamic activation user model with highest gain strategy ($\rho = 0.62$, $p < 0.01$). This again reflects the rank bias of real users (dynamic thresholds) and supports the model underlying the session-nDCG metric (prefer-first).

7.2 Cost Usage

Figure 2 shows the distribution of the cost spent by the TREC user as a portion of the "maximum cost," the cost needed to click on all relevant documents in a session (including scanning all previous snippets). On average, the TREC user used 71 % of the maximum cost; for half of the sessions the user invested 61 % of the maximum cost reflecting the satisficing theory we briefly discussed in the thresholding part. However, in 19 % of the sessions, the TREC user invests even more effort than necessary; mostly in sessions where few relevant results are found but more are clicked.

In order to compare how the user models use the cost budget, we also analyze the interactions for which some cost is spent. All user models spend the most cost on clicking but the ideal user and the median user invest approximately equal amounts for the different interactions; they scan way more results than they click. He and Wang [23]

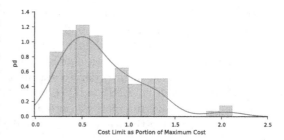

Fig. 2. Cost limits of the logged users in the TREC data.

and Tran and Fuhr [40] also suggest Markov models to investigate search behavior. A Markov model consists of a set of states and transition probabilities with the assumption that the probability of transitioning to the next state is only

Table 2. Transition probabilities between query q, click c, snippet scan s, and end e.

	TREC	Ideal	Median	Act. St.	Act. Dyn.	Click all	Pref. First	Pref. Last
$q \to s$	1.00	1.00	1.00	1.00	1.00	1.00	1.00	1.00
$s \to q$	0.03	0.09	0.12	0.07	0.01	0.02	0.03	0.03
$s \to s$	0.56	0.52	0.51	0.30	0.35	0.03	0.00	0.02
$s \to c$	0.39	0.34	0.31	0.59	0.61	0.93	0.93	0.93
$c \to s$	0.55	0.47	0.34	0.57	0.45	0.58	0.50	0.47
$c \to q$	0.25	0.28	0.35	0.25	0.31	0.23	0.28	0.28
$s \to e$	0.02	0.05	0.07	0.05	0.02	0.02	0.04	0.02
$c \to e$	0.20	0.26	0.32	0.19	0.24	0.19	0.22	0.25

dependent on the current state. The transition probability between a state a and a state b is $p(a \to b)$ given by the relative occurrence frequency.

Table 2 shows the transition probabilities of our user models and the TREC user. The user models differ the most in the probability $p(s \to s')$ of transitioning from one snippet scan to the next snippet scan and the probability $p(s \to c)$ of transitioning from a snippet scan to a click. For the ideal user, the median user, and the TREC user it is more likely to continue with the next snippet scan, for the other user models it is more likely that they will click.

7.3   Simulated Users and Evaluation

We compare the average estimated information gain of our simulated users to the "traditional" metrics session-discounted cumulative gain (sDCG), expected reciprocal rank (ERR) and MAP on the sessions of the TREC Session Track.

The behavior of our simulated user models is cost-driven such that we can describe the accumulated information gain on a search session as a function $Gain(cost_{max})$ of the cost budget. In order to give an estimate on how much information gain a user model will accumulate, we need to take into account how the users choose their cost limit. Let $f(cost_{max})$ be a probability density function that represents the likelihood of choosing a cost limit. This cost limit likelihood function is normalized such that the integral between the minimum and the maximum of the function equals 1. Smucker and Clarke [36] proposed to use such a function f in order to estimate the accumulated information gain E of a session S as $E(S) = \int_0^\infty Gain(S, cost_{max}) \cdot f(cost_{max}) \, dcost_{max}$. The probability density function we obtain is the curve in Fig. 2 approximated using a kernel density estimation. The cost budget is normalized with the maximum cost (i.e., the cost needed to click all relevant results in a session S): $maxcost(S) = cost_{scan} \cdot |D| + cost_{click} \cdot |D_{rel}| * \sum_{i=1}^{|S|} cost_{query} \cdot |q_i|$, where $|D|$ is the number of results in the session and $|D_{rel}|$ is the number of relevant results.

In order to calculate the estimated information gain for a user model and a session, we sum the gain and the likelihood of the cost budgets between 0 and an upper bound. We set this upper bound to $2.5 \cdot maxcost(S)$ since this is the highest cost limit any real user has spend in any session (cf. Fig. 2). As an increment $incr$ for the budgets we use the cost it takes to scan one snippet and perform one click. The estimated gain E of a session S then can be calculated as $E(S) = \sum_{i=0} Gain(S, incr(i)) \cdot F(i)$, where $F(i) = \int_{incr(i-1)}^{incr(i)} f(cost_{max}) \, dcost_{max}$ and $incr(i) = i \cdot (cost_{click} + cost_{scan})$. The rectangle method can be used to calculate an efficient approximation of the integral of the cost limit likelihood function F in one incrementation step i. We derive the estimated information gain of each of our seven simulated user models for the TREC Session Track 2011–2013 data and compute the correlation with the sum of the individual ERR values of the result lists, the mean of the summed average precisions of the result lists (MAP), and the session-DCG. Among the individual pairs, the highest correlation of 0.91 is between the average estimated information gain of our deterministic user simulations and session-DCG. The MAP metric correlates the least with the other metrics and our simulations (0.73). These correlations show that based on user simulations, the session-DCG metric is very reasonable. An interesting future metric could be formed by the difference of the ideal user to the more average median or activation users. If system A has better ideal user gains than system B but lower average/activation user gains, real users behaving more "average" and probably using snippet activation of some kind in their click decisions would prefer system B—which also is another argument for working on highly informative snippets giving a clue on actual result relevance.

8   Conclusion

We propose a framework to simulate deterministic user models with different stopping strategies and click behaviors. The goal is to use the simulations to better understand and evaluate user behavior in search sessions or query suggestion scenarios without requiring a huge online user population. We measure the effort of a simulated user by assigning costs to every interaction and contrast that with the achieved information gain. One of models is the ideal user with optimal click behavior and a high-information gain stopping strategy representing the perfect trade-off between cost and gain (i.e., the highest information gain possible for a given cost budget). More "average" variants are the median user with decisions towards achieving a median possible gain or the more cognitive activation-based users whose click behavior employs the spreading activation model during snippet scans. Comparing the deterministic simulations to real TREC users (interaction logs of the TREC Session Track), the "real" TREC user achieved only about half of the gain the ideal user would manage with the same cost budget. The TREC user correlates the most with an activation user having a dynamic click threshold. Using Markov model analysis, we show that the TREC users and our user models with optimal click behavior click less than other models. The estimated average gain of the simulated users correlates very well with the session-DCG metric. Though all proposed models are deterministic, our framework allows to include probabilistic decisions as well. An interesting application could be estimating the information gain with the help of large populations of simulated users in scenarios where no huge logs of millions of users are available (e.g., enterprise search). A metric based on simulation would be very transparent since for every instance of a user the achieved information gain is reproducible. The effect of changes in the ranking or the UI (that also influences cost) can be directly tested on different instances of simulated users. Different cost models also form a promising future direction since costs heavily influence search behavior [6]. Scanning a list of ten results is more costly on a phone than on a desktop while talking to a device could make queries cheaper. With variable costs, different environments can be simulated. Finally, a very important addition would be the extension of our framework such that also query (re-)formulations are simulated. Possible steps could be simulating known-item queries (clicked documents as the known item) or query simulation based on anchor texts [5,19]. This would allow the simulation of complete sessions based on a search task description without relying on the queries of the TREC Session tracks or similar datasets.

Acknowledgment. Working on simulated ideal and average users was very much inspired by many discussions the first author had with Leif Azzopardi, Charlie Clarke, Gianmaria Silvello, and Robert Villa in the "User simulation" working group of the Dagstuhl seminar 13441, organized by Maristella Agosti, Norbert Fuhr, Elaine Toms, and Pertti Vakkari.

References

1. Anderson, J.R., Matessa, M., Lebiere, C.: ACT-R: a theory of higher level cognition and its relation to visual attention. Hum. Comput. Interact. **12**(4), 439–462 (1997)
2. Arif, A., Stuerzlinger, W.: Analysis of text entry performance metrics. In: IEEE TIC-STH 2009, pp. 100–105 (2009)
3. Azzopardi, L.: Modelling interaction with economic models of search. In: SIGIR 2014, pp. 3–12 (2014)
4. Azzopardi, L.: The economics in interactive information retrieval. In: SIGIR 2011, pp. 15–24 (2011)
5. Azzopardi, L., de Rijke, M., Balog, K.: Building simulated queries for known-item topics: an analysis using six European languages. In: SIGIR 2007, pp. 455–462 (2007)
6. Azzopardi, L., Kelly, D., Brennan, K.: How query cost affects search behavior. In: SIGIR 2013, pp. 23–32 (2013)
7. Azzopardi, L., Zuccon, G.: An analysis of theories of search and search behavior. In: ICTIR 2015, pp. 81–90 (2015)
8. Azzopardi, L., Zuccon, G.: Two scrolls or one click: a cost model for browsing search results. In: Ferro, N., et al. (eds.) ECIR 2016. LNCS, vol. 9626, pp. 696–702. Springer, Heidelberg (2016). doi:10.1007/978-3-319-30671-1_55
9. Barker, K., Cornacchia, N.: Using noun phrase heads to extract document keyphrases. In: AI 2000, pp. 40–52 (2000)
10. Baskaya, F., Keskustalo, H., Järvelin, K.: Time drives interaction: simulating sessions in diverse searching environments. In: SIGIR 2012, pp. 105–114 (2012)
11. Baskaya, F., Keskustalo, H., Järvelin, K.: Modeling behavioral factors in interactive information retrieval. In: CIKM 2013, pp. 2297–2302 (2013)

12. Budiu, R., Royer, C., Pirolli, P.: Modeling information scent: a comparison of LSA, PMI and GLSA similarity measures on common tests and corpora. In: RIAO 2007, pp. 314–332 (2007)
13. Busemeyer, J.R., Diederich, A.: Cognitive Modeling. Sage Publications, Thousand Oaks (2009)
14. Carterette, B., Bah, A., Zengin, M.: Dynamic test collections for retrieval evaluation. In: ICTIR 2015, pp. 91–100 (2015)
15. Carterette, B., Kanoulas, E., Yilmaz, E.: Simulating simple user behavior for system effectiveness evaluation. In: CIKM 2011, pp. 611–620 (2011)
16. Chapelle, O., Metlzer, D., Zhang, Y., Grinspan, P.: Expected reciprocal rank for graded relevance. In: CIKM 2009, pp. 621–630 (2009)
17. Chapelle, O., Zhang, Y.: A dynamic Bayesian network click model for web search ranking. In: WWW 2009, pp. 1–10 (2009)
18. Church, H.: Word association norms, mutual information, and lexicography. Comput. Linguist. **16**(1), 22–29 (1990)
19. Dang, V., Croft, B.W.: Query reformulation using anchor text. In: WSDM 2010, pp. 41–50 (2010)
20. Dyer, M.E., Kayal, N., Walker, J.: A branch and bound algorithm for solving the multiple-choice knapsack problem. J. Comp. Appl. Math. **11**(2), 231–249 (1984)
21. Fu, W.-T., Pirolli, P.: SNIF-ACT: a cognitive model of user navigation on the world wide web. Hum. Comput. Interact. **22**(4), 355–412 (2007)
22. Hagen, M., Gomoll, J., Beyer, A., Stein, B.: From search session detection to search mission detection. In: OAIR 2013, pp. 85–92 (2013)
23. He, Y., Wang, K.: Inferring search behaviors using partially observable Markov model with duration (POMD). In: WSDM 2011, pp. 415–424 (2011)
24. Järvelin, K., Kekäläinen, J.: Cumulated gain-based evaluation of IR techniques. ACM TOIS **20**(4), 422–446 (2002)
25. Järvelin, K., Price, S.L., Delcambre, L.M.L., Nielsen, M.L.: Discounted cumulated gain based evaluation of multiple-query IR sessions. In: Macdonald, C., Ounis, I., Plachouras, V., Ruthven, I., White, R.W. (eds.) ECIR 2008. LNCS, vol. 4956, pp. 4–15. Springer, Heidelberg (2008)
26. Keane, M.T., O'Brien, M., Smyth, B.: Are people biased in their use of search engines? CACM **51**(2), 49–52 (2008)
27. Kellerer, H., Pferschy, U., Pisinger, D.: Knapsack Problems. Springer, Heidelberg (2004)
28. Klöckner, K., Wirschum, N., Jameson, A.: Depth- and breadth-first processing of search result lists. In: CHI 2004, p. 1539 (2004)
29. Manktelow, K.I.: Reasoning and Thinking. Psychology Press, Hove (1999)
30. Maxwell, D., Azzopardi, L., Järvelin, K., Keskustalo, H.: Searching and stopping: an analysis of stopping rules and strategies. In: CIKM 2015, pp. 313–322 (2015)
31. O'Brien, M., Keane, M.T.: Modeling user behavior using a search-engine. In: IUI 2007, pp. 357–360 (2007)
32. Pirolli, P., Card, S.: Information foraging in information access environments. In: CHI 1995, pp. 51–58 (1995)
33. Pisinger, D.: A minimal algorithm for the multiple-choice knapsack problem. Eur. J. Oper. Res. **83**(2), 394–410 (1995)
34. Sanderson, M., Paramita, M.L., Clough, P., Kanoulas, E.: Do user preferences and evaluation measures line up? In: SIGIR 2010, pp. 555–562 (2010)
35. Smucker, M.D., Clarke, C.L.A.: Modeling optimal switching behavior. In: CHIIR 2016, pp. 317–320 (2016)

36. Smucker, M.D., Clarke, C.L.A.: Modeling user variance in time-biased gain. In: HCIR 2012, paper 3 (2012)
37. Smucker, M.D., Jethani, C.P.: Human performance and retrieval precision revisited. In: SIGIR 2010, pp. 595–602 (2010)
38. Thomas, P., Moffat, A., Bailey, P., Scholer, F.: Modeling decision points in user search behavior. In: IIiX 2014, pp. 239–242 (2014)
39. Tran, V.T., Fuhr, N.: Using eye-tracking with dynamic areas of interest for analyzing interactive information retrieval. In: SIGIR 2012, pp. 1165–1166 (2012)
40. Tran, V.T., Fuhr, N.: Markov modeling for user interaction in retrieval. In: MUBE 2013, pp. 13–14 (2013)
41. Turpin, A., Scholer, F.: User performance versus precision measures for simple search tasks. In: SIGIR 2006, pp. 11–18 (2006)
42. Zhang, Y., Chen, W., Wang, D., Yang, Q.: User-click modeling for understanding and predicting search-behavior. In: KDD 2011, pp. 1388–1396 (2011)

Constraining Word Embeddings by Prior Knowledge – Application to Medical Information Retrieval

Xiaojie Liu, Jian-Yun Nie[(⊠)], and Alessandro Sordoni

DIRO, University of Montreal, CP. 6128, succursale Centre-ville,
Montreal, QC H3C3J7, Canada
{xiaojiex,nie,sordonia}@iro.umontreal.ca

Abstract. Word embedding has been used in many NLP tasks and showed some capability to capture semantic features. It has also been used in several recent studies in IR. However, word embeddings trained in unsupervised manner may fail to capture some of the semantic relations in a specific area (e.g. healthcare). In this paper, we leverage the existing knowledge (word relations) in the medical domain to constrain word embeddings using the principle that related words should have similar embeddings. The resulting constrained word embeddings are used to rerank documents, showing superior effectiveness to unsupervised word embeddings.

1 Introduction

Continuous word representations, called word embeddings, have known widespread uses in general NLP tasks [4, 6, 15, 17, 26, 27]. They offer an effective and efficient way of encoding semantic/syntactic relationships between words in semantic space, which typically relies on the distributional hypothesis that two words sharing similar contexts should be associated with similar vectors in the embedding space. Word embedding, and more generally, deep learning, has also been used in IR in recent years for different tasks: to suggest or to reformulate queries [16, 20], to extend language models [8, 24], or to determine a similarity score between queries and document titles [10, 22], questions and short answers [23] or queries and terms [29]. Although it is possible to optimize a deep network directly for the ad hoc search task as in [10, 12], this would require a large set of training data (e.g. clickthrough), which is not always available. An alternative approach is to train word embeddings on a document collection in an unsupervised manner. Word embeddings trained in this way may reflect some general syntactic or semantic relations between words in a language such as between "cat" and "kitten", but fail to capture some valid relations between words, which may have been established manually. For example, word embeddings trained on a medical collection fail to capture the strong relationship between *heart* and *cor* (a strongly related word used in prescriptions), while this relationship has been specified in the domain resource UMLS [3]. It is natural to leverage the knowledge to constrain or to adjust word embeddings so as to better fit the specific application domain.

© Springer International Publishing AG 2016
S. Ma et al. (Eds.): AIRS 2016, LNCS 9994, pp. 155–167, 2016.
DOI: 10.1007/978-3-319-48051-0_12

The principle we use in this paper to constrain word embeddings is that related words in our prior knowledge (e.g. synonyms) should have similar embeddings.

The idea of using prior knowledge to constrain word embeddings has been used in several recent studies in NLP [4, 7, 27]. In this paper, we adapt these approaches to medical IR, and evaluate them on several test collections - OHSUMED [11] and CLEF [9, 18]. The contributions of this paper are as follows: We propose modified constrained training methods for word embeddings and show that they can bring more improvements to MIR than the original word embeddings.

The rest of the paper is organized as follows. Section 2 gives an overview of word embedding. Sections 3 and 4 present our approach to constrain word embeddings and to document reranking. Section 5 describes our experiments and analyses. Section 6 goes through the related work and Sect. 7 presents the conclusion and future work.

2   Word Embedding

In this section, we describe the standard and regularized word embeddings.

2.1   Continuous Bag-of-Words (CBOW)

Proposed by Mikolov et al. [15], the word2vec models create a vector representation for a word according to the context words frequently appearing around it. In this section, we will only describe one of the word2vec models – CBOW, which minimizes the following objective loss function:

$$L = -\sum_{t=1}^{T} \log p(w_t|w_{t\pm k}),$$ (1)

where T is the total number of words in the corpus and $w_{t\pm k}$ are the words in the window of size k centered at position t and excluding w_t. The probability of a word given its context is defined as:

$$p(w_t|w_{t\pm k}) = \frac{\exp(w_t^T c)}{\sum_{v\in V} \exp(w_v^T c)}, c = \sum_{j=t-k, j\neq t}^{t+k} w_j,$$ (2)

where the context embedding c is simply the sum of the embeddings of words occurring in the text window.

2.2   Regularized Word Embedding

Several approaches have been proposed in recent years to constrain (regularize) unsupervised word embeddings, and we describe two approaches below.

Online Training Approach. Online training approaches alter the learning objective in word embedding estimation by adding a knowledge-based regularization term [4, 26–28]. We only describe the approach by Yu and Dredze [27]. The modified loss function is as follows:

$$L = -\frac{1}{T}\sum_{t=1}^{T} \log p(w_t|w_{t\pm k}) - \frac{C}{|R|} \sum_{(w_i,w_j)\in R} \log p(w_i|w_j), \tag{3}$$

where $(w_i, w_j) \in R$ means that that two words are linked in the resource R, $|R|$ is the number of links in R, and C is a hyper-parameter controlling the strength of the regularization. Similarly to Eq. (2), the probability $p(w_i|w_j)$ is proportional to the dot product between w_i and w_j. Therefore, the regularizer sums up a similarity measure over all pairs of related words in the resource.

We observe two shortcomings of this approach. First, any pair of linked words in the resource is considered to be a constraint of equal importance ($1/|R|$) in the regularization. Intuitively, however, a more frequent link (or a link between two frequently used words) should play a more important role in the regularization. Second, as the two terms in the objective function sum over different elements – words in the corpus and links in the resource, Yu and Dredze have to define two sets of separate learning parameters, one for the CBOW objective and another for the regularization, which are updated separately in turn. This means that when updating the parameters of the regularization, the context of a word (considered in the first term) is no longer taken into account. The risk of this process is that the second update could undo the earlier update, making the update process quite random at the end. In this paper, we propose a solution to these problems.

Offline (Retrofitting) Approach. Offline approaches (also called retrofitting) [7] adjust word embeddings outside the original training process as follows: the new embeddings should be close to the original embeddings and respect the constraints of the external resource, i.e. minimize:

$$L = \frac{1}{V}\sum_{v=1}^{V}\left|w_v' - w_v\right|^2 + \frac{\beta}{|R|} \sum_{(w_i,w_j)\in R} \left|w_i' - w_j'\right|^2, \tag{4}$$

where w_v and w_v' are the original and the new embeddings and β a parameter.

3 Constrained Word Embedding

We propose modifications to solve the problems discussed above. A tighter regularization is used in the online method: the original CBOW cost function is combined with the requirement that if a word can be well generated from a given context, its related word should also be well generated from the same context, i.e.:

$$L = \sum_{t=1}^{T} \frac{1}{|R_t|} \sum_{w_s:(w_t,w_s)\in R} [\log p(w_t|w_{t\pm k}) - \log p(w_s|w_{t\pm k})]^2 \qquad (5)$$

where $|R_t|$ is the number of words related to w_t in the resource.

A possible drawback of the above formulation is that every related word is attributed an equal weight $(1/|R_t|)$. To solve this problem, we weigh each related word w_s by its relative frequency in the document collection as follows:

$$wt(w_s|w_t) = f(w_s) \bigg/ \sum_{(w_t,w)\in R} f(w). \qquad (6)$$

where $f(w_s)$ is the frequency of w_s in the collection. The final loss function is defined as follows:

$$L = -\sum_{t=1}^{T} [\log p(w_t|w_{t\pm k}) - \alpha \sum_{w_s:(w_t,w_s)\in R} wt(w_s|w_t)[\log p(w_t|w_{t\pm k}) - \log p(w_s|w_{t\pm k})]^2] \qquad (7)$$

where α is a weighting parameter.

The above loss function solves both problems of [27]: the collection frequency of words in a relation is taken into account naturally, and the embeddings for related words are tightly related to their contexts.

We also propose a slightly modified version of retrofitting method by adding term weighting in it:

$$L = \sum_{v=1}^{V} [|w_v' - w_v|^2 + \beta \sum_{(w_v,w_s)\in R} wt(w_s|w_v)|w_v' - w_s'|^2] \qquad (8)$$

As we will see in our experiments, our modified models can outperform the original regularized embeddings in MIR.

4 Using Constrained Embeddings for MIR

Many resources exist in the medical domain. In this paper, we use UMLS Metathe-saurus [3], which is the largest resource in this area. It integrates hundreds of thesauri of different sub-domains in a uniform framework. Each concept (identified by a CUI – Concept Unique Identifier) in UMLS contains a set of expressions, which we use as synonyms. For example, the CUI C0018681 contains the expressions: {*heart, cor, hearts, cardiac, heart nos, heart structure*}. There are more types of relations defined in UMLS, but we only use synonymy relations in this paper. In addition, we only consider single-word concept expressions (i.e. *heart, cor, hearts, cardiac*), and leave multi-word expressions to future work. This results in 302,323 synonymy relations between single words from UMLS.

Once word embeddings are trained, one faces the problem of building a representation for the whole document or query. We use a simple approach commonly used in this area, by summing up all the word embeddings in the document or the query. Cosine similarity is used to measure the similarity between the document and query embeddings. This approach is similar to that used in [15, 23, 24]. We notice, however, that a simple sum will make the global embedding of a document tuned towards frequent words which are not discriminative for IR. Therefore, we use the traditional IDF weighting to weight the embedding of a word.

Word embeddings are too noisy to be used alone to rank documents. In this paper, we use them in a re-ranking approach: we first retrieve a set of 1000 documents using a traditional baseline method (BM25 or language model); then, the results are re-ranked by the following re-ranking function:

$$s(Q,D) = \gamma BOW(Q,D) + (1 - \gamma)Cosine(Q,D) \qquad (9)$$

where γ is a hyper-parameter of our model, BOW is the score of a bag-of-word method such as BM25 or LM (language model); and $Cosine$ is the cosine similarity between the query and the document embeddings. Both BOW and $Cosine$ scores are normalized as follows:

$$NormScore = (Score - MinScore)/(MaxScore - MinScore) \qquad (10)$$

where $MaxScore$, $MinScore$ are the maximum and minimum scores in the list, $Score$ and $NormScore$ are the non-normalized and normalized scores of a document.

5   Experiments

5.1   Test Collections

The experiments are performed on the following test collections: OHSUMED [11] and CLEF-eHealth 2014 [9] and 2015 [18]. We use short queries (title field). Table 1 shows some statistics of the collections.

Table 1. Statistics of test collections

Corpus	Number of queries	Number of documents	Size
OHSUMED	106	348,566	294M
CLEF2014	50	1,095,082	6.5G
CLEF2015	66	1,095,082	6.5G

We use P@10 as the main performance indicator, and MAP and NDCG@10, which are often used on these collections, as the second indicators for OHUMED and CLEF. Two-tailed t-test ($p < 0.05$) is performed for statistical significance.

5.2 Word Embedding Training

In our experiments, we use CBOW model and negative sampling [15] to train the basic word embeddings. The CBOW program is then modified to incorporate the constraints as in Eq. (7). For all the methods tested, we set the dimension of embedding to 300, the context window size (k) to 5. This setting is common in word embedding [15] and has been shown to be reasonable in [30]. We choose 10 negative samples and we filter out words appearing less than 5 times in the collection. The collections are not preprocessed before embedding training, i.e. no stemming and stopword removal. Our intuition is that stopwords could provide useful context information for word embeddings. However, this remains to be confirmed. After training, our embedding vocabulary size is 164,434 for OHSUMED and 3,989,059 for CLEF.

5.3 Retrieval Results

BM25 (with the default setting) and LM (language model with Dirichlet smoothing with $\mu = 2000$) are used as the basic retrieval methods to retrieve 1000 candidates for reranking. In order to test the effectiveness of CBOW, we also use the standard CBOW model alone (i.e. γ in Eq (9) is set to 0). The original and modified online and offline constrained word embeddings are used to rerank the documents as in Eq. (9). We use 2-fold cross-validation to set hyper-parameters (α, β, γ) of the models for each collection. We report the performance of different methods in Table 2.

We observe that the traditional CBOW alone (line **c**) leads to poor retrieval effectiveness. This could be explained by the noisy nature of word embedding for a whole document. However, when it is combined with a traditional IR method (**d** and **e**), we observe significant improvements. Similar observations have been made in [30].

Next, we observe that our online method (lines **g** and **i**) outperforms significantly CBOW and Yu's method when combined with BM25 or LM. This confirms that the constraints imposed by UMLS relations are helpful in training better word embeddings for MIR. We also see that the method of Yu does not always produce better results than CBOW, and the differences between Yu and CBOW are not statistically significant. This result could be explained by our earlier observation that the loosely tied constraint used by Yu does not necessarily lead to better word embeddings.

Retrofitting has shown better performance in several NLP tasks [7] than the method of Yu and Dredze. This is also confirmed in our results (lines **j** and **l** vs. lines **f** and **h**). However, the differences are not statistically significant. Our modified offline method (**k** and **m**) makes larger improvements. The differences with the original CBOW are statistically significant on CLEF collections. The only change between the original retrofitting (**Faruqui**) and our modified version (**Offline**) is the weighting of embeddings we added. This suggests the usefulness of embedding weighting in IR.

Table 2. Retrieval results of different methods (Significant difference with a method is marked by a letter corresponding to that method)

	OHSUMED		CLEF2014		CLEF2015	
	P@10	MAP	P@10	DCG@10	P@10	NDCG@10
(a) BM25	0.4390	0.2922	0.6720	0.6876	0.3561	0.3217
(b) LM	0.3752	0.2325	0.7280	0.7200	0.3712	0.3276
(c) CBOW ($\gamma = 0$)	0.1631	0.0401	0.0490	0.0596	0.0530	0.0616
(d) CBOW + BM25	0.4610^a	0.2986	0.7056^a	0.7085^a	0.3727^a	0.3461^a
(e) CBOW + LM	0.4438^b	0.2745^b	0.7470^b	0.7327^b	0.3909^b	0.3560^b
(f) Yu + BM25	0.4600	0.2990	0.7120	0.7060	0.3682	0.3460
(g) Online + BM25	**0.4771^{df}**	**0.3005**	0.7315^{df}	0.7320^{df}	0.3864^{df}	0.3647^{df}
(h) Yu + LM	0.4467	0.2778	0.7490	0.7340	0.3909	0.3557
(i) Online + LM	0.4581^{eh}	0.2793	0.7580^{eh}	0.7460^{eh}	0.4086^{eh}	0.3682^{eh}
(j) Faruqui + BM25	0.4695	0.3001	0.7200	0.7250	0.3818	0.3593
(k) Offline + BM25	0.4715^d	0.3001	0.7296^{dj}	0.7300^d	0.3848^d	0.3596^d
(l) Faruqui + LM	0.4470	0.2778	0.7520	0.7420	0.3955	0.3665
(m) Offline + LM	0.4486	0.2781	0.7530^e	0.7440^e	0.3970^e	0.3666^e

The online and offline constraint methods lead to similar results, with a slight advantage (not statistically significant) to the online method. This suggests that both constrained methods could be reasonably used to incorporate prior knowledge.

The above comparison shows the benefit of constrained word embeddings. To better understand the effect of constraining embeddings, we analyze a specific example of word "heart", a common medical term. The most similar words, based on word embeddings trained on OHSUMED with different methods, are shown in Table 3.

We can first observe that CBOW is able to find some strongly related words without using UMLS: *hearts, cardiovascular, cardiorespiratory*. The words *synergist, acyanotic* and *ventricular* are also concepts often used in association with *heart*. However, *ouvrier* (name of an author) and *thrive* are not strongly related to *heart*.

Table 3. The most similar words to "heart".

CBOW		Online		Offline	
Cardiac	0.4891	Cardiac	0.5205	Cardiac	0.7960
Synergist	0.4494	Hearts	0.5030	Cor	0.6957
Hearts	0.4276	Cor	0.4939	Synergist	0.5030
Cardiovascular	0.4096	Synergist	0.4690	Hearts	0.4738
Acyanotic	0.3987	Cardiovascular	0.4156	Biventricular	0.4721
Ouvrier	0.3934	Cerebrovascular	0.4149	Cyanotic	0.4720
Multiorgan	0.3931	Acyanotic	0.3985	Cardiorespiratory	0.4714
Ventricular	0.3837	Ventricular	0.3979	Ventricular	0.4651
Cardiorespiratory	0.3829	Cardiorespiratory	0.3969	Acyanotic	0.4585
Thrive	0.3766	Biventricular	0.3831	Circulatory	0.4552

UMLS contains three synonym words to *heart*: *hearts*, *cor* and *cardiac*, which are incorporated in the constrained embeddings. As we can see, these words have been added or promoted (with higher similarities) in the list using constrained methods. First, we observe that CBOW is unable to discover alone the similar word *cor*, which is often used in prescriptions for heart diseases. The prior domain knowledge provides complementary means to link this word. This is part of the benefit we expected from using prior knowledge for embedding training.

Second, we can also observe that in addition to the synonyms, other strongly related words such as *biventricular* and *cyanotic* have also been promoted in the constrained embeddings. In fact, requiring synonym embeddings to be closer also makes the embeddings of their related words closer. In this specific example, even if we do not expect to find the word *cor* in the relevant documents to *heart* in OHSUMED, the words related to *cor* such as *cyanotic* could be found in them. This indirect constraint effect can affect many more words than just synonyms.

We do not see clear differences between the lists of the Online and Offline methods. Both method are capable of finding some strongly related words.

5.4   Parameter Sensitivity

The methods we propose contain some hyper-parameters $(\alpha, \beta, \Upsilon)$, which we set by cross validation in the previous results. In this section, we examine the sensitivity of retrieval effectiveness to these parameters. We will show the variation of P@10 on OHUSMED and CLEF2015 (CLEF2014 is very similar to CLEF2015).

Figures 1, 2, 3, and 4 show that the retrieval effectiveness (P@10) varies depending on the setting of α and β. The impact of parameters depends on the test collection (OHSUMED and CLEF), and on the basic retrieval model used (BM25 and LM). Globally, the setting of parameters α and β tends to have a larger impact on CLEF than on OHSUMED. This can be explained by the nature of documents in the collections: OHSUMED contains documents written by professionals while CLEF contains web pages crawled from the Web. The domain knowledge is naturally better encoded in OHSUMED than in CLEF. So, using domain knowledge as constraint will make smaller impact on word embeddings in OHSUMED than in CLEF.

We can also see that it is preferable to set these parameter to smaller values when combined with LM than with BM25. This could indicate that less regularization is preferred with LM. Further analyses are needed to understand the reason.

It is difficult to compare directly the parameters α and β because they are used in different constraint processes. We can still observe the general trend that β is preferably set to a large value than α. This may mean that the offline method may need a larger regularization than the online method to adjust word embeddings.

On the parameter Υ (Fig. 5), we observe more consistent behavior on different collections (the variations on other collections and retrieval models are very similar). The best setting is always around 0.5–0.6.

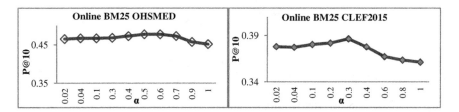

Fig. 1. Sensitivity of α in Online method (combined with BM25)

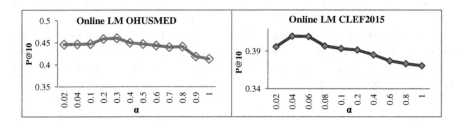

Fig. 2. Sensitivity of β in Offline method (combined with BM25)

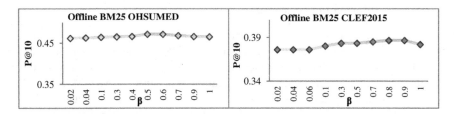

Fig. 3. Sensitivity of α in Online (combined with LM)

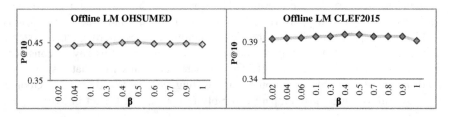

Fig. 4. Sensitivity of β in Offline method (combined with LM)

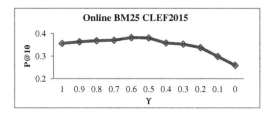

Fig. 5. Sensitivity to the re-rank parameter Υ on CLEF2015

6 Related Work

6.1 Medical Information Retrieval

A number of studies have attempted to exploit the existing resources in medical area such as UMLS. Two categories of approaches have been proposed in the literature.

The first approach is based on concepts: One first identifies concepts from documents and queries using a concept identification tool such as MetaMap [1]; then documents and queries are matched through their concepts and related concepts. Although improvements using concepts have been observed on some test collections [12, 13, 25, 31], namely in TREC Medical Record Track, which is a different IR task than the one considered in this paper, the improvements on the test collections considered in this paper have been limited and unstable [21]. An important reason lies in the relatively low accuracy of concept identification: about 70–80 % concepts identified are correct, and a number of concepts are unidentified [21].

A second method performs query expansion using the relations stored in a thesaurus [2, 14]. Typically, an additional ranking score is generated from synonyms and related terms of the query terms, and this score is combined with that of the original score. Concept phrases can also be used in this method.

In the previous experiments on the test collections we consider, query expansion approaches have been found more effective than concept-based matching [21]. All the top performing systems at CLEF 2014 and 2015 have used query expansion approaches [9].

To position our methods with respect to the existing approaches, we show the top three results in CLEF 2014 and CLEF 2015 in Table 4. For CLEF 2014, our results are comparable to those of the best team [21], which used MetaMap and all concept expressions in UMLS to perform phrase-based retrieval and query expansion. On CLEF 2015 [18], our results are clearly below the best participating system. However, this best system leveraged Google search results, and this gave a considerable advantage to the system. It is unfair to compare our results with that system. Our methods compare favorably to the other participating systems that do not use Google results. Overall, our methods compare favorably to the state of the art in MIR.

Table 4. Comparison with the best CLEF results

System	CLEF2014		CLEF2015	
	P@10	NDCG@10	P@10	NDCG@10
Best Team 1	0.7560	0.7445	*0.5394*	*0.5086*
Best Team 2	0.7540	0.7406	0.3864	0.3464
Best Team 3	0.7400	0.7301	0.3803	0.3465
Online	**0.7580**	**0.7460**	**0.4086**	**0.3682**

6.2 Word Embeddings for IR

Several studies in IR used word embeddings. [24] used word embedding in cross-language IR task. The goal was to train word embeddings in the same representation space for words in both languages. In [23], word embeddings (CBOW) are used to generate an additional feature to be embedded in a learning to rank framework to rank short answers to a question. Zuccon et al. [30] tested the effectiveness of word embeddings in IR as well as the impact of different parameters. They made similar observation that word embeddings can significantly improve IR effectiveness. De Vine et al. [5] compared several similarity measures for medical IR, and found the one based on word embeddings outperforms the others.

All the above studies showed that the semantic features captures in word embeddings are useful for IR. However, none of the above studies used constrained word embedding. In this paper, we showed that constrained word embeddings can further improve IR effectiveness.

7 Conclusion

In this paper, we explored the utilization of constrained word embedding for IR in a specialized domain. Our assumption is that constrained word embeddings can better fit the application domain and lead to better retrieval results. This is confirmed by our experiments.

Our methods to constrain word embeddings are adapted from the existing studies. In our experiments, we showed that the modifications we made lead to better retrieval results than their original versions. In particular, our modifications corrected two important problems in the original online training method and we added embedding weighting. The modifications resulted in significant changes in IR effectiveness.

We did not observe a large difference between the online and offline methods to incorporate prior knowledge. More investigations are needed to determine the best method to incorporate domain knowledge in word embeddings.

Our investigation has been limited to synonym word, while there are many other types of relation in domain resources (e.g. hierarchical relations). Such relations have been used in MIR [31] and in applications of word embedding in NLP [26]. It would be interesting to extend our study to cover more types of relation.

We only focused on single-word concepts in this study and used a very simple method to build a representation for the entire document and query. It will be interesting to investigate how an appropriate phrase embedding [6, 19], as well as a representation for the entire document and query, could be built for IR. These are some interesting topics for our future work.

Acknowledgement. This work is partly supported by an NSERC Discovery research grant.

References

1. Aronson, A.R.: Effective mapping of biomedical text to the UMLS Metathesaurus: the MetaMap program. In: Proceedings of AMIA Symposium, pp. 17–21 (2001)
2. Babashzadeh, A., Huang, J., Daoud, M.: Exploiting semantics for improving clinical information retrieval. In: SIGIR (2013)
3. Bodenreider, O.: The unified medical language system (UMLS): integrating biomedical terminology. Nucleic Acids Res. **32**, D267–D270 (2004)
4. Bian, J., Gao, B., Liu, T-Y.: Knowledge-powered deep learning for word embedding. ECML-PKDD, pp. 132–148 (2014)
5. De Vine, L., Zuccon, G., Koopman, B., Sitbon, L., Bruza, P.: Medical semantic similarity with a neural language model. In: CIKM (2014)
6. Dinu, G., Baroni, M.: How to make words with vectors: phrase generation in distributional semantics. In: Proceedings of ACL, pp. 624–633
7. Faruqui, M., Dodge, J., Jauhar, S.K., Dyer, C., Hovy, E., Smith, N.A.: Retrofitting word vectors to semantic lexicons. In: NAACL (2015)
8. Ganguly, D., Roy, D., Mitra, M., Jones, J.F.: A word embedding based generalized language model for information retrieval. In: SIGIR, pp. 795–798 (2015)
9. Goeuriot, L., Kelly, L., Li, W., Palotti, J., Pecina, P., Zuccon, G., Hanbury, A., Jones, G.J.F.: ShARe/CLEF eHealth evaluation lab 2014, task 3: user-centred health information retrieval. In: CLEF 2014 Online Working Note, pp. 43–61 (2014)
10. Huang, P.-S., He, X., Gao, J., Deng, L., Acero, A., Heck, L.: Learning deep structured semantic models for web search using clickthrough data. In: CIKM, pp. 2333–2338 (2013)
11. Hersh, W., Buckley, C., Leone, T.J., Hickam, D.: OHSUMED: an interactive retrieval evaluation and new large test collection for research. In: SIGIR, pp. 192–201 (1994)
12. Koopman, B., Zuccon, G., Bruza, P., Sitbon, L., Lawley, M.: Information retrieval as semantic inference: a graph inference model applied to medical search. Inf. Ret. **19**(1), 6–37 (2016)
13. Limsopatham, N., Macdonald, G., Ounis, I.: Inferring conceptual relationships to improve medical records search. In: Proceedings of Conference on Open Research Areas in IR, pp. 1–8 (2015)
14. Martinez, D., Otegi, A., Soroa, A., Agirre, E.: Improving search over electronic health records using UMLS-based query expansion through random walks. J. Biomed. Inf. **51**, 100–106 (2014)
15. Mikolov, T., Sutskever, I., Chen, K., Corrado, G., Dean, J.: Distributed representations of words and phrases and their compositionality. In: NIPS (2013)
16. Mitra, B.: Exploring session context using distributed representations of queries and reformulations. In: SIGIR, pp. 3–12 (2015)
17. Pennington, J., Socher, R., Manning, C.D.: Glove: global vectors for word representation. In: EMNLP, pp. 1532–1543 (2014)

18. Palotti, J., Zuccon, G., Goeuriot, L., Kelly, L., Hanbury, A., Jones, G.J.F., Lupu, M., Pecina, P.: CLEF eHealth evaluation lab 2015, task 2: retrieving information about medical symptoms. In: CLEF 2015 Online Working Notes, pp. 32–55 (2015)

19. Socher, R., Manning, C.D., Ng, A.Y.: Learning continuous phrase representations and syntactic parsing with recursive neural networks. In: Deep Learning and Unsupervised Feature Learning Workshop – NIPS (2010)

20. Sordoni, A., Bengio, Y., Vahabi, H., Lioma, C., Simonsen, J.G., Nie, J.-Y.: A hierarchical recurrent encoder-decoder for generative context-aware query suggestion. In: CIKM (2015)

21. Shen, W., Nie, J.-Y., Liu, X.-J.: An investigation of the effectiveness of concept-based approach in medical information retrieval GRIUM@CLEF2014eHealthTask3. User-centred health information retrieval. In: Proceedings of CLEF 2014 (2014)

22. Shen, Y., He, X., Gao, J., Deng, L., Mesnil, G.: A latent semantic model with convolutional-pooling structure for information retrieval. In: CIKM, pp. 101–110 (2014)

23. Severyn, A., Moschitti, A.: Learning to rank short text pairs with convolutional deep neural networks. In: SIGIR, pp. 373–382 (2015)

24. Vulic, I., Moens, M.-F.: Monolingual and cross-lingual information retrieval models based on (bilingual) word embeddings. In: SIGIR, pp. 363–372 (2015)

25. Wang, Y., Liu, X., Fang, H.: A study of concept-based weighting regularization for medical records search. In: ACL (2014)

26. Xu, C., Bai, Y., Bian, J., Gao, B., Wang, G., Liu, X., Liu, T.-Y.: RC-NET: a general framework for incorporating knowledge into word representations. In: CIKM (2014)

27. Yu, M., Dredze, M.: Improving lexical embeddings with semantic knowledge. In: ACL, pp. 545–555 (2014)

28. Zeiler, M.D., Fergus, R.: Stochastic pooling for regularization of deep convolutional neural networks. arXiv preprint arXiv:1301.3557 (2013)

29. Zheng, G., Callan, J.: Learning to reweight terms with distributed representations. In: SIGIR (2015)

30. Zuccon, G., Koopman, B., Bruza, P., Azzopardi, L.: Integrating and evaluating neural word embeddings in information retrieval. In: Proceedings of Australasian Document Computing Symposium (2015)

31. Zuccon, G., Koopman, B., Nguyen, A., Vickers, D., Butt, L.: Exploiting medical hierarchies for concept-based information retrieval. In: Proceedings of Australasian Document Computing Symposium (2012)

Personalization and Recommendation

Use of Microblog Behavior Data in a Language Modeling Framework to Enhance Web Search Personalization

Arjumand Younus[1,2(✉)]

[1] Insight Centre for Data Analytics, University College Dublin, Dublin, Ireland
arjumandyounus@gmail.com
[2] Computational Intelligence Research Group, Information Technology,
National University of Ireland, Galway, Ireland

Abstract. Diversity in users' information needs has been effectively dealt with through personalized Web search systems whereby a user's interests and preferences are taken into account within the retrieval model. A significant component of any Web search personalization model is the means with which to model a user's interests and preferences to build what is termed as a *user profile*. This work explores the use of the Twitter microblog network as a source of *user profile* construction for Web search personalization. We propose a statistical language modeling approach taking into account various aspects of a user's behavior on the Twitter network (such as Twitterers followed, mentioned and retweeted). The model also incorporates network and topical similarity measures which enables the model to be a better representation of the user's profile. The richness of the Web search personalization model leads to significant performance improvements in retrieval accuracy.

1 Introduction

Recent years have seen the emergence of personalized Web search as an effective approach to deal with the diversity present in users' information needs [15,18,23]. This diversity arises as a result of differences in users' preferences and interests and often leads to different search results satisfying different users even when the issued query is the same[1] [19]. The personalization process within Web search involves incorporation of user's preferences into the retrieval model of the search system thereby moving from a "one size fits all" approach to customization of search results for people with different information interests and goals.

Traditional retrieval models for personalized Web search utilize a user profile built from a user's search history (e.g., query logs and clickthrough data) and browsing history [7,12,17]. However, use of such history data is not feasible on account of users' privacy considerations which limit the availability of the data and furthermore, history data is more prone to noise as previous interactions with

---

[1] A query such as "Python" may refer to the programming language or the snake (Example from [17]).

© Springer International Publishing AG 2016
S. Ma et al. (Eds.): AIRS 2016, LNCS 9994, pp. 171–183, 2016.
DOI: 10.1007/978-3-319-48051-0_13

the search system are not necessarily reflective of current user needs [22]. This paper therefore argues for an alternative information source (namely, microblogs) from which to build a rich user profile.

The proliferation of Web 2.0 services has created a new form of user collaboration where users engage within a social network while at the same time generating their own content which is popularly known as user-generated content. Microblogs such as Twitter[2] are an immensely popular forum for such collaboration and we explore their worth as a source of user profile data for the Web search personalization process. Earlier research efforts that aim to exploit information from online social systems for personalized search rely mostly on social bookmarking and tagging systems [14,20]. However, Heymann et al. [9] questioned the usability of bookmarking meta-data for Web search engines by collecting a very large dataset (in fact, the largest known to the academic community) from a social bookmarking site. Heymann et al.'s findings revealed that social bookmarking lacks the size and distribution of tags necessary to make a significant impact for information retrieval at large. This is further confirmed in a user-survey based study by Younus et al. [25] revealing a very low usage of social bookmarking sites as compared to other social networking tools.

To the best of our knowledge, the use of microblogging platforms and, in particular, Twitter has not, with the exception of a few works [10,26], been explored as a source of user profile construction for personalized Web search and we undertake such a direction in this work. Our paper makes the following contributions:

1. We propose a statistical language modeling approach for user profile construction which takes into account various features of a user's Twitter network and his behavior on Twitter. This extends previous work [26] in that the model also considers the Twitterers followed by a user whereas previous work only considered the Twitterers mentioned and retweeted by a user.

2. As a further extension from previous work [26], we attempt to achieve an optimization of the model's parameters by introducing the concept of *"trust scores"* between a user and his/her Twitter network. These trust scores are derived based on various aspects of a user's Twitter's activities (i.e., mentions and retweets).

3. We propose different weighting strategies within the language model based on a user's similarity with his/her Twitter network. We extend previous work [26] by taking into account topical similarity measures instead of solely relying on network-based similarity. We additionally enhance the model through application of similarity-based weights instead of relying on a binary inclusion decision[3].

---

[2] http://twitter.com.
[3] Earlier work used a network similarity threshold based on which Twitterers not similar to the target user were excluded from the model [26].

4. We perform extensive evaluations with other personalization approaches proposed in the literature by means of an offline evaluation and a large-scale online evaluation[4]. The evaluation results show that retrieval performance substantially improves when using micoblog behavior as a source of obtaining user preferences and interests for Web search personalization.

The remainder of the paper is organized as follows. In Sect. 2, we describe works that are related to our research along with an explanation of how we differ from past work. In Sect. 3, we define the proposed methodology in sufficient detail. In Sect. 4, we discuss different variants within the parameters of the proposed personalization model based on a user's Twitter behavior and derived *trust scores*. In Sect. 5, we present the experimental evaluations. Finally in Sect. 6, we provide some conclusions with a discussion on implications of our findings.

2   Related Work

As described previously, the majority of efforts aimed at Web search personalization have attempted to model user preferences through the use of search and browse history data. These approaches are further classified based on whether the history data reflects a user's short-term [6,16] or long-term [12,17] preferences. History based approaches introduce major privacy concerns for users and based on psychological reasons the problem persists even with client-side personalization approaches [6]. The work by Teevan et al. examines a variety of sources (e.g. documents on user's hard drive, emails and so forth) in addition to history data to build a rich model of user interests. However, emails and desktop documents also contain sensitive information which users may not be willing to share with the search system.

Recent research efforts have experimented with data from online social systems as an alternative for user profile construction in the Web search personalization process with most of the approaches relying on social bookmarks and tags [2,14]. Approaches by Noll and Meinel [14] utilize the notion of frequency of occurrence for tags that users apply to resources (in this case web documents) in order to define a user-document similarity measure that re-ranks the search results. Vallet et al. [20] also utilize a user-document similarity measure based on the *term frequency-inverse document frequency (tf-idf)* scheme in which both the *tf-idf* weights in the user space and document space are calculated for computation of a joint similarity measure. A recent approach by Bouadjenek et al. [3] uses the social bookmarks assigned to documents in a collaborative filtering setting in order to take into account tags used by similar users. Other approaches that do not solely rely on social bookmarks and also take into account social networks of users remain limited to enterprise settings and hence, find limited use in Web search engines [5,22].

---

[4] Note that previous work compared our approach against a non-personalized baseline.

More recently, some efforts have attempted to undertake the process of user profile construction from Twitter data [1,13,21]. Of these approaches, some attempt to construct Twitter-based user profiles through semantic enrichment of tweet messages. As an example Abel et al. [1] apply a weighting heuristic to three types of user profiles: entity-based, topic-based and hashtag-based which is then tested in a personalized news recommendation system. Similarly, Meij et al. [13] utilize machine learning over a Wikipedia-based rich feature set to identify Wikipedia concepts in tweets. Another class of approaches aims to eliminate the information overload problem within the expanding microblogging site Twitter through recommendation of useful Twitter messages according to users' interests [11,24]. Despite the fact that the research community within text mining and information retrieval has turned its attention towards construction of Twitter-based user profiles, its use remains limited to recommendation systems and to the best of our knowledge the work by Ameni et al. is the first to utilize data collected from Twitter for Web search personalization [10].

3   Methodology

This section describes the proposed personalization model in detail. We follow a strategy in which non-personalized search results returned from a search system are re-ranked with the help of the user profile to return results that are more relevant to the user [17].

3.1   Microblog Behavior Based Language Model

We adopt a statistical language model to model various aspects of Twitter behavior. Using this model, we then define our re-ranking approach. We first present a brief overview of the Twitter-specific behaviors after which the formulations for re-ranking search results are presented.

The Twitter microblog network enables a user to follow any other user and unlike most online social networking sites the relationships of *following* and *being followed* require no reciprocation. Being a follower on Twitter enables a user to receive all the messages (called tweets) from those members the user follows. Twitter presents the opportunity to users to post 140-character long status updates about a variety of topics. Twitter also enables users to engage in conversations with each other through a feature known as *mentions* while at the same time allowing users to share a tweet written by another Twitter user with his/her followers through a feature known as *retweets*. We incorporate the *mention*, *retweet* and *follow* features of Twitter within our model with the underlying intuition that those Twitterers a particular user mentions, retweets or follows may reflect, to a large extent, the user's own preferences and interests.

For the re-ranking step, we use a language modeling approach to compute the likelihood of generating a document d from a language model estimated from a user's Twitter model as follows:

$$P(u)_{lm}(d/T) = \sum_{w \in W} P(w \mid T)^{n(w,d)} \tag{1}$$

where w is a word in the title and snippet of a document returned by a search system (i.e., d), W the set of all the words in the title and snippet of document d, $n(w,d)$ the term frequency of w in d, and u is the user for whom we want to personalize Web search results. Here, T is used to represent the uniform mixture of the user's Twitter model as follows:

$$P(w \mid T) = \lambda_o * P(w \mid T_o) + \lambda_m * P(w \mid T_{U_m}) + \lambda_r * P(w \mid T_{U_r}) + \lambda_f * P(w \mid T_{U_f}) \tag{2}$$

Let T_o denote the original tweets by the user u, T_{U_m} denotes the tweets by those Twitterers whom the user u mentions (i.e., Twitterers in set U_m), T_{U_r} denotes the tweets by those Twitterers whom the user u retweets (i.e., Twitterers in set U_r) and T_{U_f} denotes the tweets by those Twitterers whom the user u follow (i.e., Twitterers in set U_f). The individual Twitter models can be estimated as:

$$P(w \mid T_o) = \frac{1}{|T_o|} \sum_{t \in T_o} P(w \mid t) \tag{3}$$

$$P(w \mid T_{U_m}) = \frac{1}{|U_m|} \sum_{u_i \in U_m} \frac{sim(u, u_i)}{|T_{u_i}|} \sum_{t \in T_{u_i}} P(w \mid t) \tag{4}$$

$$P(w \mid T_{U_r}) = \frac{1}{|U_r|} \sum_{u_i \in U_r} \frac{sim(u, u_i)}{|T_{u_i}|} \sum_{t \in T_{u_i}} P(w \mid t) \tag{5}$$

$$P(w \mid T_{U_f}) = \frac{1}{|U_f|} \sum_{u_i \in U_f} \frac{sim(u, u_i)}{|T_{u_i}|} \sum_{t \in T_{u_i}} P(w \mid t) \tag{6}$$

i.e., a single user's Twitter model is estimated by a mixture of his own tweets, those Twitterer's tweets whom the user mentions, those Twitterers' tweets whom the user retweets and those Twitterers' tweets whom the user follows. Note that $sim(u, u_i)$ in Eqs. (4)–(6) denotes the similarity between a target user u for whom we want to personalize search results[5] and each user u_i occurring in either U_m, U_r or U_f. We explore a range of similarity measures in the next subsection.

---

[5] From this point onwards in the paper we use the phrase "target user" to refer to the user performing the search and for whom we want to personalize search results.

The constituent language models for T_o, T_{U_m}, T_{U_r} and T_{U_f} are a uniform mixture of their tweets' language models employing Dirichlet prior smoothing:

$$P(w \mid t) = \frac{n(w,t) + \mu \dfrac{n(w, coll)}{|coll|}}{|t| + \mu}$$

where $n(w,.)$ denotes the frequency of word w in (.), $coll$ is short for collection which refers to all tweets by user u (in case of Eq. (3)), all tweets by Twitterers in set U_m (in case of Eq. (4)) and all tweets by Twitterers in set U_r (in case of Eq. (5)), and $|.|$ is the overall length of the tweet or the collection.

Finally, after estimation of a user's Twitter model (using Eqs. 3–6) we use Eq. (1) to re-rank the documents returned by a search system and hence, present personalized search results to the user u.

3.2   Similarity Measure Between Users

The previous subsection presented the language modeling framework employed for the purpose of search results re-ranking which utilized the essential component of a similarity measure between the target user and the user in his/her Twitter network (more specifically, the user in mention, retweet or following network). We propose two classes of similarity measures based on the following intuitions:

– Two users are more likely to have common preferences and interests if they share many users within their Twitter network and hence, we propose network-based similarity measures.
– Two users are more likely to have common preferences and interests if they share interests in the same topics and hence, we propose topical similarity measures.

Network-Based Similarity: Previously, we defined U_m as the set of users mentioned by u, U_r as the set of users whose tweets were retweeted by user u and U_f as the set of users whose tweets were followed by user u. We present a network-based similarity measure which we then use as a weighing heuristic for a particular user in U_m, U_r or U_f.

We calculate the similarity between the current user u and each user u_i occurring in either U_m, U_r or U_f based on the heuristic that the more people u_i follows in these sets, the more likely that user's interests overlap with the user u. Furthermore, we normalise this score by the maximum of total number of users that user u_i follows or the number of users in U_m, U_r or U_f. We use the following formula to calculate the similarity score between user u and a user $u_i \in U_m$.

$$Sim(u, u_i) = \frac{|follow(u_i) \cap U_m|}{max(|follow(u_i)|, |U_m|)}$$

where $follow(u_i)$ is the set of users followed by u_i.

We also calculate similarity for all users in U_r and U_f using the same approach.

Topical Similarity: For the definition of topical similarity we make use of the Twitter-LDA model [27] in order to obtain the topics from tweets of all the users in sets U_m, U_r and U_f. It is significant to note that Twitter-LDA differs from the original LDA framework in that a single tweet is assigned a single topic instead of a distribution over topics[6]. We use the Twitter-LDA to determine the tweets' topics which are then utilized in a probabilistic model to determine topical similarity between a target user and a user in U_m, U_r or U_f as follows:

$$Sim(u, u_i) = \frac{\sum_{topic_j \in Topic_{U_m} \cap Topic_u} n(topic_j, Topic_u) + \mu \frac{n(topic_j, Topic_{U_m})}{|Topic_{U_m}|}}{|t_u| + \mu}$$

where $n(topic_j, Topic_u)$ denotes the number of tweets by the target user u related to topic j, $n(topic_j, Topic_{U_m})$ denotes the number of tweets by users in set U_m related to topic j, and t_u denotes the total number of tweets by the target user u. The topical similarity measure is essentially a weighted average of the commonality between topical distributions of a target user and the users in his/her network and is hence a good indication of shared preferences and interests.

We also calculate similarity for all users in U_r and U_f using the same approach.

4   Using Twitter Behavior for Parameter Setting

This section outlines the parameter setting heuristics that are derived based on the behavior of a target user on the Twitter social network. More specifically, we apply a Page-Rank like intuition over the network of users followed by the target user. Furthermore, depending on the amount of mentions and retweets within the tweets of a user we determine *"trust scores"* which are used as parameters for the model (λ_o, λ_m and λ_r in Eq. (2)).

4.1   Random Surfer Behavior on Twitter Network

As explained in Sect. 3.1 the proposed model takes into account those Twitterers' tweets whom the user mentions, whom the user retweets and whom the user follows. However, the likelihood that the Twitterers followed by a target user reflect his/her preferences is less unless the target user mentions or retweets the Twitterer followed[7]. The model already incorporates the mention and retweet network and hence, the likelihood that the target user is interested in a followed

---

[6] This is more suited to the task at hand as tweets are short and in general related to a single topic.

[7] It is often the case that random acquaintances are also followed on Twitter.

Twitterer mimics the *"random surfer model"* where the random surfer gets bored after several mentions and retweets and switches to a random followed Twitterer. Based on this intuition, we propose the following parameterization for λ_o, λ_m, λ_r and λ_f

$$\alpha = \lambda_o + \lambda_m + \lambda_r \tag{7}$$

$$1 - \alpha = \lambda_f \tag{8}$$

Here, α represents the damping factor which is basically the probability of the target user's interests being reflected by Twitterers mentioned or retweeted. We set the damping factor to 0.85 which is the standard value used by the PageRank algorithm giving the value for λ_f of 0.15.

4.2   Trust Scores Based on Tweeting Activities

Users differ in their behavior on Twitter in that some actively engage in conversations through the mention feature while others diffuse information in the form of retweets [4]. These differences in behavior form the basis for *"trust scores"* within our model. The *"trust scores"* measure the proportion of the target user's own tweets, tweets in which he or she engages in the mention activity and tweets in which he or she engages in the retweet activity. More precisely we set the parameters λ_o, λ_m, λ_r as follows

$$\lambda_o = \frac{|t_o|}{|t_m| + |t_r| + |t_o|} * \alpha \tag{9}$$

$$\lambda_m = \frac{|t_m|}{|t_m| + |t_r| + |t_o|} * \alpha \tag{10}$$

$$\lambda_r = \frac{|t_r|}{|t_m| + |t_r| + |t_o|} * \alpha \tag{11}$$

where, t_o represents original tweets by the target user, t_m represents those tweets by the target user in which he/she engages in the mention activity and t_r represents those tweets by the target user in which he/she engages in the retweet activity.

5   Experimental Evaluations

In this section we describe our experimental evaluations that demonstrate the effectiveness of our proposed approach. In the first step we perform an offline evaluation using user-defined relevance judgements, and in the second step we perform interleaved evaluations which as demonstrated by Matthijs and Radlinski [12] is an effective method for evaluation of real user workload on search systems. We test various variations of our system (i.e., the two different similarity measures of Sect. 3.2 and the different parameter settings) expressed by short-hand notation of Table 1.

Table 1. Variants of proposed personalization model

Description of model variant	Notation
Language model with uniform weighting and network similarity	$LM_{u,n}$
Language model with trust scores and network similarity	$LM_{tr,n}$
Language model with uniform weighting and topical similarity	$LM_{u,to}$
Language model with trust scores and topical similarity	$LM_{tr,to}$

As baseline personalization systems we use the approach by Teevan et al. [18] in addition to the approach by Matthijs and Radlinski [12]. However, we replace the search and browsing history data of these approaches by tweets of the target user and his/her network due to the limitation of such history data not being available.

5.1 Experimental Setup

We recruited 84 active Twitter users and used their Twitter data for the purpose of experimental evaluations. We obtained the search queries and underlying corpus (i.e., search documents' collection) from a publicly available dataset called *"CiteData"* by Harpale et al. [8]. *CiteData* comprises 81,432 academic articles and 41 queries. The dataset also contains relevance judgements which we do not use on account of them not being truly reflective of personalized relevance judgements and furthermore, the relevance judgements are not graded which makes it impossible to calculate normalized discounted cumulative gain.

We asked each user who participated in our user-study to select a subset of the queries that were similar to a search query that he/she had issued at some point. Each user was asked to select 12 queries from the 41 queries of the dataset and the re-ranked results were graded as highly relevant (2), relevant (1) and non-relevant (0). We re-rank the top-50 search results obtained through a non-personalized BM25 retrieval model.

The second set of experiments involves a large-scale online interleaved evaluation and this is to estimate the performance of our system on real users with real information needs so as to ensure that the results of offline evaluation do not overfit to the dataset. A browser plugin was developed and 16 out of the 84 users who participated in the offline evaluation agreed to participate in the online evaluation. We performed the interleaved evaluation over a two-week period for the 16 users and we follow the approach similar to Matthijs and Radlinski [12]. Search results from Google were re-ranked and the two rankings i.e., the original one from Google and the one produced after re-ranking by our system were interleaved to ensure that a click at random would be equally likely to be on a result from either ranking.

5.2   Experimental Results

Once we obtain relevance judgements and clickthrough data for both set of experiments, we evaluate the performance of our proposed personalization model using the evaluation metrics of mean average precision (MAP), precision at top 10 documents (P@10) and normalized discounted cumulative gain at top 10 documents (NDCG@10) which respectively measure the system's overall retrieval accuracy, its performance for those documents that are most viewed and the overall ranking positions of relevant/highly relevant documents. Table 2 shows the experimental results for the offline evaluation i.e. MAP, P@10 and NDCG@10 values for the various variants of our approach (using notations of Table 1), and other personalization approaches (denoted by *Teevan* and *Matthijs*)[8]. We report the results together across the queries and judgements for all 84 users who took part in the offline evaluation.

Table 2. Comparison of retrieval performance for variants of our proposed personalization model with other personalization models

Chosen Algo	Measures		
	MAP	$P@10$	$NDCG@10$
$LM_{u,n}$	0.564*	0.582	0.461
$LM_{tr,n}$	0.597	0.551**	0.493***
$LM_{u,to}$	0.612*	0.556	0.513**
$LM_{tr,to}$	0.651**	0.583	0.538***
Teevan	0.541***	0.472	0.420*
Matthijs	0.589	0.564**	0.488*

Note *$p < .05$, **$p < .01$, ***$p < .001$
$LM_{u,n}$ is language model with uniform weighting and network similarity
$LM_{tr,n}$ is language model with trust scores and network similarity
$LM_{u,to}$ is language model with uniform weighting and topical similarity
$LM_{tr,to}$ is language model with trust scores and topical similarity

The results for the offline evaluation show clearly, the benefits of using Twitter data to personalize search results for users. Furthermore, the MAP, P@10 and NDCG@10 scores for the personalized results corresponding to $LM_{tr,to}$ show the best performance effectively implying that topical similarity between the target

---

[8] Note that we treat the tweets' data as equivalent to history and user documents' data; furthermore, the technique by Matthijs and Radlinski utilized various segments of a web page (such as title, web page metadata which we could not utilize and hence, we use all terms in tweets except for stopwords).

user and his/her network is likely to lead to greater user satisfaction during the information-seeking process. Additionally, the incorporation of Twitter behavior in the form of *trust scores* outperforms the uniform weighting schemes.

Finally, for the online interleaved evaluation we obtained a total of 518 queries and of these 489 queries received a click on a search result. Of these 489 queries, 302 (61.8 %) queries received higher votes across our personalization model while the remaining 187 (38.2 %) received higher votes across the original Google rankings. This again demonstrates the potential for search personalization based on Twitter data to improve the search experience. Note that we only personalize using the variant that performs best in the offline evaluation (i.e., the one denoted by $LM_{tr,to}$).

6   Conclusions and Future Work

The main conclusion is that exploiting evidence available from a person's microblog behaviour to allow personalization can improve the accuracy of a system. We adopt a language modeling approach and show that including a similarity measure based on shared topical interests from a user's Twitter network provides the best performance. Moreover, taking into account a user's social network behavior leads to a rich model that dynamically adjusts parameters optimally. Future work will involve a combination of search history data and browsing history data to enable us to merge the social sources of evidence with more richer evidences of user profile information.

References

1. Abel, F., Gao, Q., Houben, G.-J., Tao, K.: Semantic enrichment of twitter posts for user profile construction on the social web. In: Antoniou, G., Grobelnik, M., Simperl, E., Parsia, B., Plexousakis, D., De Leenheer, P., Pan, J. (eds.) ESWC 2011, Part II. LNCS, vol. 6644, pp. 375–389. Springer, Heidelberg (2011)
2. Bouadjenek, M.R., Hacid, H., Bouzeghoub, M., Laicos: An open source platform for personalized social web search. In: Proceedings of the 19th ACM SIGKDD International Conference on Knowledge Discovery and Data Mining, KDD 2013, pp. 1446–1449. ACM, New York (2013)
3. Bouadjenek, M.R., Hacid, H., Bouzeghoub, M., Vakali, A.: Using social annotations to enhance document representation for personalized search. In: Proceedings of the 36th International ACM SIGIR Conference on Research and Development in Information Retrieval, SIGIR 2013, pp. 1049–1052. ACM, New York (2013)
4. Boyd, D., Golder, S., Lotan, G.: Tweet, tweet, retweet: conversational aspects of retweeting on twitter. In: 43rd Hawaii International Conference on System Sciences (HICSS), pp. 1–10. IEEE (2010)
5. Carmel, D., Zwerdling, N., Guy, I., Ofek-Koifman, S., Har'el, N., Ronen, I., Uziel, E., Yogev, S., Chernov S.: Personalized social search based on the user's social network. In: Proceedings of the 18th ACM Conference on Information and Knowledge Management, CIKM 2009, pp. 1227–1236. ACM, New York (2009)
6. Daoud, M., Tamine-Lechani, L., Boughanem, M., Chebaro, B.: A session based personalized search using an ontological user profile. In: SAC 2009, pp. 1732–1736 (2009)

7. Dou, Z., Song, R., Wen, J.-R.: A large-scale evaluation and analysis of personalized search strategies. In: WWW 2007, pp. 581–590 (2007)
8. Harpale, A., Yang, Y., Gopal, S., He, D., Yue, Z.: Citedata: a new multi-faceted dataset for evaluating personalized search performance. In: CIKM 2010, pp. 549–558 (2010)
9. Heymann, P., Koutrika, G., Garcia-Molina, H.: Can social bookmarking improve web search? In: WSDM 2008, pp. 195–206 (2008)
10. Kacem, A., Boughanem, M., Faiz, R.: Time-sensitive user profile for optimizing search personlization. In: Dimitrova, V., Kuflik, T., Chin, D., Ricci, F., Dolog, P., Houben, G.-J. (eds.) UMAP 2014. LNCS, vol. 8538, pp. 111–121. Springer, Heidelberg (2014)
11. Lu, X., Li, P., Ma, H., Wang, S., Xu, A., Wang, B.: Computing and applying topic-level user interactions in microblog recommendation. In: Proceedings of the 37th International ACM SIGIR Conference on Research & #38; Development in Information Retrieval, SIGIR 2014, pp. 843–846. ACM, New York (2014)
12. Matthijs, N., Radlinski, F.: Personalizing web search using long term browsing history. In: WSDM 2011, pp. 25–34 (2011)
13. Meij, E., Weerkamp, W., de Rijke, M.: Adding semantics to microblog posts. In: Proceedings of the Fifth ACM International Conference on Web Search and Data Mining, WSDM 2012, pp. 563–572. ACM, New York (2012)
14. Noll, M.G., Meinel, C.: Web search personalization via social bookmarking and tagging. In: Aberer, K., Choi, K.-S., Noy, N., Allemang, D., Lee, K.-I., Nixon, L.J.B., Golbeck, J., Mika, P., Maynard, D., Mizoguchi, R., Schreiber, G., Cudré-Mauroux, P. (eds.) ASWC 2007 and ISWC 2007. LNCS, vol. 4825, pp. 367–380. Springer, Heidelberg (2007)
15. Sontag, D., Collins-Thompson, K., Bennett, P.N., White, R.W., Dumais, S., Billerbeck,. B.: Probabilistic models for personalizing web search. In: Proceedings of the Fifth ACM International Conference on Web Search and Data Mining, pp. 433–442. ACM (2012)
16. Sriram, S., Shen, X., Zhai, C.: A session-based search engine. In: SIGIR 2004, pp. 492–493. ACM (2004)
17. Tan, B., Shen, X., Zhai, C.: Mining long-term search history to improve search accuracy. In: KDD 2006, pp. 718–723 (2006)
18. Teevan, J., Dumais, S.T., Horvitz, E.: Personalizing search via automated analysis of interests and activities. In: SIGIR 2005, pp. 449–456 (2005)
19. Teevan, J., Dumais, S.T., Horvitz, E.: Potential for personalization. ACM Trans. Comput.-Hum. Interact. **17**(1), 4:1–4:31 (2010)
20. Vallet, D., Cantador, I., Jose, J.M.: Personalizing web search with folksonomy-based user and document profiles. In: Gurrin, C., He, Y., Kazai, G., Kruschwitz, U., Little, S., Roelleke, T., Rüger, S., van Rijsbergen, K. (eds.) ECIR 2010. LNCS, vol. 5993, pp. 420–431. Springer, Heidelberg (2010)
21. Viejo, A., SáNchez, D., Castellí-Roca, J.: Preventing automatic user profiling in web 2.0 applications. Knowl. Based Syst. **36**, 191–205 (2012)
22. Wang, Q., Jin, H.: Exploring online social activities for adaptive search personalization, CIKM 2010, pp. 999–1008 (2010)
23. White, R.W., Chu, W., Hassan, A., He, X., Song, Y., Wang, H.: Enhancing personalized search by mining and modeling task behavior. In: Proceedings of the 22nd International Conference on World Wide Web, pp. 1411–1420. International World Wide Web Conferences Steering Committee (2013)

24. White, R.W., Chu, W., Hassan, A., He, X., Song, Y., Wang, H.: Enhancing personalized search by mining and modeling task behavior. In: Proceedings of the 22nd International Conference on World Wide Web, pp. 1411–1420. International World Wide Web Conferences Steering Committee (2013)
25. Younus, A., O'Riordan, C., Pasi, G.: Predictors of users' willingness to personalize web search. In: Larsen, H.L., Martin-Bautista, M.J., Vila, M.A., Andreasen, T., Christiansen, H. (eds.) FQAS 2013. LNCS, vol. 8132, pp. 459–470. Springer, Heidelberg (2013)
26. Younus, A., O'Riordan, C., Pasi, G.: A language modeling approach to personalized search based on users' microblog behavior. In: de Rijke, M., Kenter, T., de Vries, A.P., Zhai, C.X., de Jong, F., Radinsky, K., Hofmann, K. (eds.) ECIR 2014. LNCS, vol. 8416, pp. 727–732. Springer, Heidelberg (2014)
27. Zhao, W.X., Jiang, J., Weng, J., He, J., Lim, E.-P., Yan, H., Li, X.: Comparing twitter and traditional media using topic models. In: Clough, P., Foley, C., Gurrin, C., Jones, G.J.F., Kraaij, W., Lee, H., Mudoch, V. (eds.) ECIR 2011. LNCS, vol. 6611, pp. 338–349. Springer, Heidelberg (2011)

A Joint Framework for Collaborative Filtering and Metric Learning

Tak-Lam Wong[1(✉)], Wai Lam[2], Haoran Xie[1], and Fu Lee Wang[3]

[1] The Education University of Hong Kong, Tai Po, Hong Kong
{tlwong,hxie}@eduhk.hk
[2] Department of Systems Engineering and Engineering Management,
The Chinese University of Hong Kong, Shatin, Hong Kong
wlam@se.cuhk.edu.hk
[3] Caritas Institute of Higher Education, Tseung Kwan O, Hong Kong
pwang@cihe.edu.hk

Abstract. We have developed a framework for jointly conducting collaborative filtering and distance metric learning based on regularized singular value decomposition (RSVD), which discovers the user matrix and item matrix in the low rank space. Our approach is able to solve RSVD and simultaneously learn the parameters of Mahalanobis distance considering the ratings given by similar users and dissimilar users. One characteristic of our approach is that the learned model can be effectively applied to rating prediction and other relevant applications such as trust prediction, resulting in a solution which is coherent and optimal to both tasks. Another characteristic is that social community information and similarity information can be easily considered in our framework. We have conducted extensive experiments on rating prediction using real-world datasets to evaluate our framework. We have also compared our framework with other existing works to illustrate the effectiveness. Experimental results show that our framework achieves a promising prediction performance and outperforms the existing works.

Keywords: Collaborative filtering · Metric learning · Mahalanobis distance

1 Introduction

Collaborative Filtering (CF) have been extensively investigated due to the fact that there is massive volume of information available on the Web and CF it is readily applicable to real-world applications such as recommendation systems. For example, a number of recommendation systems have been developed to predict the movie rating given by users in the Netflix dataset[1], accomplishing very high accuracy. CF discovers the association of user-item ratings and predict the rating to a previously unseen item given by a user. One of the challenges

---

[1] http://www.netflixprize.com/.

© Springer International Publishing AG 2016
S. Ma et al. (Eds.): AIRS 2016, LNCS 9994, pp. 184–196, 2016.
DOI: 10.1007/978-3-319-48051-0_14

in CF is to handle the sparsity and high dimensionality of the user-item rating matrix. Due to this, the similarity between users are difficult to be computed directly.

Low rank matrix factorization, which identifies the latent factors of the user-item rating matrix, is one of the most common techniques used in CF. By treating the user-item rating matrix as the target matrix, the objective of matrix factorization is to discover the user matrix and item matrix, whose dot-product can approximate the target matrix. Each column of the user matrix and item matrix essentially represent a user and an item respectively. The user and item matrices are normally of lower rank to address the sparsity and the dimensionality problem and improve the efficiency in the prediction of unknown rating. However, one major limitation of low rank matrix factorization is that the similarity between two users will be unavoidably distorted because the column vectors in the user and item matrices corresponding to the smallest eigenvalues will be discarded and only a few significant columns will be retained. For example, let

$$R = \begin{pmatrix} 3 & 4 & 1 \\ 3 & 4 & 2 \\ 2 & 4 & 1 \end{pmatrix}$$

be the user-rating matrix where (i, j)-th entry corresponds to the rating given by user i to item j. The Euclidean distances between users 1 and 2, users 1 and 3, and users 2 and 3 are 1, 1, and 1.4142 respectively. If we apply low rank matrix factorization and set the rank $k = 2$ to solve $R \approx U'\Sigma V$ where $U, V \in \mathbb{R}^{3 \times 2}$ and $\Sigma \in \mathbb{R}^{2 \times 22}$. The results are

$$U = \begin{pmatrix} -0.5863 & -0.1738 \\ -0.6176 & 0.7279 \\ -0.5242 & -0.6633 \end{pmatrix}, \quad V = \begin{pmatrix} -0.5381 & 0.4055 \\ -0.7982 & -0.5269 \\ -0.2709 & 0.7470 \end{pmatrix}, \quad \text{and } \Sigma = \begin{pmatrix} 8.6604 & 0 \\ 0 & 0.8284 \end{pmatrix}.$$

If we set $\hat{R} = U'\Sigma V$ and compute the Euclidean distance according to $\hat{R}$, the distances between users 1 and 2, users 1 and 3, and users 2 and 3 are 0.7946, 0.6738, and 1.4081 respectively. As a result, low rank matrix factorization does not consider the distance between users/items in the learned low rank space. As we can observe in the above example, the Euclidean distance between users 1 and 2 is reduced from 1 to 0.7946, while the Euclidean distance between users 1 and 3 is reduced from 1 to 0.6738. The relative changes of the distance from the original space to the new space are different, even though the two distances are the same in the original space. More importantly, such changes completely depend on the user-item rating matrix and do not consider other useful information in a social network. For example, the distance between users 1 and 2 in the new space should be smaller than the distance between users 1 and 3 if users 1 and 2 are "friends" while users 1 and 3 are not in a social network.

---

[2] In CF, sometimes we directly solve $R \approx U'V$ in which Σ is embedded in U and V.

Regularized Singular Value Decomposition (RSVD) is a common technique used to solve the low rank matrix factorization problem and identify the low-rank user matrix and item matrix. Regularization is originally applied in the model to tackle the problem of model complexity and over-fitting. Several approaches have been proposed to use different regularizers to incorporate additional or prior information in learning the model. For example, Ma proposed to consider the user similarity and item similarity in the regularizer [1]. Essentially, it imposes soft constraints that given a pair of similar users, the two column vectors of the user matrix representing the two users are required to be close to each other. Similarly, the two column vectors of the item matrix representing the two items are required to be close to each other. Empirical results illustrate that prior information in the form of regularizer can substantially improve the performance in prediction. One limitation of this approach is that the closeness of two users/items is represented by the Frobenius norm of the difference between two column vectors. In other words, the distance metric is needed to be designed in advance. More importantly, the distance metric chosen does not take the data collected and the goal of the task into account.

We have developed a framework for jointly conducting collaborative filtering and distance metric learning, aiming at simultaneously discovering the user and item matrices for predicting unknown ratings, and learning the distance metric for other applications, in the new low rank space. Unlike existing works which only address the CF problem, or apply the pre-defined similarity measures to represent the closeness between users/items in the learned model, our approach can automatically discover the similarity metric when computing the user and item matrices when solving RSVD. The major idea of our approach is that given an item, a pair of similar users should give similar rating to this item. Moreover, from the discriminative perspective, the distance between them should be as close as possible in the low rank space. On the contrary, the distance between dissimilar users should be as far as possible in the new space. To achieve this, we have incorporated the parameterized Mahalanobis distance, which essentially is a linear transformation of the distance from the original space to a new space, into the regularizer of RSVD. When solving the RSVD, the user matrix, item matrix, and the parameters of the Mahalanobis distance will be learned jointly in our model. In our designed regularizer, we can easily incorporate the similarity information in the original space in our model. For example, trust information is commonly available in social networks. Trusted users can be considered to similar, while untrusted users can be considered to be dissimilar. With this trust information, the solution will naturally consider both user-item rating information and trust information. As a result, the learned user matrix, item matrix, and the parameters of Mahalanobis distance can be applied to coherently tackle both rating prediction and trust prediction problems, reducing possible conflict between the two tasks. Another characteristic of our approach is that collaborative filtering and distance metric learning serve as regularization to each other, leading to the smoothing effect and reducing overfitting.

The contribution of our work is summarized as follows:

1. We have developed a framework for jointly learning the user and item matrices in low rank space, as well as the distance metric in collaborative filtering. Unlike existing works which depend on the pre-defined distance metrics, out framework can learn the distance metric from the collected data. This is accomplished by incorporating Mahalanobis distance to the regulizers when solving RSVD.
2. Our model can easily incorporate the prior social network information such as trust or community information. This allows our model to consider multiple goals of the tasks and be applied to simultaneously solve different problems.
3. We showed that in our model derived from RSVD, collaborative filtering and distance metric learning serve as regularization to each other. As a result, overfitting can be reduced in both tasks naturally.
4. We have conducted extensive experiments to evaluate our framework and compared it existing works. Empirical results in collaborative filtering demonstrate that our approach significantly outperforms the existing works and achieves promising performance.

2   Related Work

Recommendation systems have been extensively investigated by researchers [2]. Memory-based methods aims at measure the user-user similarity based on the user profile or historical record to predict the rating of items given by a user [3–6]. However, one common shortcoming is the sparsity problem of the raw data. Normally, a user may only rate a relatively small number of the items, out of hundreds or thousands. Given two users, the number of items that are commonly rated is very small. Model-based methods aim at train a model for prediction [7–9]. For example, Zhang and Koren proposed Bayesian hierarchical linear model to tackle the CF problem [10]. In this model, the profile of each user is modeled by a linear model, whose parameters are drawn from a prior distribution. The rating to an item given by a user is then predicted by applying the model with relevant input. Xue et al. proposed a clustering-based method, which first generates clusters of similar users using K-means algorithm [8]. These generated clusters are then exploited to smooth the unknown rating, and hence improve the prediction performance for each individual user. ListCF predicts the ranking of items by a user by measuring the user-user similarity based on the Kullback-Leibler divergence between users' probability distributions over permutations of commonly rated items [11].

Matrix factorization is another commonly used model in CF [12]. The objective of matrix factorization is to discover the user matrix and the item matrix in a low-rank space, such that the dot-product can approximate the original user-item ratings. To address the sparsity problem, regularized singular value decomposition (RSVD) is applied [12,13]. Empirical results have also demonstrated that matrix factorization methods achieved promising performance.

For example, Srebro and Jaakola proposed an approximation method to discover the low rank matrices using EM algorithm and applied in CF [14]. Srebro et al. then proposed another matrix factorization method based on maximum margin principal [15]. This method imposes constraints on the norm of the factorized matrices. Salakhutdinov and Mnih developed different probabilistic matrix factorization models [6,16]. These two models consider the uncertainty involved in the user-item ratings. Instead of predicting the rating, Liu and Yang proposed a method to predict the ranking of items by a user [17].

A number of methods aiming at incorporating additional information in the learned model have been proposed [18,19]. One common method to consider the additional information is to make use of the regularizer in RSVD. For example, Noel et al. proposed to incorporate different forms of regularizer such as feature social regularizer and co-preference regularizer into the objective function when solving RSVD [20]. Ma et al. proposed two regularization models, namely, average-based regularization and individual-based regularization, and applied different similarity measures to consider the social information [21]. Later, Ma developed another method to incorporate the user-user similarity and item-item similarity [1]. Szummer and Yilmaz proposed a method to consider preference regularization to tackle the learning to rank problem in a semi-supervised setting [22].

3   Matrix Factorization

In matrix factorization, there are m users and n items. User i gives item j a rating $r_{ij} = 1, 2, \ldots, r_{max}$, where r_{max} is the maximum value for a rating. Let $R \in \mathbb{R}^{m \times n}$ be the rating matrix where the (i, j)-th entry is equal to r_{ij} if user i has rated item j and 0 otherwise. Note that a user may only rate a few items, hence R is very sparse. Let $\mathcal{E} \equiv \{r_{ij}\}$ for some pairs of i and j be the set of training examples consisting of ratings that user i has rated item j. CF aims at predicting the value of unknown ratings by making use of $\mathcal{E}$. Let $U \in \mathbb{R}^{d \times m}$ and $V \in \mathbb{R}^{d \times n}$, where $d \ll min(m, n)$, be the user matrix and item matrix. We denote $\mathbf{u}_i$ and $\mathbf{v}_j$ be the i-th column vector of U and j-th column vector of V respectively. Matrix factorization treats R as the target matrix and aims at computing U and V such that $R \approx U^\top V$. As a result, the unknown rating to item j given by user i can be predicted by computing $\hat{r}_{ij} = \mathbf{u}_i^\top \mathbf{v}_j$;

Regularized Singular Value Decomposition (RSVD) is a common technique applied to address the sparsity problem in matrix factorization problem. A quadratic loss function is defined as follows:

$$Loss = \frac{1}{2} \sum_{i,j:r_{ij} \in \mathcal{E}} (r_{ij} - \mathbf{u}_i^\top \mathbf{v}_j)^2 + \frac{\lambda_1}{2} \|U\|_F^2 + \frac{\lambda_2}{2} \|V\|_F^2 \tag{1}$$

where $\| \cdot \|_F$ refers to the Frobenius norm. The last two terms are regularizers. The objective of regularization is to avoid large values of U and V, and hence controlling the model complexity and reducing over-fitting. λ_1 and λ_2 are user-defined weighting parameters of the two regularizers. Training of RSVD aims at finding U and V by minimizing the loss function in Eq. 1.

The first derivatives of the loss function can be expressed as follows:

$$\frac{\partial Loss}{\partial \mathbf{u}_i} = \sum_{i,j:r_{ij}\in\mathcal{E}} (r_{ij} - \mathbf{u}_i^\top \mathbf{v}_j)\mathbf{v}_j + \lambda_1 \mathbf{u}_i \tag{2}$$

$$\frac{\partial Loss}{\partial \mathbf{v}_j} = \sum_{i,j:r_{ij}\in\mathcal{E}} (r_{ij} - \mathbf{u}_i^\top \mathbf{v}_j)\mathbf{u}_j + \lambda_2 \mathbf{v}_j \tag{3}$$

Since R is very sparse and not of full rank, setting Eqs. 2 and 3 to zero and solving the system the linear equations is not feasible. Instead, stochastic gradient descent is a common technique for finding the nearly optimal $\mathbf{u}_i$ and $\mathbf{v}_j$. $\mathbf{u}_i$ and $\mathbf{v}_j$ are updated iteratively as follows:

$$\mathbf{u}_i^t \leftarrow \mathbf{u}_i^{(t-1)} + \gamma_1 * [(r_{ij} - \mathbf{u}_i^{(t-1)\top}\mathbf{v}_j^{(t-1)})\mathbf{v}_j^{(t-1)} + \lambda_1\mathbf{u}_i^{(t-1)}] \tag{4}$$

$$\mathbf{v}_i^t \leftarrow \mathbf{v}_i^{(t-1)} + \gamma_2 * [(r_{ij} - \mathbf{u}_i^{(t-1)\top}\mathbf{v}_j^{(t-1)})\mathbf{u}_j^{(t-1)} + \lambda_2\mathbf{v}_i^{(t-1)}] \tag{5}$$

where $\mathbf{u}_i^t$ and $\mathbf{v}_i^t$ refer to the $\mathbf{u}_i$ and $\mathbf{v}_j$ at the t-th iteration; γ_1 and γ_2 represent the learning rate of the algorithm. This updating rules are applied for each $r_{ij} \in \mathcal{E}$ until the maximum number of iterations is reached.

4 Our Approach

As mentioned in Sect. 1, one shortcoming of typical RSVD in collaborative filtering is that the distance between two users in the low rank space will be distorted. Moreover, it does not consider prior social network information when computing U and V. Though some existing social recommendation approach attempt to incorporate the similarity between users, the pre-defined distance metric cannot effectively capture the characteristics of the data and directly accomplish the goal of the task. In this section, we first discuss the idea of distance metric learning. Next, we will present our joint model for collaborative filtering and distance metric learning

4.1 Distance Metric Learning

Following the notation used above, $U \in \mathbb{R}^{d\times m}$ denotes to the user matrix where the j-th column refers to the j-th user. Mahalanobis distance, denoted by $d_A(u_i, u_j)$, between users i and j is defined as follows:

$$d_A(\mathbf{u}_i, \mathbf{u}_j) = \|\mathbf{u}_i - \mathbf{u}_j\|_A = \sqrt{(\mathbf{u}_i - \mathbf{u}_j)^\top A(\mathbf{u}_i - \mathbf{u}_j)} \tag{6}$$

where $A \in \mathbb{R}^{d\times d}$ is a semi-definite, $A \succeq 0$. In Mahalanobis distance, A refers to the covariance matrix. If we assume all users are independent, $A = I$ and $d_A(\mathbf{u}_i, \mathbf{u}_j)$ becomes the Euclidean distance between $\mathbf{u}_i$ and $\mathbf{u}_j$. Essentially, A acts as a linear transformation of the distance between $\mathbf{u}_i$ and $\mathbf{u}_j$ from the original space to a new space.

In many applications, we may collect a set of similar or dissimilar objects. For example, in social network, we may treat a pair of users who are friends as similar users. On the contrary, two users who do not know each other are dissimilar. Distance metric learning aims at automatically learning the distance function based on the collected data. In our approach, we consider A in Mahalanobis distance as parameters, which can be learned from the training examples. The objective is the discover A such that the distance between similar users can be linearly transformed to a new space such that they are as close as possible. On the contrary, the distance between dissimilar users should be linearly transformed such that their distance in the new space is as far as possible. We denote S and D be the set of pairs of similar users and dissimilar users respectively. We can formulate the distance metric problem as an constrained optimization problem as follows:

$$\min_{A} \sum_{(\mathbf{u}_i,\mathbf{u}_j)\in S} \|\mathbf{u}_i - \mathbf{u}_j\|_A^2$$
$$\text{s.t.} \sum_{(\mathbf{u}_i,\mathbf{u}_j)\in D} \|\mathbf{u}_i - \mathbf{u}_j\|_A^2 \geq 1,$$
$$A \succeq 0.$$

The first constraint ensure that the distance between dissimilar users cannot be smaller than 1; the second constraint ensure that the A needs to be semi-positive definite. Note that it is a convex optimization problem. To simplify the learning and improve the efficiency, we set A to a diagonal matrix. As a result, the problem can further be derived to an unconstrained optimization problem as follows:

$$\min_{A} \sum_{(\mathbf{u}_i,\mathbf{u}_j)\in S} \|\mathbf{u}_i - \mathbf{u}_j\|_A^2 - \log \sum_{(\mathbf{u}_i,\mathbf{u}_j)\in D} \|\mathbf{u}_i - \mathbf{u}_j\|_A^2 \qquad (7)$$

where A is a diagonal matrix. Similarly, regularization is commonly applied to avoid overfitting in learning [23].

4.2 RSVD with Distance Metric Learning

Recall that the objective of our framework is to jointly solve RSVD and distance metric learning. To achieve this, we develop a regularizer based on the aforementioned metric learning problem and integrate to RSVD. The rationale of our approach is to simultaneously solve the RSVD and distance metric learning in a single coherent model. In essence, the loss function of RSVD with distance metric learning is expressed as follows:

$$Loss^{new} = \frac{1}{2} \sum_{i,j:r_{ij}\in \mathcal{E}} (r_{ij} - \mathbf{u}_i^\top \mathbf{v}_j)^2 + \frac{\lambda_1}{2}\|U\|_F^2 + \frac{\lambda_2}{2}\|V\|_F^2$$
$$+ \frac{\lambda_3}{2}\{ \sum_{(\mathbf{u}_i,\mathbf{u}_j)\in S} \|\mathbf{u}_i - \mathbf{u}_j\|_A^2 - \frac{\lambda_4}{2}\log \sum_{(\mathbf{u}_i,\mathbf{u}_j)\in D} \|\mathbf{u}_i - \mathbf{u}_j\|_A^2 \} \qquad (8)$$

The first three terms and the fourth term on the right hand side refer to the loss function of RSVD and metric learning respectively. To solve RSVD and distance metric learning, we jointly minimize $Loss^{new}$ with respect to U, V, and A.

The first derivatives of the loss function with respect to u_i, v_j and A can be expressed as follows:

$$\frac{\partial Loss^{new}}{\partial \mathbf{u}_i} = \sum_{i,j:r_{ij}\in\mathcal{E}} (r_{ij} - \mathbf{u}_i^\top \mathbf{v}_j)\mathbf{v}_j + \lambda_1 \mathbf{u}_i$$
$$+\lambda_3 \sum_{(\mathbf{u}_i,\mathbf{u}_j)\in\mathcal{S}} A(\mathbf{u}_i - \mathbf{u}_j) - \frac{\lambda_4}{\sum_{(\mathbf{u}_i,\mathbf{u}_j)\in\mathcal{D}}\|\mathbf{u}_i-\mathbf{u}_j\|_A^2} \sum_{(\mathbf{u}_i,\mathbf{u}_j)\in\mathcal{D}} A(\mathbf{u}_i - \mathbf{u}_j)$$
$$(9)$$

$$\frac{\partial Loss^{new}}{\partial \mathbf{v}_j} = \sum_{i,j:r_{ij}\in\mathcal{E}} (r_{ij} - \mathbf{u}_i^\top \mathbf{v}_j)\mathbf{u}_j + \lambda_2 \mathbf{v}_j \tag{10}$$

$$\frac{\partial Loss^{new}}{\partial A} = \frac{\lambda_3}{2} \sum_{(\mathbf{u}_i,\mathbf{u}_j)\in\mathcal{S}} (\mathbf{u}_i - \mathbf{u}_j)(\mathbf{u}_i - \mathbf{u}_j)^\top$$
$$-\frac{\lambda_4}{2\sum_{(\mathbf{u}_i,\mathbf{u}_j)\in\mathcal{D}}\|\mathbf{u}_i-\mathbf{u}_j\|_A^2} \sum_{(\mathbf{u}_i,\mathbf{u}_j)\in\mathcal{D}} (\mathbf{u}_i - \mathbf{u}_j)(\mathbf{u}_i - \mathbf{u}_j)^\top \tag{11}$$

We can then solve the optimization problem by iterative methods like the efficient gradient descent method. One characteristic of our approach is that U, V, and A are jointly varied to optimize $Loss^{new}$. This leads to a solution optimizing both tasks of collaborative filtering and distance metric learning. On the other hand, collaborative filtering and distance metric learning serve regularization to each other resulting to the smoothing effect and reducing over-fitting.

5   Discovery of Similar Users

Recalled that our preference regularizer in Eq. 8 contains the similarity between users. In this paper we employ three different similarity measures to discover similar users.

Jaccard Similarity. Jaccard similarity mainly consider the items that both users have rated, without considering the actual ratings given to these items. Let Q_h and Q_i be the set of items that users h and i have rated respectively. Jaccard similarity is defined as follows:

$$sim(h,i) = \frac{|Q_h \cap Q_i|}{|Q_h \cup Q_i|} \tag{12}$$

Pearson Correlation Coefficient. Pearson correlation coefficient (PCC) aims at measuring the relationship between the ratings given to the items that are rated by two users. Let Q_h and Q_i be the set of items that users h and i have rated respectively. PCC is defined as follows:

$$pcc(h,i) = \frac{\sum_{j\in Q_h \cap Q_i}(r_{hj}-\bar{r_h})(r_{ij}-\bar{r_i})}{\sqrt{\sum_{j\in Q_h \cap Q_i}(r_{hj}-\bar{r_h})^2 \sum_{j\in Q_h \cap Q_i}(r_{ij}-\bar{r_i})^2}} \tag{13}$$

where $\bar{r_h}$ refers to the mean of the ratings to all items given by user h. Since $-1 \leq pcc(h,i) \leq 1$, we define our similarity as follows:

$$sim(h,i) = \frac{1+pcc(h,i)}{2} \tag{14}$$

Kendall Rank Correlation Coefficient. Unlike PCC, Kendall rank correlation coefficient, denoted as τ, is to measure the relation between the ranking of the items that are rated by two users. Let Q_h and Q_i be the set of items that users h and i have rated respectively. $\tau(h, i)$ is defined as follows:

$$\tau(h, i) = \frac{\sum\limits_{j, k \in Q_h \cap Q_i} sign((r_{hj} - r_{hk})(r_{ij} - r_{ik}))}{\frac{1}{2}|Q_h \cap Q_i|(|Q_h \cap Q_i| - 1)} \tag{15}$$

Since $-1 \leq \tau(h, i) \leq 1$, we define our simiarity as follows:

$$sim(h, i) = \frac{1 + \tau(h, i)}{2} \tag{16}$$

The computation of PCC and τ coefficient are computationally expensive. To reduce the computational time, for any pair of users, we randomly sample N items that are rated by them to compute pcc and τ coefficient. In our experiments, N is set to 10. Next, given a user i, the top-K similar users such that the similarity is greater than 0.75 are considered to be similar to user i and constitute $S(i)$ in Eq. 8.

6   Experimental Results

We have conducted experiments on two real-world datasets to evaluate the effectiveness of our framework. The first dataset we used is the MovieLens dataset[3]. This dataset consists of 100,000 ratings (between 1 and 5) from 943 users on 1,642 movies. We call this dataset *ml-100k*. Another dataset is the Epinions dataset[4]. This dataset consists of 664,823 ratings (between 1 and 5) from 49,290 users on 139,738 different items. We call this dataset *epinions*. In each dataset, we randomly divided the data into five portions, namely u1 to u5, with equal number of ratings. In each run of the experiments, we treated four portions as the set of training examples and the remaining portion as the test data. For example, we utilized u1-u4 for training and u5 for testing. As a result, we conducted 5 runs of experiments, each of which utilized different portions as testing data, for each dataset.

Three sets of experiments were conducted to evaluate our framework. In the first set of experiments, we applied the standard RSVD method on the datasets. This can be regarded as our baseline method. We call this *RSVD approach*. In the second set of experiments, we implemented the existing method described in [1] and applied it on the datasets. We implemented the SR_{i+-}^{u+-} approach described in [1]. We call this *Ma's approach*. We compared with this approach because it also aims at improving collaborative filtering via regularization. However, it only considers the closeness between users and the closeness between items in the learned model. In the third sets of experiments, we applied our framework, using different similarity measures as described above. We call this *Our approach*.

---

[3] The dataset can be freely downloaded in http://www.grouplens.org/.
[4] The dataset can be freely downloaded in http://www.trustlet.org/wiki/Downloaded_Epinions_dataset.

Table 1. The prediction performance of RSVD approach, Ma's approach, and Our approach on the dataset ml-100k.

Testing data	RSVD approach	Ma's approach	Our approach		
			Jaccard	PCC	τ coefficient
u1	0.757	0.728	0.712	0.711	0.709
u2	0.749	0.718	0.711	0.700	0.702
u3	0.749	0.722	0.712	0.701	0.703
u4	0.750	0.725	0.710	0.710	0.709
u5	0.752	0.724	0.713	0.707	0.714
Average	0.751	0.723	0.712	0.706	0.707

Table 2. The prediction performance of RSVD approach, Ma's approach, and Our approach on the dataset epinions.

Testing data	RSVD approach	Ma's approach	Our approach		
			Jaccard	PCC	τ coefficient
u1	0.826	0.804	0.783	0.774	0.780
u2	0.824	0.803	0.780	0.779	0.783
u3	0.825	0.801	0.779	0.791	0.784
u4	0.824	0.802	0.782	0.798	0.783
u5	0.823	0.800	0.787	0.781	0.787
Average	0.824	0.802	0.782	0.785	0.783

In all these approaches, we set the dimension d in matrix factorization to 10. We also followed [1] to set the parameters λ_1, λ_2, γ_1, and γ_2 to 0.01, 0.01, 0.005, and 0.005 respectively. In our approach, we also set λ_3 to 0.01, so that all regularizers have the same weighting. The maximum iteration when running stochastic gradient descent optimization was set to 50,000. Since the ratings of the datasets we used in the experiments are discrete, we round the predicted ratings of the three approaches to the nearest integer.

We adopted the commonly used evaluation metric, namely, Mean-Absolute-Error (MAE), which is defined as follows:

$$MAE = \frac{\sum_{r_{ij} \in \mathcal{T}} |r_{ij} - \hat{r}_{ij}|}{|\mathcal{T}|} \qquad (17)$$

where $\mathcal{T}$ refers to the set of testing data.

Table 1 shows the prediction performance on the dataset ml-100k. Each row of the table refers to a run of the experiments. The first column of the table refers to the portion of the dataset used as testing data in this run. The second and third columns contains the prediction performance of RSVD approach and Ma's approach respectively. The fourth column is divided into three sub-columns, each of which contains the prediction performance of our approach using different similarity measures. The first, second, and third sub-columns refer to the Jaccard similarity, Pearson correlation coefficient (PCC), and Kendall rank correlation coefficient (τ) respectively. The last row of the table shows the average prediction performance. The average MAE of our approach using Jacaard similarity, PCC, and τ coefficient are 0.712, 0.706, and 0.707 respectively. They outperform RSVD approach and Ma's approach, whose average MAE are 0.751 and 0.723 respectively. Among the three different similar measure, our approach achieves similar prediction performance. Table 2 shows the prediction performance of different approaches on the dataset epinions. The format of Table 2 is the same as that of Table 1. Similarly, our approach achieves the best performance, with average MAE of 0.782, 0.785, and 0.783 for Jaccard similarity, PCC, and τ coefficient respectively.

7   Conclusions and Future Work

We have developed a framework for improving rating prediction in collaborative filtering by making use of preference regularization. Our framework is designed based on the idea that similar users should retain the distance in the low-rank space after RSVD. One characteristic of our framework is that collaborative filtering and distance metric learning serve as regularization to each other and naturally reduce overfitting to both. Another characteristic is that social community information and similarity information can be easily considered in our framework. We have conducted several sets of experiments on two real-world datasets to evaluate our framework. We have compared our framework with exiting works. The experimental results show that our framework achieves a very promising performance.

Acknowledgments. The work described in this paper is substantially supported by grants from the Education University of Hong Kong (Project Codes: RG 30/2014-2015R and RG 18/2015-2016R).

References

1. Ma, H.: An experimental study on implicit social recommendation. In: Proceedings of the Thirty-sixth international ACM SIGIR Conference on Research and Development in Information Retrieval, pp. 73–82 (2013)
2. Bobadilla, J., Ortega, F., Hernando, A., Gutiérrez, A.: Recommender systems survey. Knowl. Based Syst. **46**, 109–132 (2013)
3. Deshpande, M., Karypis, G.: Item-based top-n recommendation. ACM Trans. Inf. Syst. **22**(1), 143–177 (2004)
4. Jin, R., Chai, J.Y., Si, L.: An automatic weighting scheme for collaborative filtering. In: Proceedings of the Twenty-Seventh Annual International ACM SIGIR Conference on Research and Development in Information Retrieval, pp. 337–344 (2004)

5. Linden, G., Smith, B., York, J.: Amazon.com recommendations: item-to-item collaborative filtering. IEEE Internet Comput. **7**(1), 76–80 (2003)
6. Salakhutdinov, R., Mnih, A.: Probabilistic matrix factorization. In: Proceedings of Advances in Neural Information Processing Systems, pp. 1257–1264 (2007)
7. Hofmann, T.: Latent semantic models for collaborative filtering. ACM Trans. Inf. Syst. **22**(1), 89–115 (2004)
8. Xue, G.R., Lin, C., Yang, Q., Xi, W., Zeng, H.J., Yu, Y., Chen, Z.: Scalable collaborative filtering using cluster-based smoothing. In: Proceedings of the Twenty-Eighth Annual International ACM SIGIR Conference on Research and Development in Information Retrieval, pp. 114–121 (2005)
9. Si, L., Jin, R.: Flexible mixture model for collaborative filtering. In: Proceedings of the Twentieth International Conference on Machine Learning, pp. 704–711 (2003)
10. Zhang, Y., Koren, J.: Efficient bayesian hierarchical user modeling for recommendation system. In: Proceedings of the 30th Annual International ACM SIGIR Conference on Research and Development in Information Retrieval, pp. 47–54 (2007)
11. Huang, S., Wang, S., Liu, T.Y., Ma, J., Chen, Z., Veijalainen, J.: Listwise collaborative filtering. In: Proceedings of the 38th International ACM SIGIR Conference on Research and Development in Information Retrieval, pp. 343–352 (2015)
12. Koren, Y., Bell, R.M., Volinsky, C.: Matrix factorization techniques for recommender systems. IEEE Comput. **42**(8), 30–37 (2009)
13. Paterek, A.: Improving regularized singular value decomposition for collaborative filtering. In: Proceedings of the KDD Cup Workshop at SIGKDD 2007, pp. 39–42 (2007)
14. Srebro, N., Jaakkola, T.: Weighted low-rank approximations. In: Proceedings of the 20th International Conference on Machine Learning, pp. 720–727 (2003)
15. Srebro, N., Rennie, J.D.M., Jaakkola, T.: Maximum-margin matrix factorization. In: Proceedings of Advances in Neural Information Processing Systems, pp. 1329–1336 (2004)
16. Salakhutdinov, R., Mnih, A.: Bayesian probabilistic matrix factorization using markov chain monte carlo. In: Proceedings of the Twenty-Fifth International Conference on Machine Learning, pp. 880–887 (2008)
17. Liu, N.N., Yang, Q.: Eigenrank: a ranking-oriented approach to collaborative filtering. In: Proceedings of the Thirty-First Annual International ACM SIGIR Conference on Research and Development in Information Retrieval, pp. 83–90 (2008)
18. Koren, Y.: Factorization meets the neighborhood: a multifaceted collaborative filtering model. In: Proceedings of the Fourteenth ACM SIGKDD International Conference on Knowledge Discovery and Data Mining, pp. 426–434 (2008)
19. Koren, Y.: Collaborative filtering with temporal dynamics. In: Proceedings of the Fifteenth ACM SIGKDD International Conference on Knowledge Discovery and Data Mining, pp. 447–456 (2009)
20. Noel, J., Sanner, S., Tran, K.N., Christen, P., Xie, L., Bonilla, E.V., Abbasnejad, E., Nicolas, D.P.: New objective functions for social collaborative filtering. In: Proceedings of the Twenty-First International Conference on World Wide Web, pp. 859–868 (2012)
21. Ma, H., Zhou, D., Liu, C., Lyu, M.R., King, I.: Recommender systems with social regularization. In: Proceedings of the Fourth ACM International Conference on Web search and Data Mining, pp. 287–296 (2011)
22. Szummer, M., Yilmaz, E.: Semi-supervised learning to rank with preference regularization. In: Proceedings of the Twentieth ACM International Conference on Information and Knowledge Management, pp. 269–278 (2011)

23. Jin, R., Wang, S., Zhou, Y.: Regularized distance metric learning: theory and algorithm. In: Advances in Neural Information Processing Systems 22, Neural Information Processing Systems, pp. 862–870 (2009)
24. Yu, K., Schwaighofer, A., Tresp, V., Ma, W.Y., Zhang, H.: Collaborative ensemble learning: combining collaborative and content-based information filtering via hierarchical bayes. In: Proceedings of the Nineteenth Conference on Uncertainty in Artificial Intelligence, pp. 616–623 (2003)
25. Tang, J., Gao, H., Hu, X., Liu, H.: Exploiting homophily effect for trust prediction. In: Proceedings of the Sixth ACM International Conference on Web Search and Data Mining, pp. 53–62 (2013)

Scrutinizing Mobile App Recommendation: Identifying Important App-Related Indicators

Jovian Lin[1], Kazunari Sugiyama[1(✉)], Min-Yen Kan[1,2], and Tat-Seng Chua[1,2]

[1] School of Computing, National University of Singapore, Singapore, Singapore
jovian.lin@gmail.com, {sugiyama,kanmy,chuats}@comp.nus.edu.sg
[2] Interactive and Digital Media Institute, National University of Singapore,
Singapore, Singapore

Abstract. Among several traditional and novel mobile app recommender techniques that utilize a diverse set of app-related features (such as an app's Twitter followers, various version instances, *etc.*), which app-related features are the most important indicators for app recommendation? In this paper, we develop a hybrid app recommender framework that integrates a variety of app-related features and recommendation techniques, and then identify the most important indicators for the app recommendation task. Our results reveal an interesting correlation with data from third-party app analytics companies; and suggest that, in the context of mobile app recommendation, more focus could be placed in user and trend analysis via social networks.

Keywords: Recommender systems · Mobile apps · Gradient tree boosting

1 Introduction

Traditional recommendation approaches either learn a user's preference from their ratings (*i.e.*, collaborative filtering) or the contents of previously-consumed items (*i.e.*, content-based filtering). Despite the pervasive use of collaborative filtering in several domains such as books, movies, and music, its effectiveness is hindered by insufficient ratings, particularly towards newly-released items — a problem that is commonly known as the "cold-start." Moreover, due to noisy and unreliable descriptions of apps, content-based filtering does not work well in the app domain [10].

With the widespread interest and pervasiveness of mobile apps, several novel recommendation techniques that take advantage of the unique characteristics of the app domain have emerged. The first type focuses on collecting additional internal information from the user's mobile device, which analyzes the usage behavior of individual apps via anonymized network data from cellular carriers [18] as well as usage patterns of users via their in-house recommender systems [1,6,19]. The second type makes use of external information such as spatial data from GPS sensors to provide context-aware app recommendations [7,22].

© Springer International Publishing AG 2016
S. Ma et al. (Eds.): AIRS 2016, LNCS 9994, pp. 197–211, 2016.
DOI: 10.1007/978-3-319-48051-0_15

These two types, however, rely on data that is generally difficult to obtain, causing the secondary problem of data-sparsity. On the contrary, the third type consists of works that capitalize on more unique characteristics of the app domain that may not be applicable to other domains. For instance, "follower" information of an app's Twitter account was used to substitute missing user ratings [10], which proved to be useful in cold-start situations. Another work tried to find the likelihood of which a current app would be replaced by another [20]. Alternatively, by taking the fact that apps change and evolve with every new version update, a "version-sensitive" recommendation technique was constructed to identify desired functionalities (from various version descriptions of apps) that users are looking for [11].

With a variety of app recommendation techniques utilizing different sources of information, of which some may be available while others are not (*e.g.*, not all apps have user ratings), we explore the advantages of a hybrid app recommendation framework that combines traditional and novel techniques. More importantly, through the hybrid framework, we seek to identify the most important app-related indicators for the recommendation task.

The steps are as follows: First, using gradient tree boosting (GTB) [8], several recommendation techniques and their information sources are integrated to form a hybrid app recommender framework. After that, we further look into each component of the feature set to find the most significant features in the hybrid framework. Our findings show an interesting correlation with data from third-party app analytics companies, and suggest that, in the context of mobile app recommendation, more focus could be placed in user and trend analysis via social networks.

2   Related Work

2.1   Mobile App Retrieval

Chen *et al.* [5] proposed a framework for detecting similar apps by constructing kernel functions based on multi-modal heterogeneous data of each app (description text, images, user reviews, and so on) and learning optimal weights for the kernels. They also applied this approach to mobile app tagging [4]. While Chen *et al.*'s work utilized different modalities of an app, Park *et al.* [14] exclusively leveraged text information such as reviews and descriptions (written by users and developers, respectively) and designed a topic model that can bridge vocabulary gap between them to improve app retrieval. Zhang *et al.* [21] developed a mobile query auto-completion model that exploits installed app and recently opened app. In addition, Martin *et al.* [13] has published a nice survey on app store analysis that identifies some directions for software engineering such as requirements engineering, release planning, software design, testing, and so on.

2.2   Mobile App Recommendation

In order to deal with the recent rise in the number of apps, works on mobile app recommendation are emerging. Some of these works focus on collecting additional information from the mobile device to improve recommendation accuracy. Xu *et al.* [18] investigated the diverse usage behaviors of individual apps by using anonymized network data from a tier-1 cellular carrier in the United States. While Yan and Chen [19], Costa-Montenegro *et al.* [6], and Baeza-Yates *et al.* [1] analyzed internal information such as the usage patterns of each user to construct app recommendation system, Zheng *et al.* [22] and Davidsson and Moritz [7] utilized external information such as GPS sensor information to provide context-aware app recommendation. Lin *et al.* [10] utilized app-related information on Twitter to improve app recommendation in cold-start situations. Their subsequent work focused on app's uniqueness of version update, and then proposed an app recommendation system that leverages version features such as textual description of the changes in a version, version metadata [11]. These two works are compiled into [9]. Yin *et al.* [20] considered behavioral factors that invoke a user to replace an old app with a new one, and introduced the notion of "actual value" (satisfactory value of the app after the user used it) and "tempting value" (the estimated satisfactory value that the app may have), thereby regarding app recommendation as a result of the contest between these two values. Zhu *et al.* [23] and Liu *et al.* [12] incorporated both each user's interest and privacy preferences to provide app recommendation as apps could have privileges to access the user's sensitive personal information such as locations, contacts, and messages. While the aforementioned works recommend apps that are relevant to each user's interests, Bhandari *et al.* [2] proposed a graph-based method for recommending serendipitous apps.

3   Methodology

3.1   Feature Set

Inspired by Wang *et al.*'s work [17], the features that we use can be categorized into the following three distinct groups:

1. the app's underline{m}arketing-related underline{m}etadata ($\mathbb{M}$),
2. the user's underline{h}istory-related information ($\mathbb{H}$), and
3. the underline{r}ecommendation scores of different recommender systems ($\mathbb{R}$).

As illustrated in Fig. 1, every candidate app's feature vector $X_{u,a}$ is composed of all three groups of information: $X_{u,a} = \{X_a^M, X_{u,a}^H, X_{u,a}^R\}$ where $X_{u,a}$ represents the feature vector of the app a for user u, while M, H, and R represent the features from the users' history, apps' metadata, and recommendation scores from various recommendation techniques, respectively.

3.1.1   App's Marketing-Related Metadata ($\mathbb{M}$)

The features here pertain to the app's metadata or marketing-related information. We include most of the components of an app's official metadata from the iTunes App Store, such as the various genres that the app is assigned to, its price, average ratings, *etc.* We also include external information, particularly ubiquitous data from social networks, such as the number of versions an app has, the number of Facebook "likes" it has (zero if the app has no Facebook handle), and the number of Twitter followers it has (zero if the app has no Twitter handle). The *blue* components in Fig. 1 show all the information of an app's marketing-related features.

3.1.2   User's History-Related Information ($\mathbb{H}$)

User history is primarily extracted from the rating history of users, and it is a crucial component for the purpose of providing *personalized* recommendations. In addition, inspired by Wang *et al.*'s method [17] for generating additional user metadata by scrutinizing the genres of items that users have consumed, we also consider the user's preference of each app genre g. For instance, a user might be a loyal consumer of the "games" genre, yet not in the "food & drink" genre. We thus include the number of times (*i.e.*, the "count") that apps in genre g were consumed by user u (represented in *green* in Fig. 1).

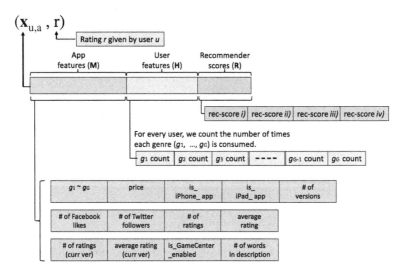

Fig. 1. An app's feature vector $(X_{u,a}, r)$, which contains app features, user features, the various recommendation scores, and the user's rating. As described in Sect. 3.1.3, "Recommender scores ($\mathbb{R}$)" are generated by *(i)* collaborative filtering, *(ii)* content-based filtering, *(iii)* "Twitter-follower-based app recommendation" (TWF) [10], and *(iv)* "version-sensitive recommendation" (VSR) [11]. (Color figure online)

3.1.3   Recommendation Scores from Different Recommender Techniques ($\mathbb{R}$)

We also include the recommendation scores generated from four recommendation techniques: *(i)* collaborative filtering, *(ii)* content-based filtering, *(iii)* "Twitter-follower-based app recommendation" (TWF) [10], and *(iv)* "version-sensitive recommendation" (VSR) [11]. These are represented by the red components in Fig. 1.

We employ probabilistic matrix factorization (PMF) [15] to implement collaborative filtering as it is a state-of-the art technique that models the user-item ratings matrix as a product of two lower-rank user and item matrices, and it has been used in many previous recommendation works due to its highly flexibility and extendability. We also employ latent Dirichlet allocation (LDA) [3] to implement content-based filtering (on apps' textual descriptions) as it effectively provides an interpretable and low-dimensional representation of the items. In addition, we select TWF and VSR due to their ability to make use of ubiquitous information from Twitter's API and version data from third-party app analytics companies, respectively. With the hybrid app recommendation that is modeled by gradient tree boosting (GTB) [8], we further look into each component of the feature set (*i.e.*, M, H, and R) in the hybrid model based on relative influence[1].

3.2 Combining App Features

Inspired by BellKor's winning solution for the Netflix Prize[2], we turn to Gradient Tree Boosting (GTB), a machine learning algorithm that iteratively constructs an ensemble of weak decision tree learners through boosting [8]. It produces an accurate and effective off-the-shelf procedure for data mining that can be directly applied to the data without requiring a great deal of time-consuming data preprocessing or careful tuning of the learning procedure.

To generate recommendations, the learned GTB predicts the rating that a user may give to an app. After which, it ranks all recommended apps in descending order of rating to produce a ranked list for each user. Here, we use a popular Python machine learning package from scikit-learn[3] to implement GTB.

4   Experimental Setup

We construct our experimental dataset by crawling the information on Apple's iTunes App Store[4] (app metadata, users, and ratings), App Annie[5] (version information of apps), Twitter (for the Twitter followers of apps), and Facebook (for the "likes" information of apps). Our dataset includes 33,802 apps, 16,450 users, and 3,106,759 ratings after we retain only unique users who give at least 30 ratings. Among the 33,802 apps, 7,124 (21.1 %) have Twitter accounts, 9,288 (27.5 %) have Facebook accounts, and 10,520 (31.1 %) have at least five versions.

---

[1] Friedman [8] proposed the relative influence for boosted estimates to reflect each feature's contribution of reducing the loss by splitting on the feature.

[2] Y. Koren: "The BellKor Solution to the Netflix Grand Prize," http://www.stat.osu.edu/~dmsl/GrandPrize2009_BPC_BellKor.pdf.

[3] http://scikit-learn.org/stable/modules/ensemble.html (Ver 0.15.0).

[4] https://itunes.apple.com/us/genre/ios/id36?mt=8.

[5] https://www.appannie.com/.

Note that 678 (2.0 %) apps have both Twitter and Facebook accounts. We perform 5-fold cross validation, where in each fold, we take the first 80 % of the apps (chronologically) as training data for the individual recommendation techniques, use the following 10 % as the training data for the unified model (*i.e.*, the probe set of GTB), and use the remaining 10 % for testing.

4.1   Comparative Recommender Systems

We compare two types of recommender systems: individual and hybrid. For individual systems which are baselines, we implement the four state-of-the-art recommender algorithms mentioned in Sect. 3.1.3, namely, collaborative filtering (PMF) [15], content-based filtering (LDA) [3], TWF [10], and VSR [11]. For the hybrid systems, we create three subsets of the GTB framework using a smaller set of features. That is, on top of our gradient boosting hybrid framework GTB($\mathbb{M}$, $\mathbb{H}$, $\mathbb{R}$), we create three more hybrid systems: GTB($\mathbb{R}$), GTB($\mathbb{H}$, $\mathbb{R}$), and GTB($\mathbb{M}$, $\mathbb{R}$), where "$\mathbb{M}$", "$\mathbb{H}$", and "$\mathbb{R}$" represent the various information $\boldsymbol{X}_a^M$, $\boldsymbol{X}_{u,a}^H$, and $\boldsymbol{X}_{u,a}^R$ mentioned in Sect. 3.1, respectively.

Table 1 shows the details of the various recommendation techniques and their feature set. For the individual recommender systems, the feature set contains the user's history-related features ($\boldsymbol{X}_{u,a}^H$) that are generated from the user's previous ratings history as well as the app data. The hybrid models further integrate the product's marketing-related metadata ($\boldsymbol{X}_a^M$) and the recommender scores generated by the individual recommender systems ($\boldsymbol{X}_{u,a}^R$).

4.2   Evaluation Metric

Our system ranks the recommended apps based on the probability in which a user is likely to download the app. This methodology leads to two possible evaluation metrics: precision and recall. However, a missing rating in the training set is ambiguous as it may either mean that the user is not interested in the app, or that the user does not know about the app (*i.e.*, truly missing). This makes it difficult to accurately compute precision [16]. But since the known ratings are true positives, recall is a more pertinent measure as it only considers the positively rated apps within the top M, namely, a high recall with a lower M will be a better system. We thus chose Recall@M (especially, $M = 50$) as our primary evaluation metric.

5   Experimental Results

5.1   Individual Recommender Techniques

Figure 2 shows Recall@50 obtained by different recommender systems. Among the individual recommender techniques (*i.e.*, the first four bars from the left), content-based filtering (LDA) achieves the best performance, *i.e.*, it outperforms collaborative filtering (PMF), TWF, and VSR. At first, it is surprising

Table 1. Various recommendation techniques.

Technique	Feature set
PMF [15]	Collaborative filtering with $\boldsymbol{X}_{u,a} = \{\boldsymbol{X}_{u,a}^{H}\}$
LDA [3]	Content-based filtering with $\boldsymbol{X}_{u,a} = \{\boldsymbol{X}_{u,a}^{H}\}$
TWF [10]	Twitter-follower recommender with $\boldsymbol{X}_{u,a} = \{\boldsymbol{X}_{u,a}^{H}\}$
VSR [11]	Version-sensitive recommendation with $\boldsymbol{X}_{u,a} = \{\boldsymbol{X}_{u,a}^{H}\}$
GTB($\mathbb{R}$)	$\boldsymbol{X}_{u,a} = \{\boldsymbol{X}_{u,a}^{R}\}$
GTB($\mathbb{H}, \mathbb{R}$)	$\boldsymbol{X}_{u,a} = \{\boldsymbol{X}_{u,a}^{H}, \boldsymbol{X}_{u,a}^{R}\}$
GTB(M, $\mathbb{R}$)	$\boldsymbol{X}_{u,a} = \{\boldsymbol{X}_{a}^{M}, \boldsymbol{X}_{u,a}^{R}\}$
GTB(M, $\mathbb{H}, \mathbb{R}$)	$\boldsymbol{X}_{u,a} = \{\boldsymbol{X}_{a}^{M}, \boldsymbol{X}_{u,a}^{H}, \boldsymbol{X}_{u,a}^{R}\}$

that content-based filtering (LDA) is the best individual technique among the other individual algorithms, especially against state-of-the-art ones. But given that the dataset contains some apps that: *(i)* do not have enough ratings for collaborative filtering, *(ii)* do not have Twitter accounts (78.9 %), and *(iii)* do not have sufficient version information (68.9 %), it is reasonable that these techniques underperform due to the lack of sufficient information for every app, whereas content-based filtering (LDA) works better because apps always have app descriptions to construct a recommendation model. In other words, in general and practical situations where there are a variety of apps that have and do not have ratings, Twitter accounts, and version information, content-based filtering is the more reliable technique.

5.2 Hybrid Recommender Techniques

Next, we explore the GTB models in Fig. 2 (the last four bars). All of our GTB models outperform the individual techniques described in Sect. 5.1. This is expected as many other works that use GTB, particularly those involved in the Netflix prize, have also reported improvements against individual baselines. We also observe a general improvement in recall when we incorporate more components into the feature set. For example, GTB(M,$\mathbb{R}$) and GTB(M,$\mathbb{H}$,$\mathbb{R}$) outperform GTB($\mathbb{R}$) and GTB(M,$\mathbb{R}$), respectively. We observe an interesting small anomaly, in which GTB($\mathbb{H}$,$\mathbb{R}$) slightly underperforms GTB($\mathbb{R}$), whereas GTB(M,$\mathbb{R}$) significantly outperforms both GTB($\mathbb{R}$) and GTB($\mathbb{H}$,$\mathbb{R}$). In other words, the recommendation scores ($\mathbb{R}$) is more effective when it is combined with app metadata (M) than when it is combined with user features ($\mathbb{H}$). This suggests that app metadata (M) complements the feature of recommendation scores ($\mathbb{R}$) — which actually makes sense as, given the assortment of app metadata (M) that coincides with recommendation scores ($\mathbb{R}$), a correlation pattern can be better identified. For example, the app metadata of Twitter followers would complement the recommendation score provided by TWF, while the number of versions would complement the recommendation score generated by VSR;

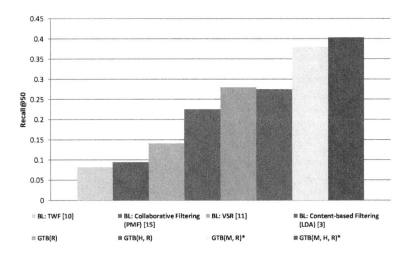

Fig. 2. Recall@50 obtained by individual and hybrid recommender systems. "BL" stands for "baselines." "*" denotes the difference between combined techniques (GTB(M,$\mathbb{R}$) and GTB(M,$\mathbb{H}$,$\mathbb{R}$)) and the best baseline (content-based filtering (LDA)) is statistically significant for $p < 0.01$.

likewise, the number of ratings would complement the recommendation score given by collaborative filtering. On the contrary, as features from user history ($\mathbb{H}$) mainly consists of the number of times each genre is consumed, it has less obvious correlations.

5.3   Ablation Testing

5.3.1   Ablation Testing for Hybrid Recommendation Techniques

The experimental results described in Sect. 5.2 show the overall effectiveness of all four combined recommendation techniques as well as user features and app information. To gain a deeper understanding of the individual recommendation techniques, we further perform ablation testing by excluding one of the four recommendation techniques from GTB(M,$\mathbb{H}$,$\mathbb{R}$), while at the same time, using the user features and app metadata, $X_{u,a}^{H}$ and X_{a}^{M}.

Table 2 shows recall@50 obtained by the ablation testing in which we ablate one recommendation technique out of the four. We observe the followings from Table 2:

- Content-based filtering (LDA), which achieves the best recall among all individual baselines, also causes the largest dip in recall when we ablate it from the unifying model. That is, "GTB($\mathbb{H}$,M,$\mathbb{R}$) excluding content-based filtering" has the lowest score (0.237) among the four ablation baselines. This is unsurprising as it is expected when we omit the strongest individual predictor.
- Although VSR individually outperforms collaborative filtering (0.141 against 0.094), ablating it from the unifying model does not have very much impact;

Table 2. Recall@50 obtained by ablation testing.

Feature	Recall@50
GTB(M, H, R)	**0.403**
GTB(M, H, R), excluding TWF [10]	0.363
GTB(M, H, R), excluding VSR [11]	0.346
GTB(M, H, R), excluding Collaborative filtering (PMF) [15]	0.292
GTB(M, H, R), excluding Content-based filtering (LDA) [3]	0.237
TWF [10]	0.082
VSR [11]	0.141
Collaborative Filtering (PMF) [15]	0.094
Content-based filtering (LDA) [3]	0.225

in fact, ablating collaborative filtering (PMF) has more impact than ablating VSR.
- It would seem that, from this initial ablation study, both of the traditional recommendation techniques, collaborative filtering (PMF) and content-based filtering (LDA) are more effective than VSR and TWF as the two traditional techniques bring about the two biggest dips in recall when we ablate them.
- However, we should not let this relative ablation comparison undermine the improvements that VSR and TWF have brought about. In fact, VSR and TWF improve recall by 16.5 % and 11.0 %, respectively. More importantly, by utilizing these unique and less obvious signals in the app domain (compared with other traditional domains in recommender systems), we have gained significant improvements for general app recommendation[6]. In other words, different pieces of evidences (*e.g.*, Twitter followers and versions) that, when present, can be utilized sufficiently to create a discernible improvement in recommendation quality.

Still, this initial ablation testing does not paint a full picture, especially regarding VSR and TWF, as 68.9 % of apps do not have sufficient version information while 78.9 % of apps do not have Twitter accounts (see Sect. 4). Therefore, the lack of information does not provide a well grounded conclusion. In order to investigate the real utility of VSR and TWF, we further scrutinize our data by utilizing a subset of data that has sufficient Twitter and version information in the unifying model.

5.3.2 Ablation Testing Using Sufficient Twiter Information
Similar to Sect. 5.3.1, we also perform ablation testing using a dataset with full Twitter information. Table 3 shows recall@50 obtained by this study where

---

[6] In fact, on 21 September 2009, the grand prize of US$1,000,000 was given to the BellKor's Pragmatic Chaos team which bested Netflix's own algorithm for predicting ratings by 10.06 %. That is, US$1M for an improvement of 10.06 %.

$GTB_{TWF}(...)$ represents the model that uses full Twitter information in our controlled ablation testing. Table 3 indicates the followings:

- Under a dataset with full Twitter information, we observe a reordering of recommendation techniques whereby TWF becomes consequential — ablating it causes the largest dip in recall scores (0.338) for the unifying model.
- Not only does this justify TWF's effectiveness but more importantly, it indicates that when certain evidence is available (here, Twitter followers information), this changes the signals that are used in the unifying model, allowing TWF to displace the traditional, well-established recommendation techniques.

5.3.3   Ablation Testing Using Sufficient Version Information

Furthermore, we perform another ablation testing using a dataset with full version information. Table 4 shows the recall@50 obtained by this study where $GTB_{VSR}(...)$ represents the model that uses full version information in our controlled ablation testing. According to Table 4, we observe the followings:

- Similar to our ablation testing with TWF in Sect. 5.3.2, under a dataset with full version information, we observe a reordering of recommendation techniques.
- Even though VSR does not displace collaborative filtering in this ablation testing, it still results in the second largest dip in recall scores (0.344) when we ablate it from the unifying model. In addition, under this dataset, improvement in recall obtained by VSR increases from 16.5 % (in Table 2) to 22 %.
- This further substantiates that when certain evidence is accessible, it changes the way signals are used in the unifying model, which the reordering of recommendation techniques in our ablation study suggests.

The ablation studies on the two controlled datasets (pertaining to full Twitter and version information) clearly demonstrate the importance of TWF and VSR in app recommendation, without which we would not have been able to capture Twitter and version signals for the purpose of improving recommendation quality.

Table 3. Recall@50 obtained by controlled ablation testing using sufficient **Twitter** information.

Feature	Recall@50
$GTB_{TWF}(\mathbb{M}, \mathbb{H}, \mathbb{R})$	**0.446**
$GTB_{TWF}(\mathbb{M}, \mathbb{H}, \mathbb{R})$, excluding VSR [11]	0.412
$GTB_{TWF}(\mathbb{M}, \mathbb{H}, \mathbb{R})$, excluding Collaborative filtering (PMF) [15]	0.390
$GTB_{TWF}(\mathbb{M}, \mathbb{H}, \mathbb{R})$, excluding Content-based filtering (LDA) [3]	0.386
$GTB_{TWF}(\mathbb{M}, \mathbb{H}, \mathbb{R})$, excluding TWF [10]	0.338

Table 4. Recall@50 obtained by controlled ablation testing using sufficient **version** information.

Feature	Recall@50
$GTB_{VSR}(\mathbb{M}, \mathbb{H}, \mathbb{R})$	**0.418**
$GTB_{VSR}(\mathbb{M}, \mathbb{H}, \mathbb{R})$, excluding TWF [10]	0.396
$GTB_{VSR}(\mathbb{M}, \mathbb{H}, \mathbb{R})$, excluding Collaborative filtering (PMF) [15]	0.361
$GTB_{VSR}(\mathbb{M}, \mathbb{H}, \mathbb{R})$, excluding VSR [11]	0.344
$GTB_{VSR}(\mathbb{M}, \mathbb{H}, \mathbb{R})$, excluding Content-based filtering (LDA) [3]	0.335

5.4   Feature Importance in GTB

We further analyze each component of the feature set in Fig. 1 of the $GTB(\mathbb{M}, \mathbb{H}, \mathbb{R})$ model based on the relative influence. GTB allows us to measure the importance of each component feature. Basically, the more often a feature is used in the split points of a tree, the more important the feature is. Feature importance is essential because the input features are seldom equally relevant. While only a few of them often have substantial influence on the response, the vast majority are irrelevant and could just as well have not been included. Thus, it is helpful to learn the relative importance or contribution of each input feature in predicting the response. Figure 3 shows the relative importances of the top features and gives the following insights (starting with the most important feature):

- Not surprisingly, the *average rating (all versions)* is the most important factor as, when the average rating is high, it is natural for users to download the app because of its positive ratings. Therefore, this feature can be used as a strong signal in the unifying framework to make a split in the decision tree. This reasoning is also similar for the *average rating (current version)*.
- *Price* (*i.e.*, free vs paid) is also an important factor, and this evidence coincides with the trend that apps in the app store are heading towards the freemium model — with the proportion of free apps taking up 90 % of the app store. Therefore, the price of an app could be a strong signal for a split in the decision tree.
- The *number of ratings* is also a strong indicator, as the more ratings an app has garnered, the clearer the sign that it is popular and hence, likely to be consumed. It is also a clear sign that the collaborative filtering technique can be employed.
- Not only the *number of Twitter followers* to the app's Twitter handle is an indicator of a strong social reach, but also the availability of additional Twitter-followers information is an indicator that our Twitter-followers based recommendation technique can be utilized. Additionally, on a related note, the same reasoning could be used to explain why the *number of Facebook likes* is also one of the top features, as this indicator from Facebook is also a hint of the app's social presence on the popular social networking site.

- The *number of versions* also plays an important role as this is a sign that our version-sensitive recommendation technique (VSR) may be employed. Given that this feature is one of the top features of GTB, it suggests that the version-sensitive recommendation technique [11] is useful here.
- We also observe that some app genres fall under the top features, notably "games," "entertainment," and "social networking" — with "games" having a much more significant influence score. The three genres are consistent with alternate findings by Flurry Analytics[7] whereby they discovered that people spend most of their time in apps in the "games," "social networking," and "entertainment" genres across iOS and Android devices.

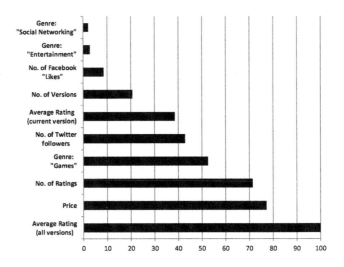

Fig. 3. Top features with the highest relative influence.

Finally, we also observe that our results of the top GTB features in Fig. 3 coincide with another set of findings from Flurry Analytics, ComScore, and Net-MarketShare[8]. For instance, the significant chunks that relate to genres (*i.e.*, "games," "entertainment," and "social messaging") coincide with our genre labels shown in Fig. 3. Additionally, the "Facebook" and "Twitter" chunks also coincide with the "# of Facebook likes" and "# of Twitter followers" features in Fig. 3, which suggests that apps with a strong presence on these two popular social networks have a tendency to be spotted and subsequently consumed, making them popular candidates to be recommended. The data from the alternate user studies (See footnotes 7 and 8) demonstrates a strong correlation with

---

[7] Flurry Analytics; "iOS & Android Smart Device Time Spent per App Category"; http://cl.ly/image/3m0P0g2r3f2C.

[8] Flurry Analytics, ComScore, NetMarketShare; "Time Spent on iOS and Android Connected Devices"; http://cl.ly/image/201x2H1Q1j3H.

our GTB feature component analysis shown in Fig. 3. It indicates how two disciplines (*i.e.*, user studies and GTB feature component analysis) from two different sources of opinions and quantitive angles managed to arrive at similar findings. This further suggests a future direction in mobile app recommendation whereby more focus could be placed in user and trend analysis through social networks — a direction that deviates from traditional research in recommender systems.

6   Conclusion

Given that different recommendation techniques work in different settings, we need to evaluate a method for integrating the various sources of information into a hybrid model that can recommend a set of apps to a target user. To achieve this, we have proposed incorporating the user's prior history, app metadata, and the recommendation scores of various individual recommendation techniques into a hybrid recommendation model for app recommendation. We then used gradient tree boosting (GTB) as the core of the unifying framework to integrate the recommendation scores by using user features and app metadata as additional features for the decision tree. Experimental results show that the unifying framework achieves the best performance against individual state-of-the-art baselines. We also performed a series of in-depth analysis through ablation studies, and demonstrated how different pieces of evidences (such as Twitter and version information) that, when available, could be utilized sufficiently, and how the unifying model dynamically alters the recommendation based on available signals. Finally, we discovered an interesting correlation between important feature components in our unifying framework and user analysis from third-party data analytics companies, which further suggests a future direction in mobile app recommendation, where more focus could be placed in user and trend analysis via social networks.

References

1. Baeza-Yates, R., Jiang, D., Silvestri, F., Harrison, B.: Predicting the next app. that you are going to use. In: Proceedings of the 8th ACM International Conference on Web Search and Data Mining (WSDM 2015), pp. 285–294 (2015)
2. Bhandari, U., Sugiyama, K., Datta, A., Jindal, R.: Serendipitous recommendation for mobile apps using item-item similarity graph. In: Banchs, R.E., Silvestri, F., Liu, T.-Y., Zhang, M., Gao, S., Lang, J. (eds.) AIRS 2013. LNCS, vol. 8281, pp. 440–451. Springer, Heidelberg (2013)
3. Blei, D.M., Ng, A.Y., Jordan, M.I.: Latent dirichlet allocation. J. Mach. Learn. Res. (JMLR) **3**, 993–1022 (2003)
4. Chen, N., Hoi, S.C.H., Li, S., Xiao, X.: Mobile app tagging. In: Proceedings of the 9th ACM International Conference on Web Search and Data Mining (WSDM 2016), pp. 63–72 (2016)
5. Chen, N., Hoi, S.C.H., Xiao, X. SimApp: a framework for detecting similar mobile applications by online kernel learning. In: Proceedings of the 8th ACM International Conference on Web Search and Data Mining (WSDM 2015), pp. 305–314 (2015)

6. Costa-Montenegro, E., Barragáns-Martínez, A.B., Rey-López, M.: Which app? a recommender system of applications in markets: implementation of the service for monitoring users' interaction. Expert Syst. Appl. **39**(10), 9367–9375 (2012)
7. Davidsson, C., Moritz, S.: Utilizing implicit feedback and context to recommend mobile applications from first use. In: Proceedings of the Workshop on Context-Awareness in Retrieval and Recommendation (CaRR 2011), pp. 19–22 (2011)
8. Friedman, J.H.: Greedy function approximation: a gradient boosting machine. Ann. Stat. **29**, 1189–1232 (2001)
9. Lin, J.: Mobile app recommendation. Ph.D. thesis, National University of Singapore (2014)
10. Lin, J., Sugiyama, K., Kan, M.-Y., Chua, T.-S.: Addressing cold-start in app recommendation: latent user models constructed from twitter followers. In: Proceedings of the 36th International ACM SIGIR Conference on Research and Development in Information Retrieval (SIGIR 2013), pp. 283–292 (2013)
11. Lin, J., Sugiyama, K., Kan, M.-Y., Chua, T.-S.: New and improved: modeling versions to improve app recommendation. In: Proceedings of the 37th International ACM SIGIR Conference on Research and Development in Information Retrieval (SIGIR 2014), pp. 647–656 (2014)
12. Liu, B., Kong, D., Cen, L., Gong, N.Z., Jin, H., Xiong, H.: Personalized mobile app recommendation: reconciling app functionality and user privacy preference. In: Proceedings of the 8th ACM International Conference on Web Search and Data Mining (WSDM 2015), pp. 315–324 (2015)
13. Martin, W., Sarro, F., Jia, Y., Zhang, Y., Harman, M.: A survey of app store analysis for software engineering. Technical report RN/16/02, University College London (2016)
14. Park, D.H., Liu, M., Zhai, C., Wang, H.: Leveraging user reviews to improve accuracy for mobile app retrieval. In: Proceedings of the 38th International ACM SIGIR Conference on Research and Development in Information Retrieval (SIGIR 2015), pp. 533–542 (2015)
15. Salakhutdinov, R., Mnih, A.: Bayesian probabilistic matrix factorization using markov chain monte carlo. In: Proceedings of the 25th International Conference on Machine Learning (ICML 2008), pp. 880–887 (2008)
16. Wang, C., Blei, D.M.: Collaborative topic modeling for recommending scientific articles. In: Proceedings of the 17th ACM SIGKDD International Conference on Knowledge Discovery and Data Mining (KDD 2011), pp. 448–456 (2011)
17. Wang, J., Zhang, Y., Chen, T.: Unified recommendation and search in e-commerce. In: Hou, Y., Nie, J.-Y., Sun, L., Wang, B., Zhang, P. (eds.) AIRS 2012. LNCS, vol. 7675, pp. 296–305. Springer, Heidelberg (2012)
18. Xu, Q., Erman, J., Gerber, A., Mao, Z., Pang, J., Venkataraman, S.: Identifying diverse usage behaviors of smartphone apps. In: Proceedings of the ACM SIG-COMM Conference on Internet Measurement Conference (IMC 2011), pp. 329–344 (2011)
19. Yan, B., Chen, G.: AppJoy: personalized mobile application discovery. In: Proceedings of the 9th International Conference on Mobile Systems, Applications, and Services (MobiSys 2011), pp. 113–126 (2011)
20. Yin, P., Luo, P., Lee, W.-C., Wang, M.: App recommendation: a contest between satisfaction and temptation. In: Proceedings of the 6th International Conference on Web Search and Data Mining (WSDM 2013), pp. 395–404 (2013)

21. Zhang, A., Goyal, A., Baeza-Yates, R., Chang, Y., Han, J., Gunter, C.A., Deng, H.: Auto-completion, towards mobile query : an efficient mobile application-aware approach. In: Proceedings of the 25th International World Wide Web Conference (WWW), pp. 579–590 (2016)

22. Zheng, V.W., Cao, B., Zheng, Y., Xie, X., Yang, Q.: Recommendation, collaborative filtering meets mobile : a user-centered approach. In: Proceedings of the 24th AAAI Conference on Artificial Intelligence (AAAI 2010), pp. 236–241 (2010)

23. Zhu, H., Xiong, H., Ge, Y., Chen, E.: Mobile app recommendations with security and privacy awareness. In: Proceedings of the 20th ACM SIGKDD International Conference on Knowledge Discovery and Data Mining (KDD 2014), pp. 951–960 (2014)

User Model Enrichment for Venue Recommendation

Mohammad Aliannejadi$^{(\boxtimes)}$, Ida Mele, and Fabio Crestani

Faculty of Informatics, Università della Svizzera Italiana, Lugano, Switzerland
{mohammad.alian.nejadi,ida.mele,fabio.crestani}@usi.ch

Abstract. An important task in recommender systems is suggesting relevant venues in a city to a user. These suggestions are usually created by exploiting the user's history of preferences, which are, for example, collected in previously visited cities. In this paper, we first introduce a user model based on venues' categories and their descriptive keywords extracted from Foursquare tips. Then, we propose an enriched user model which leverages the users' reviews from Yelp. Our participation in the TREC 2015 Contextual Suggestion track, confirmed that our model outperforms other approaches by a significant margin.

1 Introduction

Recent years have witnessed an increasing use of location-based social networks (LBSNs) such as Yelp, TripAdvisor, and Foursquare. These social networks collect valuable information about users' mobility records, which often consist of their check-in data and may also include users' ratings and reviews. Therefore, being able to recommend personalized venues to users plays a key role in satisfying the user needs on such social networks.

One of the challenges in recommending venues is to model the user based on her profile (e.g., the ratings of previously visited venues). In the past, researchers proposed to make recommendations based on the similarity between the users' preferences and the venues' description and categories [12]. Others leveraged the opinions of users about a given place, which are, for example, extracted from the users' online reviews [14]. We believe that these two techniques should be used together to get better recommendations.

Recent research has focused on recommending venues using collaborative-filtering technique [8,15], where the system recommends venues based on users whose preferences are similar to those of the target user (i.e., the user who receives the recommendations). Collaborative-filtering approaches are very effective, but they suffer from the cold-start (i.e., they need to collect enough information about a user for making recommendations) and the data-sparseness problems. Furthermore, these approaches mostly rely on check-in data to learn the preferences of users, and such information is insufficient to get a complete picture of what the user likes or dislikes in a specific venue (e.g., the food, the view). In order to overcome this limitation, we model the users by applying deeper analysis on users' past

© Springer International Publishing AG 2016
S. Ma et al. (Eds.): AIRS 2016, LNCS 9994, pp. 212–223, 2016.
DOI: 10.1007/978-3-319-48051-0_16

ratings as well as their reviews. In addition, following the principle of collaborative filtering, we exploit the reviews from different users with similar preferences.

In this paper we present a novel approach for suggesting venues to users, where the users are modeled based on venues' content as well as users' reviews. For the former we use the categories of the venues enriched by keywords extracted from users' online reviews, which provide a more detailed description of the venue itself. Although the venue information is valuable for inferring "what type" of places a user may like or dislike, it does not give any clue on the reasons "why" a user rated as positive or negative a particular place. We need to exploit the user's opinions in order to understand what the user may have appreciated of a place. One way to obtain these opinions is mining the users' reviews and see how much they liked the venue and, more importantly, for which reasons: was it for the quality of food, for the good service, for the cozy environment, or for the location? In cases where we lack reviews from some of the users (e.g., they have rated a venue but omitted to review it) and therefore we cannot extract opinions, we apply the collaborative-filtering principle and we use reviews from other users with similar interests and tastes. Our intuition is that a user's opinion regarding an attraction could be learned based on the opinions of others who expressed the same or similar rating for the same venue. To do this we exploit information from multiple sources and combine them to gain better performance. We report the results of our participation in the TREC Contextual Suggestion Track 2015 which show how our model outperforms all the other runs by a significant margin and is placed as the first run in the track.

The remainder of the paper is organized as follows. Section 2 reviews related work. Then, we present our methodology in Sect. 3. Section 4 describes our experiments. Finally, Sect. 5 is a short conclusion and description of future work.

2   Related Work

Recommendation systems help users to find interesting items in large collections. These systems can be employed for recommending products (e.g., books, songs), information (e.g., news and blog articles), or venues (e.g., restaurants, pubs). In recent years, due to the availability of Internet access on mobile devices, there has been a large interest in venue recommendation and contextual suggestion [6] given that the context would be easily provided by the mobile device.

Recommendation algorithms can be divided into four categories: content-based approaches, rating-based collaborative filtering, preference-based product ranking, and review-based approaches [3].

Content-based approaches build user and item profiles based on items' contents (i.e., description, meta-data, keywords) and measure the similarity between the profiles. In [12], the authors applied Part-of-Speech (POS) tagging to the venues' descriptions in order to get the most informative terms which are then expanded using the synonyms from WordNet[1]. Venues' description and categories can be helpful to infer users' preferences and, in particular, the type of

---

[1] WordNet - http://wordnet.princeton.edu.

places the user likes, but they do not enable understanding the reasons behind a positive or negative rate of the user for a venue, and this is considered one of the limitations of such approaches.

Rating-based collaborative filtering approaches are based on finding out common features among users' preferences and recommending venues to people with similar interests. These models are usually based on matrix factorization for dealing with huge collections of data, and they mostly use check-in data for recommending places [4,9]. The usual assumption is that if the user goes to a place multiple times, she probably likes it, but this does not take into account the users' ratings and their reviews. In [11], the authors utilize factorization machines to leverage the feedback of the user as well as contextual information for improving the venue recommendation. These approaches usually suffer from the data-sparsity problem.

Preference-based product ranking is applied when an item (venue) can be described as a set of attributes such as price, view, staff, etc. A user can be modeled as a weighted combination of all the attributes that represent how much a particular user cares about a specific attribute and/or how it affects a user's opinion about an item (venue) [3]. Unfortunately, due to the lack of such data, such techniques cannot be applied to the venue-recommendation scenario.

Review-based approaches aim to build enhanced user profiles using their reviews. When a user writes a review about a venue, there is a wealth of information which reveals the reasons why that particular user is interested in a venue or not. Chen et al. [3] state three main reasons for which the reviews can be beneficial for a recommender system: (1) extra information that can be extracted from reviews enables a system to deal with large data sparsity problem; (2) reviews have been proven to be helpful to deal with the cold-start problem; (3) even in cases when the data is dense, they can be used to determine the quality of the ratings or to extract user's contextual information. As an example, Hariri et al. [10] tried to predict user's context from their reviews about venues by learning a Labeled Latent Dirichlet Allocation model on a dataset from TripAdvisor and using the predicted contextual information to measure the relevance of a venue to a user.

Researchers, observing the effectiveness of reviews for recommending venues [2,7], have been motivated to model the users based on reviews. To overcome the problem that for some users there might be no reviews, Yang et al. [14] demonstrated how it is possible to get improved recommendations by modeling a user with the reviews of other users' whose tastes are similar to the ones of the target user. In particular, they modeled users by extracting positive and negative reviews to create positive and negative profiles for users and venues. The recommendation is then made by measuring and combining the similarity scores between all pairs of profiles. Inspired by their work, we also use reviews of users with similar tastes, but instead of applying a simple similarity measure between venues and users, we use a binary classification.

In this paper, we propose a novel method for recommending venues that builds up on two models: a category-based user model to answer the question "what kind of places would a user like?" and a review-based user model, which answers the question "why would a user like a place?" Differently from other works, we combine the venues' content and the users' reviews to get better recommendations. Moreover, to overcome the problem of the lack of users' reviews from which it is possible to extract users' opinions we rely on the basic assumption of collaborative-filtering [13] and we assume that similar users tend to share similar ratings for the same venue.

3   User Modeling

In this section we firstly describe the user model based on the venues' categories and keywords extracted from Foursquare's tastes. Then, we present how to model the user with opinions extracted from reviews.

3.1   Content-Based User Model

Categories of venues represent a valuable source of information that can be used to infer the types of places a user may like or dislike. Moreover, in cases where users do not provide any reviews or these are not sufficient to model their preferences, categories represent the only resource we can leverage to make venue recommendations.

To model the user's interests using venue categories, we adopt a frequency-based approach. For simplicity we describe how we model the positive categories, while the negative-category model is built similarly. We design Algorithm 1 to calculate the category frequencies (cf_{pos}) and build a positive category model for each user. Let $V = \{v_1, \ldots, v_M\}$ be the set of venues which were positively rated in the user's history. Each place is associated with a set of categories, $C(v) = \{c_1, \ldots, c_z\}$. We assume that if the user rated a venue positively, she also liked the corresponding categories. So, let CM_{pos} be the set of positively rated categories, which is made of all the categories of the venues belonging to V. We compute the frequency of these categories by counting the number of times a user rated the category positively: $\mathrm{count}(c_j) = \sum_{v_s \in V} \sum_{c_k \in C(v_s)} \delta(c_j, c_k)$, where

$$\delta(c_j, c_k) = \begin{cases} 1 & \text{if } c_j = c_k \\ 0 & \text{if } c_j \neq c_k. \end{cases}$$

Each category frequency in the positive (negative) category model is normalized in order to have a score between 0 and 1. Note that the users may have rated the same category with different scores depending on the venues they liked or disliked.

Given a user u and a venue v, the category-based similarity score $S_{CM}(u, v)$ between them is calculated as follows:

$$S_{CM}(u, v) = \sum_{c_i \in C(v)} \mathrm{cf}_{pos}(c_i) - \mathrm{cf}_{neg}(c_i), \tag{1}$$

Algorithm 1. User Positive Category Modeling

---

for all $v_i \in V$ **do**

    for all $c_j \in C(v_i)$ **do**

        if $c_j \notin \mathrm{CM}_{pos}$ **then**

            $\mathrm{CM}_{pos} \leftarrow \mathrm{CM}_{pos} \cup c_j$

            $\mathrm{count}(c_j) = \sum_{v_s \in V} \sum_{c_k \in C(v_s)} \delta(c_j, c_k)$

            $N = \sum_{v_s \in V} \sum_{c_k \in C(v_s)} 1$

            $\mathrm{cf}_{pos}(c_j) = \mathrm{count}(c_j)/N$

        end if

    end for

end for

---

where cf_{pos} and cf_{neg} are respectively the positive and negative categories' frequencies.

Foursquare's Taste Keywords. The previous model can be enriched by using special terms extracted from users' reviews about a venue. Foursquare provides a list of keywords, also known as "tastes" to better describe a venue. As an example, 'Central Park' in 'New York City' is described by these taste terms: *picnics, biking, trails, park, scenic views,* etc. Such keywords are very informative, since they often express characteristics of a venue, and they can be considered as a complementary source of information for venue categories.

Table 1 shows all taste keywords and categories for a sample restaurant on Foursquare. As we can see, the taste keywords represent much more detailed information about the venue compared to categories. The average number of taste keywords for venues (8.73) is much higher than the average number of categories for venues (2.8). It suggests that these keywords could describe a venue in more details compared to categories.

Consequently, we consider these keywords as a complementary source of information for categories and use the same frequency-based approach to further enrich the user model. Given a user u and a place to recommend v we compute the similarity score with this category-based model enriched with Foursquare's taste terms as we did for the simple category-based model (see Eq. 1) and we call it $S_{TM}(u, v)$.

3.2 Review-Based User Model

We believe that modeling a user solely based on content of venues she visited or liked is very general and would not allow to understand the specific reasons for which a user liked or disliked a place. For example, consider a user who rated two venues belonging to the same categories *Restaurant, Italian,* and *Pizza,* with a positive and a negative rating, respectively. Looking only at the category and at the rates, we cannot know if the user does not like Italian restaurants and pizza places in general, or if she did not appreciated the second venue for some other reasons (e.g., food quality, service). In order to understand why the user liked or disliked a venue, we need to determine the reasons behind a positive

Table 1. A sample of taste keywords and categories for a restaurant.

Taste keywords	pizza, lively, cozy, good for dates, authentic, casual, pasta, desserts good for a late night, family-friendly, good for groups, ravioli, lasagna, salads, wine, vodka, tagliatelle, cocktails, bruschetta
Categories	pizza place, italian restaurant

or negative rating. This is only possible if reviews are available. In particular, analyzing the text of the reviews, we can observe that the user rated positively the first venue, because she appreciated the food and the kind service, while she did not like the second venue because of the food quality and the location.

So, to figure out for which reasons the user expressed an opinion we need to know the user's reviews about the rated venues. Unfortunately, there is often a lack of explicit reviews from the users, so we tried to overcome this problem by using opinions expressed by other users who rated the venue with a similar score. Lacking any other information, our intuition is that a user liked/disliked a place for the same reasons that others liked/disliked that place. Although this assumption might not be perfect and might not always be valid, it provides the best way to model users in case we lack other information.

Binary Classification. For each user, we build a model by training a binary classifier with the positive and negative reviews of previously visited venues. We decided to use a binary classification, because we assume that a user, before planning a trip or trying a new venue, would read the online reviews of other users to have an insight on the places of interest. Suppose that the user would like to try a restaurant and, in order to decide whether it is worth to go or not, she checks the online reviews of other customers. The user may have a positive or negative idea about the restaurant depending on the ratings and comments of other people.

Subsequently, if the user rates the restaurant positively, we can assume that her judgment after reading positive reviews about the venue was positive, so she tried it and expressed an opinion similar to the other customers. An alternative to binary classification would be a regression model, but we decided not to adopt it for two reasons. First, as explained before, when users make their minds reading online reviews they have to take a binary decision: like or dislike that same place. Secondly, due to the sparsity of the dataset, a binary discrimination of venues and reviews helps our system to model users more accurately.

Support Vector Machine. SVM was first introduced by Cortes and Vapnik [5], and it is considered one of the most powerful supervised classifiers in machine learning. The SVM classifier model deals with binary-classification problems in which the training data is supposed to be divided into two classes using a hyperplane which is defined by a number of support vectors. The underlying idea behind supervised learning approaches is to learn from training examples. SVM finds optimal separated hyperplanes for a binary classification problem through mapping of the input vectors into a high-dimensional feature space in a nonlinear manner. It constructs a linear model for estimating the decision function based on the support vectors. In case the training data is linearly separable, SVM results in an optimal hyperplane with maximum margin between the hyperplane and the training samples which are closest to the hyperplane, namely, the support vectors.

Our problem can be easily mapped to a binary-classification problem, as a user either likes or dislikes a venue, so we can apply successfully the SVM classifier. We separate relevant and non-relevant suggestions for each user into two classes, $y_i \in \{-1, 1\}$, and the number of labeled training examples is N. Therefore, the training examples are $(\mathbf{x}_1, y_1), \ldots, (\mathbf{x}_N, y_N), \mathbf{x} \in R^d$ where d is the number of features for each instance. The decision function without using a kernel for linearly separable training data is:

$$y_j = \text{sign}(\mathbf{w}^* \bullet \mathbf{x}_j - b),$$

where x_j is an unknown vector, $\bullet$ represents the dot product, and $\mathbf{w}^*$ is:

$$\mathbf{w}^* = \sum_{i=1}^{r} \alpha_i y_i \mathbf{x}_i,$$

where r is the number of nonzero α's.

In order to find the optimal discriminant hyperplane, one needs to find the optimal weight vector $\mathbf{w}^*$ such that $\|\mathbf{w}^*\|$ is the minimum. This operation can be done using Lagrangian Multipliers.

Our preliminary experiments show that among all possible kernels for SVM, linear kernel exhibits the best performance, so we choose linear kernel to train SVM classifier.

Training the Classifier. As we will explain in Sect. 4.1, for the training we used example suggestions, basically venues rated by users. In particular, positive training samples are extracted from positive reviews of positive example suggestions, while negative samples are from negative reviews of negative example suggestions. Note that we ignore the middle rate, which corresponds to a neutral opinion. We ignore negative reviews of positive example suggestions and positive reviews of negative example suggestions since they are not supposed to contain any useful information as they do not share the same perspective about a particular place.

As classifiers we used Support Vector Machine (SVM) and Naïve Bayes classifier. We consider the TF-IDF score for each term as our feature vector, since it indicates the importance of each term to the users. Moreover, it provides a good means to filter out off-topic and noisy terms from reviews. In short, given a user u and candidate suggestion p, the similarity score between them, $S_{BM}(u, p)$, is the value of the decision function of the SVM classifier or the confidence score of the Naïve Bayes classifier.

3.3 Venue Ranking

To rank venues for each user, we combine all scores described above. We calculate a linear combination of all the scores for each $\langle user, venue \rangle$ pair. The similarity score between a user, u, and a venue, v, is calculated as follows:

$$SIM(u,v) = \alpha \times S_{CM}^{Yelp}(u,v) + \beta \times S_{CM}^{TAdvisor}(u,v)$$
$$+\eta \times S_{TM}(u,v) + \gamma \times S_{BM}(u,v), \tag{2}$$

where $S_{CM}^{Yelp}(u,v)$ and $S_{CM}^{TAdvisor}(u,v)$ are the scores based on the categories from Yelp and TripAdvisor, respectively. $S_{TM}(u,v)$ is the score achieved with Foursquare's taste keywords, and $S_{BM}(u,v)$ is the score computed using reviews of users (see Sect. 3.2). The weights α, β, η, and γ are assigned to the scores to balance the impact of each of them in the final similarity. Finally, for each user u the venues are ranked based on $SIM(u,v)$ similarity score.

4   Experiments

This section describes the dataset, the experimental setup for assessing the performance of our methodology, and the experimental results.

4.1   Dataset and Experimental Setup

Our experiments were conducted on the collection provided by the Text REtrieval Conference (TREC) for the Batch Experiments of the 2015 Contextual Suggestion Track[2]. This track was originally introduced by the National Institute of Standards and Technology (NIST) in 2012 to provide a common evaluation framework for participants that are interested in dealing with the challenging problem of contextual suggestions and venue recommendation.

In short, given a set of example places as user's preferences (profile) and contextual information (e.g., the *city* where the venues should be recommended), the task consists in returning a ranked list of 30 candidate places which match the user's profile. Regarding the user context, it may contain the following information: trip type (business, holiday, or other), trip duration (night out, day trip, weekend trip, or longer), group type (alone, friends, family, or other), and season (winter, summer, autumn, or spring). Moreover, user's age and gender may also be included. While the user profiles consist of a list of venues a particular user has already rated. The ratings range between 0 (very uninterested) and 4 (very interested).

The collection, provided by TREC, consists of a total $9K$ distinct venues and 211 users. For each user, the contextual information plus a history of 60 previously rated attractions are provided. Additionally, for our experiments, we gathered information about the venues and their corresponding reviews from three LBSNs. In particular, we extracted the venues' categories from Yelp and TripAdvisor, the taste keywords from Foursquare, and the reviews from Yelp.

Given a user and a list of 30 candidate suggestions, the recommendation system ranks them. Such generated ranking is then evaluated using relevance assessments, which provide information about whether a given candidate suggestion is relevant to a user or not.

---

[2] https://sites.google.com/site/treccontext/trec-2015.

Table 2. Results for our methods compared with other competitors and TREC median scores. *CatRev-SVM* denotes our submitted system which uses SVM classifier and *CatRev-NB* denotes our submitted system which uses Naïve Bayes classifier.

Approach	P@5 Rank	P@5	MRR
CatRev-SVM	1	**0.5858**	**0.7404**
CatRev-NB	7	0.5450	0.6991
BASE1	2	0.5706	0.7190
BASE2	3	0.5583	0.6815
TREC Median		0.5090	0.6716

Our ranking of recommendations is done as described in Sect. 3.3. In order to find the optimum setting for the weights associated with each score of Eq. 2, we conducted a 5-fold cross validation that leads to the following setting: $\alpha = 1.0$, $\beta = 0.3$, $\eta = 0.3$, and $\gamma = 1.0$. As we can see from the values of the weights, Yelp dataset is more significant than TripAdvisor for the categories, and the opinion-based model has a bigger impact on the score, as well.

4.2   Results and Discussions

We demonstrate the effectiveness of our model by reporting and analyzing in details the official results of the TREC 2015 Contextual Suggestion Track [6]. We report the performance of our models as well as the two top ranked models reported in the track, briefly comparing the approaches.

The first model is an approach based on collaborative filtering (BASE1) presented in [11]. More specifically, they use factorization machine for venue recommendation. The instances which are fed into the factorization machine are composed of three blocks representing user, context, and venue features. The second one is a similarity-based approach (BASE2) presented in [14]. They create profiles for users and venues using reviews and measure the similarity between the profile pairs to rank the venues. We also compare our results with the median performance of all submitted runs to TREC (TREC Median). In Table 2 we report the values of P@5 (precision-at-5) and MRR (Mean Reciprocal Rank) for our two classifiers: Support Vector Machine (CatRev-SVM) and Naïve Bayes (CatRev-NB), and for the competitors. We run t-test for CatRev-SVM and CatRev-NB and the results were statistically significant at $p < 0.001$. Note that we could not carry out the t-test for the BASE1 and BASE2 approaches, since we do not have the rankings from the other competitors.

Results in Table 2 demonstrate that both our models perform well compared to TREC median. Specifically, the methodology which utilizes SVM classifier to model a user based on reviews performs best compared to all other submitted runs to TREC and is ranked as top 1 [1,6]. It confirms that our approach of modeling user with reviews from similar users using a machine learning classification algorithm and combining it with other content-based scores is effective for

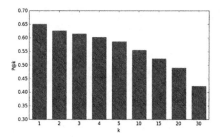

Fig. 1. CatRev-SVM: Precision for different values of k = 1, 2, 3, ..., 30.

venue recommendation. Better results, however, can be achieved by SVM classifier, since it is more suitable for text classification, which is a linear problem and feature vectors are high dimensional with weights. Moreover, the advantage of linear SVM is that the execution time is very low and there are very few parameters to tune.

It is also worth noting that in several cases there is a lack of negative reviews about venues and the sizes of the positive and negative sets differ significantly. Most of the classification algorithms do not perform well with unbalanced sets, because they tend to correctly classify the class with the larger number of training samples and lower down the overall error rate. However, SVM does not suffer from this, since it does not try to directly minimize the error rate but instead tries to separate the two classes using a hyperplane maximizing the margin. This makes SVM relatively intolerant of the relative size of each class.

In Fig. 1 we show the behavior of precision for CatRev-SVM at different $k = 1, 2, 3, \ldots, 30$. As we can see, the higher precision is achieved with lower k values, and this is desirable since users on their mobiles are more likely to select a venue on the top of the list.

We report in Fig. 2 the distribution of venues over 30 of the most liked types of venues in the dataset. As we can see, the most visited places are *American Restaurant* (10 % of the dataset), *Park* (6 % of the dataset), followed by *Bar* (5 % of the dataset). The figure also shows the number of suggested venues that are liked by the user (the lighter bar). Note that the bars are ordered by their number of likes from left to right.

Following our previous work [12], we calculate a *liked rate* for each type of venue. It is the percentage of suggested venues that are liked by all the users. This percentage is shown on the top of each bar. We could observe that the *Plaza* category is the one with the highest *liked rate* (75 %), followed by *Beach* (73 %) and *Trail* (71 %). Frequently visited categories, such as *American Restaurant* and *Park*, have a *liked rate* equal to 50 % and 61 %, respectively. The least categories in term of *liked rate* are *Sandwich Place* (30 %) and *Café* (39 %). It is also worth noting that according to this figure, the number of users who liked *American Restaurant* is more than *Park*; however, *Park* category has a significantly higher *liked rate* than *American Restaurant*. Note that we cut the long-tail categories, namely, the categories that are not frequently liked, and we did this study only

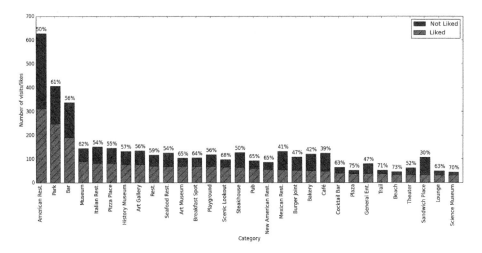

Fig. 2. Distribution of the number of suggestions the users Liked and Not Liked for the 30 different types of venues. The categories are ordered by the number of liked venues belonging to that particular category. The percentage of liked venues (*liked rate*) of all suggested venues for each category is written on top of their corresponding bar.

on top 30 liked categories. This study suggests that using a prior probability over categories could potentially benefit a recommender system, and we plan to further explore this direction in a future work.

5 Conclusions and Future Work

In this paper we proposed a simple but novel approach for recommending venues. We used frequency-based scores in order to model users' interest and venues, and we enriched the model using users' opinions extracted from reviews written by similar users. Experimental results corroborated the effectiveness of our approach and, although simple, our system managed to outperform all other submitted systems in the TREC 2015 Contextual Suggestion track. This proves the effectiveness of our model compared to state-of-the-art systems under exactly the same settings.

As future work, we would like to propose new scores for other contextual signals that are available in the dataset, such as the *trip type* and *duration*, *group type* and *season*. Furthermore, we would like to enrich the model by including the preference tags that a user indicates when she rates a venue. One possible way to include them is to find a mapping between them and Foursquare's taste keywords using an iterative algorithm. Finally, it would be interesting to try different Learning to Rank approaches for combining different scores.

References

1. Aliannejadi, M., Bahrainian, S.A., Giachanou, A., Crestani, F., University of lugano at TREC 2015: Contextual suggestion and temporal summarization tracks. In: TREC, Gaithersburg, Maryland, USA, November 2015
2. Chen, G., Chen, L.: Recommendation based on contextual opinions. In: Dimitrova, V., Kuflik, T., Chin, D., Ricci, F., Dolog, P., Houben, G.-J. (eds.) UMAP 2014. LNCS, vol. 8538, pp. 61–73. Springer, Heidelberg (2014)
3. Chen, L., Chen, G., Wang, F.: Recommender systems based on user reviews: the state of the art. User Model. User-Adap. Inter. **25**(2), 99–154 (2015)
4. Cheng, C., Yang, H., King, I., Lyu, M.R.: Fused matrix factorization with geographical and social influence in location-based social networks. In: AAAI, Toronto, Canada, July 2012
5. Cortes, C., Vapnik, V.: Support-vector networks. Mach. Learn. **20**(3), 273–297 (1995)
6. Dean-Hall, A., Clarke, C.L.A., Kamps, J., Kiseleva, J., Voorhees, E.M.: Overview of the TREC 2015 contextual suggestion track. In: TREC, Gaithersburg, USA, November 2015
7. Esparza, S.G., O'Mahony, M.P., Smyth, B.: On the real-time web as a source of recommendation knowledge. In: RecSys, Barcelona, Spain, pp. 305–308 September 2010
8. Gao, H., Tang, J., Hu, X., Liu, H.: Exploring temporal effects for location recommendation on location-based social networks. In: RecSys, Hong Kong, China, pp. 93–100, October 2013
9. Griesner, J., Abdessalem, T., Naacke, H.: POI recommendation: towards fused matrix factorization with geographical and temporal influences. In: RecSys, Vienna, Austria, pp. 301–304, September 2015
10. Hariri, N., Zheng, Y., Mobasher, B., Burke, R.: Context-aware recommendation based on review mining. In: Proceedings of the 9th Workshop on Intelligent Techniques for Web Personalization and Recommender Systems (2011)
11. McCreadie, R., Vargas, S., MacDonald, C., Ounis, I., Mackie, S., Manotumruksa, J., McDonald, G.: University of glasgow at TREC 2015: experiments with terrier in contextual suggestion, temporal summarisation and dynamic domain tracks. In: TREC, Gaithersburg, USA, November 2015
12. Rikitianskii, A., Harvey, M., Crestani, F.: A personalised recommendation system for context-aware suggestions. In: de Rijke, M., Kenter, T., de Vries, A.P., Zhai, C.X., de Jong, F., Radinsky, K., Hofmann, K. (eds.) ECIR 2014. LNCS, vol. 8416, pp. 63–74. Springer, Heidelberg (2014)
13. Su, X., Khoshgoftaar, T.M.: A survey of collaborative filtering techniques. Adv. Artif. Intell. **2009**, 421425:1–421425:19 (2009)
14. Yang, P., Fang, H.: University of delaware at TREC 2015: combining opinion profile modeling with complex context filtering for contextual suggestion. In: TREC, Gaithersburg, USA, November 2015
15. Yuan, Q., Cong, G., Ma, Z., Sun, A., Magnenat-Thalmann, N.: Time-aware point-of-interest recommendation. In: SIGIR, Dublin, Ireland, pp. 363–372, July 2013

Learning Distributed Representations for Recommender Systems with a Network Embedding Approach

Wayne Xin Zhao[1,2(✉)], Jin Huang[1,2], and Ji-Rong Wen[1,2]

[1] School of Information, Renmin University of China, Beijing, China
{batmanfly,jin.huang,jrwen}@ruc.edu.cn
[2] Beijing Key Laboratory of Big Data Management and Analysis Methods,
Beijing, China

Abstract. In this paper, we present a novel perspective to address recommendation tasks by utilizing the network representation learning techniques. Our idea is based on the observation that the input of typical recommendation tasks can be formulated as graphs. Thus, we propose to use the k-partite adoption graph to characterize various kinds of information in recommendation tasks. Once the historical adoption records have been transformed into a graph, we can apply the network embedding approach to learn vertex embeddings on the k-partite adoption network. Embeddings for different kinds of information are projected into the same latent space, where we can easily measure the relatedness between multiple vertices on the graph using some similarity measurements. In this way, the recommendation task has been casted into a similarity evaluation process using embedding vectors. The proposed approach is both general and scalable. To evaluate the effectiveness of the proposed approach, we construct extensive experiments on two different recommendation tasks using real-world datasets. The experimental results have shown the superiority of our approach. To the best of our knowledge, it is the first time that a network representation learning approach has been applied to recommendation tasks.

Keywords: Recommender systems · Network embedding · Item recommendation · Tag recommendation

1 Introduction

In recent years, recommender systems have played an important role in helping match users with information resources [4]. Various recommendation algorithms have been developed in the past years [1], including collaborative filtering methods, content-based methods, and hybrid methods. Collaborative filtering methods build a model from a user's past behaviors as well as decisions made by other similar users. Content-based methods extract a set of important features of an item in order to recommend new items with similar features. These two

© Springer International Publishing AG 2016
S. Ma et al. (Eds.): AIRS 2016, LNCS 9994, pp. 224–236, 2016.
DOI: 10.1007/978-3-319-48051-0_17

types of methods are often combined in practical systems to form the hybrid methods. Although previous methods have been shown to be effective to some extent, there exist several problems with these approaches. First, these methods are usually task-oriented and cannot serve as a general solution to multiple recommendation settings. It may not be easy for existing methods to adapt to a different recommendation setting. Second, existing recommendation algorithms may not be scalable to large datasets. For example, the efficiency of classic item-based KNN recommendation algorithms is largely limited by the construction of the KNN graph [4]; matrix factorization involves eigen-decomposition of the data matrix which is expensive and usually with approximation calculation [13]. Thus, how to balance generality and scalability has become an important problem in practice. The main research focus of our paper is to develop a general and scalable recommendation framework.

To address this issue, our intuition is based on the observation that the input of typical recommendation tasks can be formulated as graphs. For example, we have presented two illustrative examples for top-N item recommendation and tag recommendation respectively in Fig. 1. For item recommendation, we have two sets of vertices, namely users and items; While for tag recommendation, we have three sets of vertices, namely users, items and tags. Once we have built the graph representation, the recommendation task can be considered as a relatedness or relevance evaluation problem: given a or more query vertices, we would like to identify the most related vertices. For example, for item recommendation, the query vertex can be set to a specific user, while for tag recommendation, the query vertices can be set to a combination of a user and an item. With such a formulation, the difficulty lies in how to develop an effective way to evaluate the relatedness on the graph.

Our approach is inspired by the recent progress in network representation learning and deep learning [8,10,18]. Network representation learning characterizes a vertex in a graph with a low-dimensional dense vector, $a.k.a.$, embedding vector. Embedding representations provide a promising way to represent and extract structural patterns in the networks, and several pioneering works

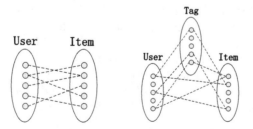

(a) User-item bipartite network.   (b) User-item-tag tripartite network.

Fig. 1. Illustrative examples of bipartite and tripartite networks for recommendation tasks.

have shown the effectiveness of network embedding models [10, 18]. Especially, vertex similarity or relatedness can be well measured with the embedding vectors. Following this point, our general solution to recommendation tasks has been developed as a three-step procedure. In the first step, we build a k-partite adoption network which is constructed with all the historical adoption records (*e.g.*, product purchase records). Here, the term of "adoption" has been used because the recommendation task can be considered as modeling the adoption process of a user. Second, we apply the network embedding techniques to learn the embeddings for the vertices on the k-partite adoption network. Embeddings from different vertices are projected into the same latent space, where similarity measurements can be used to evaluate the relatedness between vertices (*e.g.*, cosine similarity). Finally, the recommendation task will be casted into a similarity evaluation process. For example, given a user, we can directly rank the candidate items by the cosine similarity values between the user and item embeddings.

The proposed approach is both general and scalable. On one hand, our formulation (*i.e.*, k-partite adoption network) can be used to characterize multiple recommendation settings with rich contextual information. When new contextual information is needed to consider, we can simply discretize the contextual information into discrete variables and represent them as new vertices. On the other hand, our approach utilizes the neural network models to derive the vertex embedding representations. In this case, the model is designed to optimize within local neighborhoods instead of performing global computations. This allows us to develop scalable algorithms such as stochastic gradient descent with weight sampling [18]. To evaluate the effectiveness of the proposed approach, we construct extensive experiments on two different recommendation tasks using real-world datasets. The experimental results have shown the superiority of our approach. To the best of our knowledge, it is the first time that recommendation task has been addressed by a network representation learning approach.

2   Prelimenaries

Comparing to traditional recommendation algorithm, our goal is to provide a unified approach to multiple recommendation tasks. In a general sense, the input of the recommendation task corresponds to the adoption behaviors of users. The main idea is to represent the adoption records using a k-partite network, and then embedding representations are used to represent vertices on the graph. With such representations, we can fulfill the recommendation task using simple similarity measurements. Next we introduce the preliminaries for this paper.

Definition 1. k-**Tuple Adoption Record**. *An adoption record can be modeled as a k-tuple:* $\langle e_1, \ldots, e_j, \ldots, e_k \rangle$*, where each entry* e_j *($1 \leq j \leq k$) refers to the value for the j-th feature (*a.k.a. *attribute) in an adoption record. Here we require that the value of each entry must be a positive discrete value.*

Such a formulation is general to model different recommendation tasks, even with rich context information. For example, in top-N item recommendation, a pair $\langle u, i \rangle$ is used to represent the record that user u has adopted item j. While in top-N tag recommendation, a triplet $\langle u, i, t \rangle$ is used to represent the record that tag t has been given by user u on item i. It is similar to the feature coding in context-aware recommendation models such as SVDFeature [3] and libFM [11].

Definition 2. k-Partite Adoption Network. *A k-partite adoption network is a graph whose vertices are or can be partitioned into k different independent sets: edges only exist in vertices from the same set. The edge weight is a real number which indicates the importance of the corresponding link.*

Given a set of k-tuple adoption records, it is easy to construct a k-partite network. The values for the j-th attribute are characterized by the j-th vertex set in the k-partite adoption network. With loss of generality, we next present the construction of adoption graph with the settings of $k = 2$ and $k = 3$. These two cases correspond to two classic and widely studied tasks, top-N item and tag recommendation. Other cases with large values for k can be solved in a similar way, and we leave it as future work. We assume that the k-partite network is an undirected graph.

Definition 3. *Bipartite User-Item (UI) Network.* *Let $\mathcal{U}$ denote the set of all the users, and $\mathcal{I}$ denote the set of all the items. A bipartite user-item network can be denoted by $\mathcal{G}^{(bi)} = (\mathcal{V}, \mathcal{E}, \mathbf{W})$, where the vertex set $\mathcal{V} = \mathcal{U} \cup \mathcal{I}$, the edge set $\mathcal{E} \subset \mathcal{U} \times \mathcal{I}$, the weight matrix $\mathbf{W}$ stores the edge weights, and $W_{u,i}$ denote the link weight between a user u and an item i.*

Definition 4. *Tripartite User-Item-Tag (UIT) Network.* *Let $\mathcal{U}$ denote the set of all the users, $\mathcal{I}$ denote the set of all the items, and $\mathcal{T}$ denote the set of all the tags. A tripartite user-item-tag network can be denoted by $\mathcal{G}^{(tri)} = (\mathcal{V}, \mathcal{E}, \mathbf{W})$, where the vertex set $\mathcal{V} = \mathcal{U} \cup \mathcal{I} \cup \mathcal{T}$, the edge set $\mathcal{E} \subset \big((\mathcal{U} \times \mathcal{I}) \cup (\mathcal{U} \times \mathcal{T}) \cup (\mathcal{I} \times \mathcal{T}) \big)$, and the weight matrix $\mathbf{W}$ stores the edge weights.*

In a bipartite UI network, we set the weight to the number that a user has adopted an item. Different from a UI network, in a UIT network, there can be three different types of edges consisting of vertices from two out of the three vertex sets, which correspond to the edge weights $W_{u,i}$, $W_{u,t}$ and $W_{i,t}$. To set these weights, we use a simple counting method, W_{e_1,e_2} is equal to the number that e_1 and e_2 occur in all the adoption records.

With the above definitions, we can perform the recommendation task by evaluating the relatedness between query vertices and candidate vertices. For example, in top-N item recommendation, a given user will be treated as the query vertex, and we search over the candidate item vertices to find out the most related ones. Next, we introduce a network embedding approach.

3   A Network Embedding Approach to Recommendation Tasks

Recently, networking representation learning is widely studied [10,18], and it provides an effective way to explore the networking structure patterns using low-dimensional embedding vectors. Not limited to discover structure patterns, network representations have been shown to be effective to serve as important features in many network-independent tasks, such as text classification [17]. In our current task, we aim to learn low-dimensional representations for the vertices on the k-partite adoption network. The embedding representation should encode important topological information and the similarity can be evaluated by these embedding vectors.

3.1   The General Network Embedding Model

Formally, we use a d-dimensional embedding vector $\boldsymbol{v}_e \in \mathbb{R}^d$ to denote the embedding representation for a vertex e on the k-partite adoption network. We first describe a general network embedding model.

Let us start with studying how to model the generative probability for an undirected edge between two vertices e_s and e_t, formally denoted as $P(e_s, e_t)$. The main intuition is if two vertices v_i and v_j form a link on the network, their networking representations should be similar. In other words, the inner product $\boldsymbol{v}_{e_s}^\top \cdot \boldsymbol{v}_{e_t}$ between the corresponding two networking representations will yield a large similarity value for two linked vertices. We define the probability of a link (e_s, e_t) by using a sigmoid function as follows

$$P(e_s, e_t) = \sigma(\boldsymbol{v}_{e_s}^\top \cdot \boldsymbol{v}_{e_t}) = \frac{1}{1 + \exp(-\boldsymbol{v}_{e_s}^\top \cdot \boldsymbol{v}_{e_t})}. \tag{1}$$

The probability $P(e_s, e_t)$ indicates that the link strength between two vertices e_s and e_t. Recall that we also have the real weights for edges, i.e., the weight matrix $\mathbf{W}$. We can also derive an empirical estimation $\hat{P}(e_s, e_t)$ as follows

$$\hat{P}(e_s, e_t) = \frac{W_{e_s, e_t}}{\sum_{(e_{s'}, e_{t'}) \in \mathcal{E}} W_{e_{s'}, e_{t'}}}. \tag{2}$$

Following previous study on networking representation learning [18], we minimize the KL-divergence of two probability distributions

$$L(\mathcal{G}) = D_{KL}(\hat{P}(\cdot, \cdot) \| P(\cdot, \cdot)) \propto \sum_{(e_s, e_t) \in \mathcal{E}} W_{e_s, e_t} \log P(e_s, e_t). \tag{3}$$

This essentially models first-order proximity modeled in the LINE model [18]. Although we can also model the second-order proximity as in LINE, the empirical results showed that the second-order did not perform well on tag recommendation. A possible reason will be that the second-order proximity mainly captures the shared contexts between vertices, and such an indirect modeling method does not work well on recommendation task. Thus, we only model the first-order proximity of the k-partite adoption network in this work. We follow the learning method in [18] to optimize Eq. 3, which applies stochastic gradient descent with negative sampling and weight sampling. The running complexity is about $\mathcal{O}(n \cdot d \cdot \#neg \cdot M)$, where n is the iteration number, d is the number of latent factors, $\#neg$ is the number of negative samples and M is the number of training instances.

3.2 Utilizing Embedding Representations for Recommendations

In the above, we have presented how to learn networking embeddings on the k-partite adoption graph. After parameter learning, we can obtain the embeddings for each vertex on the graph. Next, we will study how to make recommendations with these embeddings. We consider two tasks, namely the top-N item recommendation and the top-N tag recommendation.

Top-N Item Recommendation with Bipartite Network Embedding. Given a user u, the first task aims to produce a candidate list of N items based on her adoption history. In this task, we have two kinds of entities in the recommendation setting, namely users and items. We follow Definition 3 to construct the bipartite user-item network $\mathcal{G}^{(bi)}$ (See Fig. 1(a)). Then we run the network embedding model shown in Sect. 3.1, and derive the embedding representations for both users and items, denoted by $\boldsymbol{v}_u$ and $\boldsymbol{v}_i$ respectively. Given a query vertex, *i.e.*, a user, we would like to identify the most related item vertices. Formally, the task can be fulfilled using the following ranking funciton

$$\text{score}(u, i) = \boldsymbol{v}_u^\top \cdot \boldsymbol{v}_i. \tag{4}$$

Given a user u, we can rank the items using Eq. 4 to generate the recommendations. Here, we do not consider repetitive adoption behaviors of users, it will be easy to adapt to the case where repetitive adoptions are considered.

Top-N Tag Recommendation with Tripartite Network Embedding. Given a user u and an adopted item i, the second task aims to produce a candidate list of N tags based on the tagging history. In this task, we have three kinds of entities in the recommendation setting, namely users, items and tags. We first follow Definition 4 to construct the tripartite user-item-tag network $\mathcal{G}^{(tri)}$ (See Fig. 1(b)). Then run the network embedding model (in Sect. 3.1) on the tripartite network, and derive the embedding representations for both users, items and tags, denoted by $\boldsymbol{v}_u$, $\boldsymbol{v}_i$ and $\boldsymbol{v}_t$ respectively. Formally, the task can be fulfilled using the following ranking function

$$\text{score}(u, i, t) = \boldsymbol{v}_u^\top \cdot \boldsymbol{v}_t + \boldsymbol{v}_i^\top \cdot \boldsymbol{v}_t. \tag{5}$$

Our ranking function follows the idea in [13], which models the three-way data by using pairwise interaction factorization.

High-Order Recommendation with Network Embedding. In more complex tasks, we can have multiple kinds of entities (*i.e.*, attributes or features) to consider. Our approach is quite general to incorporate arbitrary types of discrete features into the recommendation setting. The procedure can be described as follows. We first construct the k-partite adoption graph, and then learn the embedding representations for each vertex on the network. After obtaining the embedding representations, we can define the score functions, such as Eqs. 4 and 5, to rank candidate vertices for recommendation.

In what follows, we will call *top-N item recommendation* as *item recommendation*, and call *top-N tag recommendation* as *tag recommendation* for short.[1] We refer to our method as *Network Embedding based Recommendation Model (NERM)*.

4   Experiments

In this section, we conduct extensive experiments to evaluate the effectiveness of the proposed approach in two tasks, namely item and tag recommendation.

Table 1. Summary statistics of the two datasets for item recommendation.

Datasets	#Users	#Items	#Records	Sparsity
JD	94,440	46,573	2,767,366	0.063 %
MovieLens	198,155	17,505	22,290,822	0.6426 %

4.1   Evaluation on Top-N Item Recommendation

Dataset. We use two shared datasets for the evaluation of item recommendation: the *JD* dataset in [24] and the *MovieLens* dataset[2]. JD dataset is a large product purchase collection, in which each adoption record consists of a user ID, a product ID and an adoption timestamp. MovieLens dataset is a large movie rating collection, in which each adoption record consists of a user ID, a movie ID and an adoption timestamp.[3] Table 1 summarizes the basic statistics of the two datasets. We select these two datasets because they are large and represent different data applications.

Evaluation Metrics. For item recommendation, we adopt four widely used evaluation metrics, including Precision@K, Recall@K, Mean Average Precision (MAP) and Mean Reciprocal Rank (MRR). Usually, only top ranked recommendations are important to consider, thus we set K to 10 in the experiments.

---

[1] A tag itself can be treated as an item, too. Here we follow the conventions in tag recommendation which distinguishes between an item and a tag.

[2] http://grouplens.org/datasets/movielens.

[3] The dataset was originally used for rating prediction, and we use it for item recommendation.

Table 2. Performance comparisons of the proposed method and baselines on item recommendation.

Methods	JD				MovieLens			
	P@10	R@10	MAP	MRR	P@10	R@10	MAP	MRR
BPR	0.171	0.360	0.337	0.564	0.097	0.169	0.148	0.195
DeepWalk	0.259	0.443	0.502	0.806	0.203	0.243	0.249	0.358
NERM	**0.275**	**0.477**	**0.528**	**0.819**	**0.206**	**0.256**	**0.258**	**0.368**

Experimental Setting. Since each adoption record is attached with a timestamp, we consider a time-sensitive evaluation. We split the entire dataset by timestamps: the first 80 % data is used as training data while the rest 20 % data is used as test data.[4]

Baselines. We consider using the following methods as the comparison baselines

- **BPR** [12]. BPR is a Bayesian personalized ranking method for learning with implicit feedback. It adopts a pairwise loss function which assumes that an adopted item should be more weighted compared with an unadopted item.
- **DeepWalk** [10]. DeepWalk is a recently proposed network embedding method. It first generates multiple random paths based on a social network, and further employs the word2vec [8] to deal with vertex sequences.

The BPR method was originally proposed to solve the item recommendation task, representing state-of-the-art. DeepWalk is also a network embedding method and we incorporate it as a comparison. There can be several parameters to tune in baselines and our method. We hold out 10 % of training data as the development set for parameter optimization. For BPR, the number of latent factors is set to 256, and the number of negative samples is set to 300. For DeepWalk, the number of embedding dimensions is set to 1024 and we use the hierarchal softmax algorithm to learn the parameters. For our method NERM, the number of embedding dimensions is set to 1024, and the number of negative samples is set to 8.

Results and Analysis. Table 2 presents the experimental results of the compared methods on the task of item recommendation. Overall, we have made the following observations. First, both network embedding methods are much better than the competitive baseline BPR. Second, NERM is slightly better than DeepWalk. These results indicate the effectiveness of the network embedding approach. BPR employs a pairwise ranking function to learn the preference order using implicit feedback, and each training case is a pair consisting a positive item and a negative item. DeepWalk generates truncated random vertex sequences,

---

[4] The number of items in both datasets is large, and it will be quite time-consuming to consider all the unadopted items as candidate recommendations. We follow [24] to pair each adopted item with 50 negative unadopted items to form the candidate recommendation list.

and derive the embeddings by using a hierarchical softmax. Compared to these two methods, NERM directly optimizes each edge (*i.e.*, each user-item adoption record) in the network and adopt the negative sampling as the optimization method.

4.2   Evaluation on Top-N Tag Recommendation

Dataset. We use two shared datasets in [13] for the evaluation of tag recommendation. These two datasets have been widely used for tag recommendation. Different from [13], we do not perform p-core filtering. Table 3 summarizes the basic statistics of the two datasets.

Table 3. Summary statistics of the two datasets for tag recommendation.

Datasets	#Users	#Items	#Tags	#Records
Last.fm	1,893	12,524	9,750	186,479
Bookmarks	1,868	69,224	40,898	437,593

Experimental Setting. For tag recommendation, we following the same experimental setting in [13] for evaluation. We adopt Precision@K, Recall@K and F@K as the evaluation metrics. For each user, we take the last annotated item together with the attached tags as the test data, while the rest tagging records are used as training data.

Baselines. We consider using the following methods as the comparison baselines

- **PITF** [13]. PITF is a factorization model for tag recommendation, that explicitly models the pairwise interactions (PITF) between users, items and tags. The model is learned with an adaption of the Bayesian personalized ranking (BPR) criterion which originally has been introduced for item recommendation.
- **DeepWalk** [10]. It is similar to that is described in previous experiments on item recommendation.

PITF represents a competitive baseline for tag recommendation[5]. As shown in [13], PITF is better than FolkRank [5] on the two datasets, thus we do not

Table 4. Performance comparisons of the proposed methods and baselines on tag recommendation.

Methods	Last.fm						Bookmarks					
	P@1	R@1	F@1	P@5	R@5	F@5	P@1	R@1	F@1	P@5	R@5	F@5
PITF	0.305	0.125	0.178	0.189	0.351	0.245	0.381	0.132	0.197	0.204	0.304	0.244
DeepWalk	0.088	0.044	0.059	0.040	0.099	0.057	0.064	0.024	0.035	0.038	0.074	0.050
NERM	**0.327**	**0.165**	**0.220**	0.182	**0.370**	0.244	**0.396**	**0.135**	**0.201**	**0.228**	**0.323**	**0.267**

---

[5] We do not compare with other methods with item contents or temporal information.

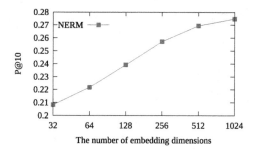

Fig. 2. Varying the number of embedding dimensions for item recommendation on JD dataset.

compare with FoldRank here. We hold out 10% of items in the training set as the development set for parameter optimization. For PITF, the number of latent factors is set to 256, and the number of negative samples is set to 200. For DeepWalk, the number of embedding dimensions is set to 128 and we use the hierarchal softmax algorithm to learn the parameters. For our method NERM, the number of embedding dimensions is set to 128, and the number of negative samples is set to 200.

Results and Analysis. Table 4 presents the experimental results of the compared methods on the task of tag recommendation. Overall, we have made the following observations. First, the proposed method NERM nearly performs best for all the entries, slightly worse than the state-of-the-art method PITF on Last.fm in terms P@5 and F@5. Second, the network embedding method DeepWalk performs poorly on the tag recommendation task. By combining the results on item recommendation, we can conclude that NERM is effective to deal with recommendation tasks as a general method. DeepWalk does not perform well on tag recommendation, a possible reason is that it may require more principled random walk methods on k-partite graphs. Currently, we follow [10] to use a uniform sampling method, however, such a method may not be suitable to k-partite graphs. For example, it is likely that vertices in some independent set cannot be well covered even with many random paths. We will investigate into it as a future work.

4.3   Parameter Tuning

In our model NERM, an important parameter to tune is the number of embedding dimensions. We vary it in the set $\{32, 64, 128, 256, 512, 1024\}$ and see how it affects the performance. The tuning results are shown in Figs. 2 and 3. As we can see that, for item recommendation, we need to set a large number; While for tag recommendation, the optimal number is set to 128. The major reason is that the two datasets used for item recommendation are much larger than those used in tag recommendation.

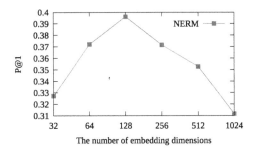

Fig. 3. Varying the number of embedding dimensions for tag recommendation on Bookmarks dataset.

5 Related Work

Recommender Systems. In the literature of recommender systems, two widely studied tasks are *rating prediction* and *top-N recommendation*. Rating prediction aims to predict the ratings from users to items, while top-N recommendation aims to generate a short list of recommendations for users [4]. Our focus in this paper is top-N recommendation. There are three typical approaches for top-N recommendation. First, rating prediction methods were directly applied where the predicted rating value was used for ranking [9,23]. Second, implicit feedback information was utilized to improve the recommendation performance, such as the weighting-based method [6]. Thirdly, specific loss function in the optimization objective was developed, including AUC-based loss function [7,12] and MAP-based loss function [14]. Tag recommendation can be considered as a special task for top-N recommendation, where tag are recommended to a user on a specific item. Various methods have been proposed for tag recommendation, including random walk methods [5], time-sensitive methods [15], higher order singular value decomposition [16], and pairwise interaction tensor factorization [13]. Recently, several context-aware models have been also proposed in order to utilize complex contextual information for rating prediction [3,11]. Different from previous studies, we aim to build a general and scalable recommendation framework for top-N recommendation, and present a new perspective by applying the network embedding techniques.

Distributed Representation Learning. Recent years have witnessed the great success of distributed representation learning and neural networks. It provides an effective way to represent and extract useful knowledge in many tasks, including text classification [17], knowledge graph mining [22] and recommender systems [21]. Especially, a promising direction is network embedding with distributed representation learning [10,18]. For example, DeepWalk [10] adapted Skip-Gram [8], a widely used language model in natural language processing area, for network representation learning on truncated random walks. LINE [18] is a scalable network embedding algorithm which modeled the first-order and second-order proximities between vertices. More recently, heterogenous network embedding [2] or focus on deep network embedding [19] have been studied. We are also aware that several works have applied distributed representation learning [20] or neural network models [21] to recommendation tasks. Our work is highly built on these studies. The novelty lies in the idea which casts the recommendation task into a network embedding task. To our knowledge, it is the first time that network embedding methods have been applied to recommendation tasks.

6   Conclusion and Future Work

In this paper, we made the first attempt that utilized the network representation learning techniques for recommendation tasks. We first transformed the adoption records into a k-partite adoption network, then learned distributed representations for the vertices, and finally calculated the embedding similarity for recommendation. To evaluate the effectiveness of the proposed approach, we constructed extensive experiments on two different recommendation tasks using real-world datasets. The experimental results have shown the superiority of our approach. To the best of our knowledge, it is the first time that a network representation learning approach has been applied to recommendation tasks. Currently, we adopt a simple network architecture for efficient parameter learning. In the future, we consider employing more complex deep neural networks [2,19] for recommender systems. We will also test how the current framework performs on context-aware recommendation which involves multiple kinds of contextual information, such as users' demographics and items' reviews.

Acknowledgements. The authors thank the anonymous reviewers for their valuable and constructive comments. The work was partially supported by National Natural Science Foundation of China under the grant number 61502502 and Beijing Natural Science Foundation under the grant number 4162032.

References

1. Bobadilla, J., Ortega, F., Hernando, A., GutiéRrez, A.: Recommender systems survey. Know. Based Syst. **46**, 109–132 (2013)
2. Chang, S., Han, W., Tang, J., Qi, G., Aggarwal, C.C., Huang, T.S.: Heterogeneous network embedding via deep architectures. In: SIGKDD, pp. 119–128 (2015)
3. Chen, T., Zhang, W., Lu, Q., Chen, K., Zheng, Z., Yu, Y.: Svdfeature: a toolkit for feature-based collaborative filtering. J. Mach. Learn. Res. **13**, 3619–3622 (2012)
4. Deshpande, M., Karypis, G.: Item-based top-n recommendation algorithms. ACM Trans. Inform. Syst. (TOIS) **22**(1), 143–177 (2004)
5. Hotho, A., Jäschke, R., Schmitz, C., Stumme, G.: Folkrank: a ranking algorithm for folksonomies. In: LWA 2006, pp. 111–114 (2006)
6. Hu, Y., Koren, Y., Volinsky, C.: Collaborative filtering for implicit feedback datasets. In: ICDM, pp. 263–272 (2008)
7. Kabbur, S., Ning, X., Karypis, G.: Fism: factored item similarity models for top-n recommender systems. In: Proceedings of the 19th ACM SIGKDD International Conference on Knowledge Discovery and Data Mining, pp. 659–667. ACM (2013)

8. Mikolov, T., Sutskever, I., Chen, K., Corrado, G.S., Dean, J.: Distributed representations of words and phrases and their compositionality. In: NIPS (2013)
9. Ning, X., Karypis, G.: Slim: sparse linear methods for top-n recommender systems. In: IEEE 11th International Conference on Data Mining, pp. 497–506. IEEE (2011)
10. Perozzi, B., Al-Rfou, R., Skiena, S.: Deepwalk: online learning of social representations. In: Proceedings of SIGKDD (2014)
11. Rendle, S.: Factorization machines with libfm. ACM TIST **3**(3), 57 (2012)
12. Rendle, S., Freudenthaler, C., Gantner, Z., Schmidt-Thieme, L.: Bpr: Bayesian personalized ranking from implicit feedback. In: UAI, pp. 452–461 (2009)
13. Rendle, S., Schmidt-Thieme, L.: Pairwise interaction tensor factorization for personalized tag recommendation. In: WSDM, pp. 81–90 (2010)
14. Shi, Y., Karatzoglou, A., Baltrunas, L., Larson, M., Hanjalic, A., Oliver, N.: Tfmap: optimizing map for top-n context-aware recommendation. In: Proceedings of the 35th International ACM SIGIR Conference on Research and Development in Information Retrieval, pp. 155–164. ACM (2012)
15. Song, Y., Zhuang, Z., Li, H., Zhao, Q., Li, J., Lee, W., Giles, C.L.: Real-time automatic tag recommendation. In: SIGIR, pp. 515–522 (2008)
16. Symeonidis, P., Nanopoulos, A., Manolopoulos, Y.: Tag recommendations based on tensor dimensionality reduction. In: RecSys (2008)
17. Tang, J., Qu, M., Mei, Q.: Pte: Predictive text embedding through large-scale heterogeneous text networks. In: SIGKDD (2015)
18. Tang, J., Qu, M., Wang, M., Zhang, M., Yan, J., Mei, Q.: Line: Large-scale information network embedding. In: WWW (2015)
19. Wang, D., Cui, P., Zhu, W.: Structural deep network embedding. In: SIGKDD (2016)
20. Wang, H., Wang, N., Yeung, D.: Collaborative deep learning for recommender systems. In: SIGKDD, pp. 1235–1244 (2015)
21. Wang, P., Guo, J., Lan, Y., Xu, J., Wan, S., Cheng, X.: Learning hierarchical representation model for nextbasket recommendation. In: SIGIR (2015)
22. Yang, C., Liu, Z., Zhao, D., Sun, M., Chang, E.Y.: Network representation learning with rich text information. In: IJCAI (2015)
23. Yang, X., Steck, H., Guo, Y., Liu, Y.: On top-k recommendation using social networks. In: Proceedings of the Sixth ACM Conference on Recommender Systems, pp. 67–74. ACM (2012)
24. Zhao, W.X., Wang, J., He, Y., Wen, J., Chang, E.Y., Li, X.: Mining product adopter information from online reviews for improving product recommendation. TKDD **10**(3), 29 (2016)

Factorizing Sequential and Historical Purchase Data for Basket Recommendation

Pengfei Wang$^{(\boxtimes)}$, Jiafeng Guo, Yanyan Lan, Jun Xu, and Xueqi Cheng

Key Laboratory of Network Data Science and Technology, CAS,
Institute of Computing Technology, Chinese Academy of Sciences, Beijing, China
wangpengfei@software.ict.ac.cn,
{guojiafeng,lanyanyan,junxu,cxq}@ict.ac.cn

Abstract. Basket recommendation is an important task in market basket analysis. Existing work on this problem can be summarized into two paradigms. One is the item-centric paradigm, where sequential patterns are mined from users' transactional data and leveraged for prediction. However, these approaches usually suffer from the data sparseness problem. The other is the user-centric paradigm, where collaborative filtering techniques have been applied on users' historical data. However, these methods ignore the sequential behaviors of users, which are often crucial for basket recommendation. In this paper, we introduce a hybrid method, namely the *Co-Factorization* model over *S*equential and *H*istorical purchase data (*CFSH* for short) for basket recommendation. Compared with existing methods, our approach enjoys the following merits: (1) By mining and factorizing global sequential patterns, we can avoid the sparseness problem in traditional item-centric methods; (2) By factorizing item-item and user-item matrices simultaneously, we can exploit both sequential and historical behaviors to learn user and item representations better; (3) Experimental results on three real-world transaction datasets demonstrated the effectiveness of our approach as compared with the existing methods.

Keywords: Basket recommendation · Sequential patterns · Recommendation

1 Introduction

Market basket analysis aims to discover meaningful patterns from massive users' transaction data [3]. It helps retailers analyze sale trends, optimize commodity placement, and comprehend users' preferences. Especially with the prevalence of mobile applications and online e-commerce systems, market basket analysis becomes even more important in stimulating the consumptions and enlarging the selling profits, by providing the key technologies for personalized next-basket recommendation.

Generally, existing methods on basket recommendation can be summarized into two paradigms. One is the item-centric paradigm, which explores the sequential transaction data by predicting the next purchase based on the last actions.

© Springer International Publishing AG 2016
S. Ma et al. (Eds.): AIRS 2016, LNCS 9994, pp. 237–248, 2016.
DOI: 10.1007/978-3-319-48051-0_18

A major advantage of this model is its ability to capture sequential behavior for good recommendations, e.g. for a user who has recently bought a mobile phone, it may recommend accessories that other users have bought after buying that phone. A number of approaches have been proposed to mine the meaningful sequential patterns from users' transactional data [11,15]. However, the directly mined sequential patterns are usually very sparse and hard to cover many long-tailed items and users.

The other is the user-centric paradigm, with the key idea as that "one is likely to buy the items favored by similar users". One of the most successful methods in this paradigm is the model based collaborative filtering [1,6,9,17]. The typical way is to represent users' historical purchase behaviors as a user-item matrix by discarding transaction information, and apply matrix factorization for recommendation. Obviously, these methods are good at modeling users' general interests, but hard to capture the sequential purchase behaviors of users which is often crucial for the basket recommendation.

A better solution for basket recommendation, therefore, is to take both item-centric and user-centric paradigms into consideration. For this purpose, we propose a hybrid method in this paper, namely *Co-F*actorization model over *S*equential and *H*istorical purchase data (CFSH for short). Specifically, on one hand, we apply sequential pattern mining methods to the massive transaction data, and aggregate all the mined patterns to form a item-item matrix. On the other hand, we construct a user-item matrix based on users' whole historical purchase data. These two matrices are then simultaneously factorized to learn the low-dimensional representations of both users and items. With the learned representations, we provide personalized basket recommendation based on both sequential behaviors and users' general interests.

Compared with existing basket recommendation methods, our approach has the following advantages:

- By mining and factorizing global sequential patterns, our approach can avoid the data sparseness problem in traditional item-centric methods;
- By factorizing the item-item matrix and user-item matrix simultaneously, our approach can exploit both sequential and historical behaviors to learn better representations of both users and items;
- By adopting a hybrid recommendation approach, our model enjoys the merits of both item-centric and user-centric paradigms, and thus achieves better performances on basket recommendation.

We conducted empirical experiments over three real-world purchase datasets: two from retailers and one from the online e-commerce website. Both the existing item-centric and user-centric methods have been taken into comparison. The results demonstrate that the proposed CFSH method can perform significantly better than the baseline methods on basket recommendation task.

2   Related Work

In this section, we briefly review the related work on basket recommendation from the following two aspects, i.e. item-centric models and user-centric models.

2.1   Item-Centric Models

The key idea lies in item-centric models is the phenomenon observed in users' purchase behavior that "buying one item leads to buying another next". Therefore, item-centric models, mostly relying on Markov chains, explore the sequential transaction data by predicting the next purchase based on the last actions. For example, in early work, different data mining methods including ApriorALL, SPADE have been designed for mining frequent sequential patterns among items. Zimdar et al. [2] investigate how to extract sequential patterns to learn the next state using probablistic decision-tree models. Mobasher et al. [16] study different sequential patterns for recommendation and find that contiguous sequential patterns are more suitable for sequential prediction task than general sequential patterns. However, the directly mined sequential patterns are usually too sparse with respect to the size of users and items. One way to tackle the data sparseness problem is to learn latent models over the sequential patterns.

2.2   User-Centric Models

User-centric models, in contrast, does not take sequential behavior into account but make recommendation based on users' whole purchase history. The key idea is collaborative filtering (CF) which can be further categorized into memory-based CF and model-based CF [19]. The memory-based CF provides recommendations by finding k-nearest-neighbor of users or items based on certain similarity measure [15]. While the model-based CF tries to factorize the user-item correlation matrix for recommendation.

For example, Lee et al. [9] treat the market basket data as a binary user-item matrix, and apply a binary logistic regression model based on principal component analysis (PCA) for recommendation. Hu et al. [8] conduct the factorization on user-item pairs with least-square optimization and use pair confidence to control the importance of observations. Rendle et al. [18] propose a different optimization criterion, namely Bayesian personalized ranking, which directly optimizes for correctly ranking over item pairs instead of scoring single items. They apply this method to matrix factorization and adaptive KNN to show its effectiveness. General speaking, these models are good at capturing users' general taste, but can hardly adapt its recommendations directly to users' recent purchases without modeling sequential behavior.

3   Our Framework

In this paper, we propose to factorize both users' sequential and historical purchase data for basket recommendation. By adopting such a hybrid recommendation method, we can enjoy the advantages of both item-centric and user-centric paradigms. In this section, we will present the proposed method, namely the *Co-Factorization* model over *Sequential* and *Historical* purchase data (CFSH for short) in detail. Specifically, we first introduce the notations used in this work. Then we describe how to factorize the sequential and historical purchase data for basket recommendation respectively. The hybrid model CFSH is then presented based on the above two recommendation paradigms. Finally, we present the optimization algorithm for the proposed hybrid recommendation model.

3.1   Notations

Let $U = \{u_1, u_2, \ldots, u_{|U|}\}$ be a set of users and $I = \{i_1, i_2, \ldots, i_{|I|}\}$ a set of items, where $|U|$ and $|I|$ denote the total number of unique users and items respectively. For each user $u \in U$, a purchase history T^u of his transactions is given: $T^u := (T_1^u, T_2^u, \ldots, T_{t_u-1}^u)$, where $T_t^u \subseteq I$, $t \in [1, t_u - 1]$. The purchase history of all users is denoted as $T := \{T^{u_1}, T^{u_2}, \ldots, T^{u_{|U|}}\}$. More formally, let $V^U = \{v_u^U \in R^n | u \in U\}$ denote all the user vectors and $V^I = \{v_i^I \in R^n | i \in I\}$ all the item vectors. Given each user's purchase history, the task is to recommend items that user u would probably buy at the next (i.e. t_u-th) visit.

3.2   Factorizing Sequential Purchase Data

The item-centric models explore the sequential patterns in transaction data and provide basket recommendation based on users' last actions. A severe problem of existing methods in this paradigm is the sparseness of sequential patterns, which is too sparse to generalize well for long-tail users and items. In our work, therefore, we propose to mine the sequential patterns in a global way, and factorize the mined patterns to learn better low dimensional representations of items for recommendation. Here we first give the definition of sequential patterns in transactional data.

Definition 1 *Sequential Pattern*. *Given the transaction set* $T^u :=$ $(T_1^u, \ldots, T_{t_u-1}^u)$ *of user* u, *Sequential Pattern is defined as a weighted pair of items* $< i_k, i_{k'}, s_{k,k'}^u >$, *where* $i_k \in T_p^u$, $i_{k'} \in T_q^u$, $p < q$, *and* $s_{k,k'}^u$ *denotes the support of the sequential pattern.*

If we restrict that the patterns are mined from consecutive transactions of each user, then the obtained patterns are called as Contiguous Sequential Pattern (CSP for short) [4]. Otherwise, they are called as Non-Contiguous Sequential Patterns (NCSP for short). Existing work finds that CSPs are more suitable for sequential prediction task. Figure 1 shows an example, where six CSPs can be mined from a user with three transactions. The mined CSPs can be represented as a matrix S^u, where the entry $S_{k,k'}^u$ corresponds to the pattern $< i_k, i_{k'}, s_{k,k'}^u >$, and the question mark denotes a missing pattern. Obviously, the CSPs capture the local dependency between user's purchase behaviors. For example, a user would probably buy a sim card in the next transaction if she bought a phone in the previous transaction. A higher support of the pattern indicates higher possibility that sequential behavior will take place.

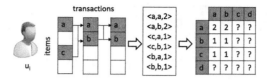

Fig. 1. Contiguous sequential patterns mined from a single user.

We conduct the above mining procedure over each user's transaction data and aggregate all the mined CSPs into one global matrix $S = \sum_u S^u$. This matrix captures all the globally salient sequential patterns, and is more dense and more robust to noise than any single user's matrix (i.e. S^u). We then factorize S to learn the low dimensional representations of items, by assuming that the support of the sequential pattern can be recovered by the representations of the two items within the pattern. The objective function of our matrix factorization is shown as follows:

$$\min\{ \ \|S - V^I V^{I^T}\|^2 + \lambda\|V^I\|^2\}$$
$$\text{s.t.} \quad V^I \geq 0. \tag{1}$$

where λ is the regularization coefficient.

With the learned low dimensional representations of items, we can then provide personalized basket recommendation given the user's last transaction. Specifically, given the user u's last transaction $T^u_{t_u-1}$, we calculate the probability of item i_l to appear in the next transaction as the aggregated support for the item

$$P(i_l \in T^u_{t_u}|T^u_{t_u-1}) \propto agg\_supp_u(i_l) = \sum_{i_k \in T^u_{t_u-1}} v^I_k \cdot v^I_l \tag{2}$$

We then sort the items according to the probability and obtain the top-K items for recommendation.

3.3 Factorizing Historical Purchase Data

For the user-centric paradigm, we follow the previous practice and apply model based collaborative filtering. Specifically, we first construct the user-item matrix H based on users' whole historical purchase data, as shown in Fig. 2. Note here all the sequential information has been discard. The entry $H_{u,k}$ denote the purchase count of item i_k for user u. The higher the count is, the more likely the user will buy the item.

We then factorize the user-item matrix H to learn the low dimensional representations of users and items with the following objective function.

$$\min\{\| \ H - V^U V^{I^T} \ \|^2 + \lambda_1 \| \ V^U \ \|^2 + \lambda_2 \| \ V^I \ \|^2\}$$
$$\text{s.t.} \quad \begin{cases} V^U \geq 0 \\ V^I \geq 0 \end{cases} \tag{3}$$

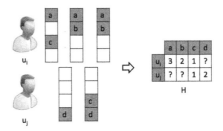

Fig. 2. Customer-item matrix mined from transactions. Users' purchase count to a certain item indicates users' general interest to it. Elements with ? indicate unobserved purchase count.

where λ is the regularization coefficient With the learned representations of users and items, we calculate the probability of item i_l to appear in user u's next transaction as the preference of the user on this item:

$$P(i_l \in T_{t_u}^u | u) \propto pref_u(i_l) = \boldsymbol{v}_u^U \cdot \boldsymbol{v}_l^I \tag{4}$$

We then sort the items according to the probability and obtain the top-K items for recommendation.

3.4   Hybrid Method

Now we have described how to provide personalized basket recommendation in two ways, i.e. one explores the correlation between items and the other relies on the correlation between users and items. The former can well capture the sequential behaviors of users while the latter can model users' general interests. Our idea is then to combine the two ways so that we can enjoy the powers of both paradigms and meanwhile complement each other to achieve better performances.

Specifically, we propose to simultaneously factorize the item-item matrix and the user-item matrix, by sharing the same low dimensional representations of items. The objective function of our hybrid method is as follows:

$$\min\{\alpha \parallel H - V^U V^{I^T} \parallel^2 + (1 - \alpha) \parallel S - V^I V^{I^T} \parallel^2 + \lambda_1 \parallel V^U \parallel^2 + \lambda_2 \parallel V^I \parallel^2\}$$

$$\text{s.t.} \quad \begin{cases} V^U \geq 0 \\ V^I \geq 0 \end{cases}$$

$$\tag{5}$$

where $\alpha \in [0,1]$ is the coefficient which balance the importance of the two paradigms, and λ_1 and λ_2 denote the regularization coefficients.

With the learned low dimensional representations of users and items, we provide personalized basket recommendation also in a hybrid way. Specifically, given user u's last transaction $T_{t_u-1}^u$, we calculate the probability of item i_l to appear in the next transaction as follows

$$P(i_l \in T_{t_u}^u | T_{t_u-1}^u) = \alpha \sum_{i_k \in T_{t_u-1}^u} \boldsymbol{v}_k^I \cdot \boldsymbol{v}_l^I + (1-\alpha)\boldsymbol{v}_u^U \cdot \boldsymbol{v}_l^I \qquad (6)$$

We then sort the items according to the probability and obtain the top-K items for recommendation.

3.5 Optimization of the Hybrid Objective

To optimize the hybrid objective mentioned above, we find there is no closed-form solution for Eq. 5. Therefore, we apply an alternative minimization algorithm to approximate the optimal result [5]. The basic idea of this algorithm is to optimize the loss function with respect to one parameter, with all the other parameters fixed. The algorithm keeps iterating until convergence or the maximum of iterations. First we fix V^I, and calculate V^U to minimize Eq. 5. The updating algorithm is shown in Algorithm 1.

Algorithm 1. Hybrid Factoring Sequential pattern and Historical data

---
Input: :Input $H, S, \alpha, \lambda_1, \lambda_2, num$
Output: V^U, V^I
  t=0
  repeat
    $t \leftarrow t+1$;
    $V^U \leftarrow V^U \sqrt{\dfrac{\alpha H V^I}{\alpha V^U V^{I^T} V^I + \lambda_1 V^U}}$;

    $V^I \leftarrow V^I \sqrt{\dfrac{\alpha H^T V^U + (1-\alpha)S^T V^I}{\alpha V^I V^{U^T} V^U + (1-\alpha)V^I V^{I^T} V^I + \lambda_2 V^I}}$;

  until converge or t>num
  return V^U, V^I;

---

4 Experiment

4.1 Data Description

We evaluate different recommendation methods based on three real-world purchase datasets, i.e. two retail datasets namely Ta-Feng and BeiRen, and one e-commerce dataset namely T-Mall.

- The Ta-Feng[1] dataset is a public dataset released by RecSys conference, which covers items from food, office supplies to furniture.
- The BeiRen dataset comes from BeiGuoRenBai[2], a large retail enterprise in China, which records its supermarket purchase history during the period from Jan. 2013 to Sept. 2013.

---

[1] http://recsyswiki.com/wiki/Grocery_shopping_datasets.
[2] http://www.brjt.cn/.

– The T-Mall[3] dataset is a public online e-commerce dataset released by Taobao[4], which records the online transactions in terms of brands.

We first conducted some pre-processes on these purchase datasets. For both Ta-Feng and BeiRen dataset, we remove all the items bought by less than 10 users and users that has bought in total less than 10 items. For the T-Mall dataset, which is relatively smaller, we remove all the items bought by less than 3 users and users that has bought in total less than 3 items. The statistics of the three datasets after pre-processing are shown in Table 1. Finally, we split all the datasets into two non overlapping set, one used for training and the other for testing. The testing set contains only the last transaction of each user, while all the remaining transactions are put into the training set.

Table 1. Statistics of the datasets used in our experiments.

Dataset	Users $|U|$	Items $|I|$	Transactions T	avg.transaction size	avg.transaction per user
Ta−Feng	9238	7982	67964	5.9	7.4
BeiRen	9321	5845	91294	5.8	9.7
T-Mall	292	191	1805	1.2	5.6

4.2   Baseline Methods

We evaluate our hybrid model by comparing with several existing methods on basket recommendation:

– TOP: The top popular items in training set are taken as recommendations for each user.
– MC: An item-centric model which mines the global CSPs from purchase data, and predict the basket based on the last transaction of the user.
– FMC: A factorized MC model, which factorizes the global CSP matrix to learn the low dimension representations of items, and predict the basket based on the last transaction of the user.
– NMF: A state-of-the-art user-centric model based on collaborative filtering [13]. Here Nonnegative Matrix Factorization is applied over the user-item matrix, which is constructed from the transaction dataset by discarding the sequential information. Obviously, this model can be viewed as a sub-model of our approach, since it is exactly the same as the user-centric model described in Sect. 3.3. For implementation, we adopt the publicly available codes from NMF:DTU Toolbox[5].

---

[3] http://102.alibaba.com/competition/addDiscovery/index.htm.
[4] http://www.taobao.com.
[5] http://cogsys.imm.dtu.dk/toolbox/nmf/.

For all the latent models, including NMF, FMC and CFSH, we run several times with random initialization by setting the dimensionality $d \in \{50, 100, 150, 200\}$ on Ta-Feng and BeiRen datasets, and $d \in \{10, 15, 20, 25\}$ on T-Mall dataset. We set all the regularization parameters to 0.01, and parameter α used in CFSH equals to 0.6. We repeat these experiments ten times, and compare the average performances of different methods in the following sections.

4.3   Evaluation Metrics

The performance is evaluated for each user u on the transaction $T_{t_u}^u$ in the testing dataset. For each recommendation method, we generate a list of K items (K=5) for each user u. We use the following quality measures to evaluate the recommendation lists against the actual bought items.

- F1-score: F1-score is the harmonic mean of precision and recall, which is a widely used measure in recommendation [7,12,14].
- Hit-Ratio: Hit-Ratio is a All-but-One measure used in recommendation [10]. If there is at least one item in the test transaction also appears in the recommendation list, we call it a *hit*. Hit-Ratio focuses on the *recall* of a recommender system, i.e. how many people can obtain at least one correct recommendation.

4.4   Performance on Basket Prediction

In this section we compare our hybrid model to state-of-the-art methods in basket recommendation.

Figure 3 shows the results on Ta-Feng, BeiRen, and T-Mall respectively. We have the following observations from the results: (1) Overall, the Top method is the weakest. However, we find that the Top method outperforms MC on the T-Mall dataset. By analyzing the dataset, we found that the filling rate of the CSP matrix of the T-Mall dataset is around 3.7 %, which is much lower than that of the Ta-Feng dataset (11.8 %) and BeiRen dataset (15.2 %). It indicates that the CSPs mined from the T-mall dataset are too sparse for MC to generate reasonable recommendations for users. (2) By either factorizing global sequential patterns, FMC can obtain better results over the MC method. Specifically, we can see obvious improvement in terms of Hit-Ratio, which indicates the factorized methods can cover more users than the original MC method. The improvement becomes larger on the T-Mall dataset which is much more sparse when only considering directly mined CSPs. (3) By factorizing the historical purchase data, the NMF method also outperforms the MC method in most cases, and its performance is between the two factorized item-centric methods. (4) By combining both item-centric and user-centric paradigms, the proposed CFSH method performs best on all the three datasets. Take Ta-Feng dataset as an example, when compared with second best performed baseline method, the relative performance improvement by CFSH is around 16.2 %, 9.8 %, 21.1 % and 16.5 % in terms of Precision, Recall, F1-Score, and Hit-Ratio respectively. All the improvement are

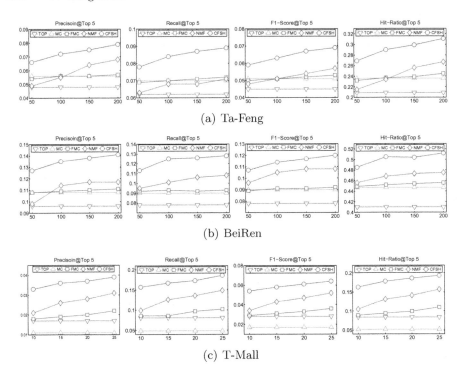

Fig. 3. Performance comparison on three datasets.

statistically significant (p-value < 0.01). The results demonstrate that by factorizing global sequential patterns and historical purchase data simultaneously, we can learn better representations of users and items and thus obtain improved performance on basket recommendation.

To further investigate the performance of different methods, we split the users into three groups (i.e. inactive, medium and active) based on their activeness and conducted the comparisons on different user groups. Take the Ta-Feng dataset as an example, a user is taken as inactive if there are less than 5 transactions in his/her purchase history, and active if there are more than 20 transactions in the purchase history. The remaining users are taken as medium. In this way, the proportions of inactive, medium and active are 40.8%, 54.5%, and 4.7% respectively. Here we only report the comparison results on Ta-Feng dataset with the dimension equals 50 due to the page limitation. In fact, similar conclusions can be drawn from other datasets. The results are shown in Table 2.

From the results we can see that, not surprisingly, the Top method is still the worst on all the groups. Furthermore, we find that MC, FMC works better than NMF on both inactive and medium users in terms of all the measures; While on active users, NMF can achieve better performance than MC, FMC. The results indicate that it is difficult for NMF to learn a good user representation with few transactions for recommendation. Finally, CFSH can achieve the

Table 2. Performance comparison of different methods on Ta-Feng over different user groups with dimensionality set as 50.

User activeness	Method	Precision	Recall	F-Measure	Hit-Ratio
Unactive (3765)	Top	0.043	0.047	0.036	0.181
	MC	0.050	0.052	0.042	0.206
	FMC	0.049	0.052	0.041	0.210
	NMF	0.047	0.039	0.037	0.198
	CFSH	**0.059**	**0.059**	**0.048**	**0.236**
Medium (5031)	Top	0.053	0.073	0.052	0.230
	MC	0.061	0.083	0.059	0.261
	FMC	0.059	0.081	0.057	0.253
	NMF	0.058	0.073	0.051	0.234
	CFSH	**0.072**	**0.097**	**0.068**	**0.269**
Active (442)	Top	0.043	0.067	0.045	0.207
	MC	0.047	0.076	0.049	0.210
	FMC	0.049	0.077	0.051	0.212
	NMF	0.057	0.079	0.056	0.223
	CFSH	**0.061**	**0.093**	**0.062**	**0.246**

best performances on all the groups in terms of all the measures. The results demonstrate that by combining both item-centric and user-centric paradigms, we can enjoy the merits of both methods and complement each other to achieve better performance.

5   Conclusions

In this paper, we proposed a new hybrid recommendation method, namely CFSH, for basket recommendation based on massive transactional data. The major purpose is to leverage the power of both item-centric and user-centric recommendation paradigms in capturing correlations between items and users for better recommendation. By conducting experiments on three real world purchase datasets, we demonstrated that our approach can produce significantly better prediction results than the state-of-the-art baseline methods.

In the future work, we would like to explore other context information, e.g. time and location, for better basket recommendation. Obviously, people's shopping behavior may be largely affected by these factors. It would be interesting to investigate how these types of information can be integrated into our proposed hybrid method.

References

1. Reutterer, T., Mild, A.: An improved collaborative filtering approach for predicting cross-category purchases based on binary market basket data. J. Retail. Consum. Serv. **10**(3), 123–133 (2003)

2. Meek, C., Zimdars, A., Chickering, D.M.: Using temporal data for making recommendations. UAI (2001)

3. Cavique, L.: A scalable algorithm for the market basket analysis. J. Retail. Consum. Serv. **14**, 400–407 (2007)

4. Chen, J., Shankar, S., Kelly, A., Gningue, S., Rajaravivarma, R.: A two stage approach for contiguous sequential pattern mining. In: Proceedings of the 10th IEEE International Conference on Information Reuse & Integration (2009)

5. Ding, C., Li, T., Peng, W., Park, H.: Orthogonal nonnegative matrix t-factorizations for clustering. In: SIGKDD (2006)

6. Gatzioura, A., Sanchez Marre, M.: A case-based recommendation approach for market basket data. IEEE Intell. Syst. **30**, 20–27 (2014)

7. Godoy, D., Amandi, A.: User profiling in personal information agents: a survey. Knowl. Eng. Rev. **20**(4), 329–361 (2005)

8. Hu, Y., Koren, Y., Volinsky, C.: Collaborative filtering for implicit feedback datasets. In: Proceedings of the Eighth IEEE International Conference on Data Mining, ICDM 2008, pp. 263–272. IEEE Computer Society, Washington, DC (2008)

9. Kim, S., Lee, J.-S., Jun, C.-H., Lee, J.: Classification-based collaborative filtering using market basket data. Expert Syst. Appl. **29**, 700–704 (2005)

10. Karypis, G.: Evaluation of item-based top-n recommendation algorithms. In: CIKM (2001)

11. Koenigstein, N., Koren, Y.: Towards scalable and accurate item-oriented recommendations. In: Proceedings of the 7th ACM Conference on Recommender Systems, RecSys 2013, pp. 419–422. ACM, New York (2013)

12. Lacerda, A., Ziviani, N.: Building user profiles to improve user experience in recommender systems. In: WSDM, WSDM 2013, pp. 759–764. ACM, New York (2013)

13. Lee, D.D., Seung, H.S.: Algorithms for non-negative matrix factorization. In: Leen, T., Dietterich, T., Tresp, V. (eds.) Advances in Neural Information Processing Systems 13, pp. 556–562. MIT Press, Cambridge (2001)

14. Lin, W., Alvarez, S.A., Ruiz, C.: Efficient adaptive-support association rule mining for recommender systems. Data Min. Knowl. Discov. **6**(1), 83–105 (2002)

15. Linden, G., Smith, B., York, J.: Amazon.com recommendations: Item-to-item collaborative filtering. IEEE Internet Comput. **7**(1), 76–80 (2003)

16. Mobasher, B.: Using sequential and non-sequential patterns in predictive web usage mining tasks. In: The IEEE International Conference on Data Mining Series (2002)

17. Forsatia, C.R., Meybodib, M.R.: Effective page recommendation algorithms based on distributed learning automata and weighted association rules. Expert Syst. Appl. **37**, 1316–1330 (2010)

18. Rendle, S., Freudenthaler, C., Gantner, Z., Schmidt-Thieme, L.: Bpr: Bayesian personalized ranking from implicit feedback. In: UAI (2009)

19. Su, X., Khoshgoftaar, T.M.: A survey of collaborative filtering techniques. Adv. Artif. Intell. **2009**, 4:2 (2009)

IR Evaluation

Search Success Evaluation
with Translation Model

Cheng Luo[✉], Yiqun Liu, Min Zhang, and Shaoping Ma

State Key Laboratory of Intelligent Technology and Systems, Tsinghua National
Laboratory for Information Science and Technology, Department of Computer
Science and Technology, Tsinghua University, Beijing 100084, China
c-luo12@mails.tsinghua.edu.cn, yiqunliu@tsinghua.edu.cn
http://www.thuir.cn

Abstract. Evaluation plays an essential way in Information Retrieval
(IR) researches. Existing Web search evaluation methodologies usually
come in two ways: offline and online methods. The benchmarks generated by offline methods (e.g. Cranfield-like ones) could be easily reused.
However, the evaluation metrics in these methods are usually based on
various user behavior assumptions (e.g. Cascade assumption) and may
not well accord with actual user behaviors. Online methods, in contrast,
can well capture users' actual preferences while the results are not usually reusable. In this paper, we focus on the evaluation problem where
users are using search engines to finish complex tasks. These tasks usually involve multiple queries in a single search session and propose challenges to both offline and online evaluation methodologies. To tackle this
problem, we propose a search success evaluation framework based on
machine translation model. In this framework, we formulate the search
success evaluation problem as a machine translation evaluation problem:
the ideal search outcome (i.e. necessary information to finish the task) is
considered as the reference while search outcome from individual users
(i.e. content that are perceived by users) as the translation. Thus, we
adopt BLEU, a long standing machine translation evaluation metric, to
evaluate the success of searchers. This framework avoids the introduction
of possibly unreliable behavior assumptions and is reusable as well. We
also tried a number of automatic methods which aim to minimize assessors' efforts based on search interaction behavior such as eye-tracking and
click-through. Experimental results indicate that the proposed evaluation
method well correlates with explicit feedback on search satisfaction from
search users. It is also suitable for search success evaluation when there
is need for quick or frequent evaluations.

Keywords: Search engine evaluation · Search success evaluation · User
behavior

This work was supported by Natural Science Foundation (61622208, 61532011,
61472206) of China and National Key Basic Research Program (2015CB358700).

S. Ma et al. (Eds.): AIRS 2016, LNCS 9994, pp. 251–266, 2016.
DOI: 10.1007/978-3-319-48051-0_19

1   Introduction

Evaluation plays a critical role in IR research as objective functions for system effectiveness optimization. Traditional evaluation paradigm focused on assessing system performance on serving "best" results for single queries. The Cranfield method proposed by Cleverdon [4] evaluates performance with a fixed document collection, a query set, and relevance judgments. The relevance judgments of the documents are used to calculate various metrics which are proposed based on different understanding of users' behavior. We refer this type of evaluation paradigm as *offline* evaluation, which is still predominant form of evaluation.

To line up the evaluation and the real user experience, *online* evaluation tries to infer users' preference from implicit feedback (A/B test [17], interleaving [14]), or explicit feedback (satisfaction [11]). Online methods naturally take user-based factors into account, but the evaluation results can hardly be reused.

Offline and Online methods have already achieved great success in promoting the development of search engine. However, offline evaluation metrics do not always reflect real users' experience [26]. The fixed user behavior assumptions (e.g. Cascade assumption) behind offline metrics may lead to failures on individual users. Consider an example in our experiment (depicted in Fig. 1), user A and B worked on the same task in one search engine and behaved in similar ways, the offline measurements should also be similar. However, the actual normalized scores given by external assessors showed that there was a relatively great difference in their success degrees.

In a typical search, according to the Interpretive Theory of Translation (ITT) [15], the search process can be modelled as three interrelated phases: (1) reading the content, (2) knowledge construction and (3) answer presentation. Inspired by this idea, we formalize the search success evaluation as a machine translation evaluation problem and propose a Search Success Evaluation framework based on Translation model (SSET). The ideal search outcome, which can be constructed manually, is considered as the *"reference"*. Meanwhile, the individual search outcome collected from a user is regarded as a *"translation"*. In this way, we can evaluate *"what degree of success the user has achieved"* by evaluating the correspondence between the ideal search outcome and individual search outcome. We investigate a number of machine translation (MT) evaluation

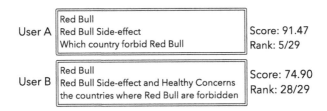

Fig. 1. An example of two user sessions with similar offline evaluation results but completely different online feedback (Scores by assessors and rank among all 29 participants)

metrics and choose BLEU (Bilingual Evaluation Understudy) for its simpleness and robustness.

To reduce the effort of manually construction of ideal search outcome, we also propose an automatic extraction method with various users' behavior data. Experiments indicate that evaluation with automated extracted outcome performs comparatively as well as with manually organized outcomes. Thus, it is possible to perform automated online evaluation including relatively large scale of users. In summary, our contribution includes: (1) A search evaluation framework based on machine translation model. To the best of our knowledge, our study is among the first to evaluate success with machine translation models. (2) An extraction method for the automatic generation of references with the help of multiple users' search interaction behavior (e.g. eye-tracking) is proposed and enables quick or frequent evaluations. (3) Experiment framework and data shared with the research community.

2   Related Work

Online/Offline Search Evaluation. Cranfield-like approaches [4] introduced a way to evaluate ranking systems with a document collection, a fixed set of queries, and relevance assessments from professional assessors. Ranking systems are evaluated with metrics, such as Precision, Recall, nDCG etc. The Cranfield framework has the advantage that relevance annotations on query-document pairs can be reused.

Beyond Cranfield framework, IR community strives to make evaluation more centred on real users' experience. The *online* evaluation methods, observing user behavior in their natural task procedures offer great promise in this regard. The *satisfaction* [11] method will ask the users to feedback their satisfaction during the search process explicitly, while the *interleaving* [14], *A/B testing* [17] methods try to infer user preference depending on implicit feedbacks, such as click-through etc. The evaluation results can hardly be reused for other systems which are not involved in the online test.

Session Search Evaluation. Beyond serving "best" results for single queries, for search sessions with multiple queries, several metrics are proposed by extending the single query metrics, i.e. the *nsDCG* based on *nDCG* [12] and *instance recall* based on *recall* [23]. Yang and Lad [31] proposed a measure of expected utility for all possible browsing paths that end in the kth reformulation. Kanoulas et al. [16] proposed two families of measures: one model-free family (for example, *session Average Precision*) that makes no assumption about the user's behavior and the other family with a simple model of user interactions over the session (*expected session Measures*).

Search Success Prediction. Previous researchers intuitively defined *search success* as the *information need fulfilled* during interactions with search engines. Hassan et al. [10] argued that relevance of Web pages for individual queries only represented a piece of the user's information need, users may have different information needs underlying the same queries. Ageev et al. [1] proposed a principled formalization of different types of "success" for informational tasks. The success model consists of four stages: *query formulation, result identification, answer extraction* and *verification of the answer*. They also presented a scalable game-like prediction framework. However, only binary classification labels are generated in their approach.

What sets our work apart from previous approaches is the emphasis on the outcomes the users gained through multiple queries. Our framework evaluates the success based on the information gained by users rather than implicit behavior signals. Ageev et al.'s definition about "success" was designed to analyze the whole process of their designed informational tasks. In our work, we simplify this definition and mainly focus on in what degree the user has gained enough information for certain search tasks.

Machine Translation Evaluation. Machine Translation models have been explored in Information Retrieval research for a long time [3,7]. However, machine translation evaluation methods have not been explored in search success evaluation problem. Several automatic metrics were accomplished by comparing the translations to references, which were expected to be efficient and correlate with human judgments. BLEU was proposed by Papineni et al. [24] to evaluate the effectiveness of machine translation systems. The scores are calculated for individual language segments (e.g. sentences) combining modified *n-gram* precision and brevity penalty. Several metrics were proposed later by extending BLEU [5,28].

Based on our definition of success, we mainly focus on whether a user has found the key information to solve the task. BLEU offers a simple but robust way to evaluate how good the users' outcome are comparing to pre-organized ideal search outcome on *n-gram* level. Other MT evaluation metrics could be adopted in this framework in a similar way as BLEU and we would like to leave them to our future work.

3   Methodology

3.1   Search Success Evaluation with Translation Model (SSET)

During a typical Web search, the user's information gathering actions can be regarded as "distilling" information gained into an organized answer to fulfill his/her information need [3]. We take the view that this distillation is a form of translation from one language to another: from documents, generated by Web page authors, to search outcome, with which the user seeks to complete the search task. Different from the standard three step translation process (understanding, deverbalization and re-expression [19,27]), "re-expression" is not always necessary in search tasks. In our framework, we retain the re-expression step by asking the participants to summarize their outcomes with the help of a predefined question so that we can measure the success of search process. The details of the framework will be represented in Sect. 4.

Comparing to previous success evaluation methodologies, we put more emphasis on user perceived information corresponding to the search task. We at first define some terminologies to introduce our framework:

Individual Search Outcome: for a specific user engaged in a search task, *search outcome* is the information gained by the user from interactions to fulfill the task's information need.

Ideal Search Outcome: for a certain topic, *ideal search outcome* refers to all possible information that can be found (by oracle) through reading the relevant documents provided by the search engine to fulfill the task's information need.

Search Success: *Search Success* is the situation that the user has collected enough information to satisfy his/her information need.

For a particular search task, a user read a sequence of words R^U. The search outcome of the user can be described by another sequence of words as S^J, while the ideal search outcome can be represented by a sequence of words as T^K. We assume users' search outcomes and the ideal search outcomes have identical vocabularies due to both of them come from the retrieved documents.

Fig. 2. Overview of search success evaluation framework with translation model

In this work, we propose a search success evaluation framework with translation model (SSET), which is presented in Fig. 2. Suppose the user's individual search outcome is a *"translation"* from examined documents in the search session, we can treat the ideal search outcome as a *"reference"*, which can be constructed manually by human assessors, or automatically based on group of users' interaction behaviors.

When evaluating a translation, the central idea behind a lot of metrics is that "the closer a machine translation is to a professional human translation, the better it is" [24]. We assume that *"the closer a user's individual search outcome is to an ideal search outcome, the more successful the search is"*. The SSET model is proposed to evaluate the search success by estimating the closeness between the individual search outcome and the ideal search outcome.

In SSET, we first compute a modified *n-gram* precision for the individual search outcome S^J, according to the ideal search outcome T^K:

$$p_n = \frac{\sum_{n\text{-}gram \in S^J} min(c(n\text{-}gram, T^K), c(n\text{-}gram, S^J))}{\sum_{n\text{-}gram\prime \in T^K} c(n\text{-}gram\prime)} \qquad (1)$$

where $c(n\text{-}gram, S)$ indicates the times of appearances of the $n\text{-}gram$ in S. In other words, one truncates each $n\text{-}gram$'s count, if necessary, to not exceed the largest count observed in the ideal search outcome. Then the brevity penalty BP is calculated by considering the length of the individual search outcome (c), and the length of the ideal search outcome (r):

$$BP = \begin{cases} 1 & if \quad c > r \\ e^{(1-r/c)} & if \quad c \leq r \end{cases} \tag{2}$$

The SSET score combine both the modified precision of $n\text{-}gram$ in different lengths and the brevity penalty, where w_n is the weight of the modified precision of $n\text{-}gram$, we use the typical value $N = 4$ in our experiment.

$$score_{SSET} = BP \cdot exp \left(\sum_{n=1}^{N} w_n \log p_n \right) \tag{3}$$

In this way, we can measure how close the individual search outcome is to ideal search outcome. The ideal search outcome organized by human assessors could be reused to evaluate other retrieval systems.

However, the generation process of ideal search outcome is still expensive and time-consuming. The individual search outcome generated by explicit feedback would also bring unnecessary effort to the users. We further explore the automation generation of *ideal search outcome* and *individual search outcome* based on multiple search interaction behaviors.

3.2   Automated Search Outcome Generation

We employ a bag-of-words approach, which has proven to be effective in many retrieval settings [9,18], to generate pseudo documents as the users' individual search outcome and ideal search outcome.

Consider a user u involving in certain task t, we can calculate a modified TF-IDF score for each $n\text{-}gram$ in the snippets and titles read by the user, the IDFs of terms are calculated on the titles and snippets of all the tasks' SERPs:

$$s_{n\text{-}gram} = \sum_{r \in ViewedBy\_u_t} (c(n\text{-}gram, r) \cdot w_r) \cdot IDF_{n\text{-}gram} \tag{4}$$

where w_r denotes the weight of documents. We can estimate w_r with the user's clicks or eye-fixations. For the score based on user clicks, $w_r = \#clicks\_on\_r$. For the score based on fixations, $w_r = \sum_{f \in fixations\_on\_r} \log(duration_f)$.

Thus, we can construct the user's individual search outcome *indiv_so* by joining the top-k $n\text{-}grams$ with greatest scores in different lengths. Note that all the $n\text{-}grams$ appearing in the task description are removed because we want to capture what extended information the user have learned from the search process. We could calculate $s'_{n\text{-}gram}$ for the $n\text{-}gram$ with group of users' clicks or eye-fixations in a similar way. The ideal search outcome could be organized by

Table 1. SSET with different outcome measurements

	Individual search outcome	Ideal search outcome
SSET1MM	Summarized by participants	Organized by assessors
SSET2MA	Summarized by participants	Generated based on Eye-fixation
SSET3AM	Generated based on Eye-fixation	Organized by assessors
SSET4AA	Generated based on Eye-fixation	Generated based on Eye-fixation

utilizing group of users' interactions, which is assumed to be a kind of "wisdom of crowds".

More fine-grained user behaviors (e.g. fixations on the words) are promising to help the generation of search outcome extraction. Due to the limit of experimental settings in our system, we would leave them to our future work.

Based on different ideal/individual search outcome extraction methods, we get several SSET models which are shown in Table 1. The performance of these models will be discussed in Sect. 5. In the experiments we find that the SSET models with eye-tracking data always outperform the models with clickthrough information, thus, we only report the performance of models with eye-tracking in the remainder of this paper.

4   Experiment Setups

We conducted an experiment to collect user behaviors and search outcomes for completing complex tasks. During the whole process, users' queries, eye fixation behaviors on Search Engine Result Pages (SERPs), clicks and mouse movements are collected.

Search Task. We selected 12 informational search tasks for the experiment. 9 of them were picked out from recent years' TREC Session Track topics. According to the TREC Session Track style, we organized 3 tasks based on the participants' culture background and the environment they live in. The criteria is that the tasks should be clearly stated and the solutions of them cannot be retrieved simply by submitting one query and clicking the top results. Each task contains three parts: an information need description, an initial query and a question for search outcome extraction. The description briefly explains the background and the information need. To compare user behavior on query level, the first query in each task was fixed. We summarized the information needs and extracted key words as relatively broad queries. People may argue that the fixed initial queries might be useless for searcher. Statistics shows that there are average 2.33 results clicked on the SERPs of initial queries. At the end of the task, the question is showed to the participants which requires them summarize the information gained in the searching process and the answers were recorded by voice.

Experimental System. We built an experimental search system to provide modified search results from a famous commercial search engine in China.

First, all ads and sponsors' links were removed. Second, we removed vertical results to reduce possible behavior biases during searching process [30]. Third, we remove all the query suggestions because we suppose that query reformulation might reflect potential interests of users. Besides these changes, the search system looks like a traditional commercial search engine. The users could issue a query, click results, switch to the landing pages and modify their queries in a usual way. All the interactions were logged by our background database, including clicks, mouse movement and eye-tracking data.

Eye-tracking. In the experiment, we recorded all participants' eye movements with a Tobii X2-30 eye-tracker. With this tracking device, we are able to record various ocular behaviors: fixations, saccades and scan paths. We focus on eye fixations since fixations and durations indicate the users' attention and reading behavior [25].

Participants. We recruited 29 undergraduate students (15 females and 14 males) from a University located in China via email, online forums, and social networks. 17 of 29 participants have perfect eyesight. For the others, we calibrated the eye-tracker carefully to make sure the tracking error was acceptable. All the participants were aged between 18 and 26. Ten students are major in human sciences, fifteen are major in engineering while the others' majors range from arts and science. All of them reported that they are familiar with basic usage of search engines.

Procedure. The experiment proceeded in following steps, as shown in Fig. 3. First the participants were instructed to read the task description carefully and they were asked to retell the information need to make sure that they had understood the purpose of the search tasks. Then the participants could perform searches in our experimental system as if they were using an ordinary search engine. We did not limit their time of searching. The participants could finish searching when they felt satisfied or desperate. After searching for information, the participants were asked to judge and rate queries/results regarding their contribution to the search task. More specifically, they were instructed to make the following three kinds of judgments in a 5-points Likert scale, from strong disagreement to strong agreement:

- For each **clicked result**, how useful it is to solve the task?
- For each **query**, how useful it is to solve the task?
- Through the search **session**, how satisfied did the participant feel?

At last, the system would present a question about the description, which usually encourage the participant to summarize their searches and extract search outcome. The answers from users would be recorded by voice. We notice that answering the question by voice-recording could not only reduce participants' effort but also give them a hint that they should be more serious about the search tasks.

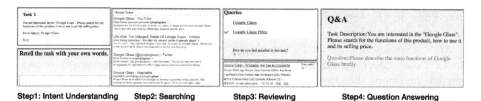

Step1: Intent Understanding   Step2: Searching   Step3: Reviewing   Step4: Question Answering

Fig. 3. Experimental procedure (Translated from original Chinese system)

5 Experimental Results and Disscussions

This section will lead to answers to 3 research questions:

RQ1: How well do the result of SSET correlate with human assessments? Can we use it as an understudy of human assessments?

RQ2: What's the relationship between SSET, and offline/online metrics?

RQ3: Does automatic methods work for SSET? Can we extract the ideal/individual search outcomes based on a single user's or group of users' behavior automatically?

5.1 Data and Assessments

In our experiment, we collected search behavior and success behavior from 29 participants on 12 unique tasks. To evaluate search success, we recruited 3 annotators to assess the degree of success based on the users' answer after each task. The assessors were instructed to make judgments with magnitude estimation (ME) [29] methods, rather than ordinal Likert scale. ME could be more precise than traditional multi-level categorical judgments and ME results were less influenced by ordering effects than multi-points scale [6]. For each task, before assessments, the assessors were represented with the task description and the question. The records of 29 participants are randomly listed on a webpage, each assessor make judgments sequentially. For each record, the assessor can listen to the record one or more times and then assign a score between 0 and 100 to the record in such a way that the score represents how successful the record is. The score was normalized according to McGee et al.'s method [22]. In this paper, we use the mean of normalized scores from three assessors as the Ground Truth of search success evaluation.

While assessing the participants' answers, we find that the question of Task 10 ("Please tell the support conditions of the Hong Kong version iphone to domestic network operators.") fails to help the participants to summarize their search outcome depending on the task description ("You want to buy a iphone6 in Hong Kong. Please find the domestic and Hong Kong price of iphone6, how to purchase iphone in Hong Kong, whether it is necessary to pay customs on bringing iphone home, whether the Hong Kong version of iphone would support domestic network operators, etc."), because the question just focuses on a detailed fact about the task. Thus, in the reminder analysis of this paper, Task 10 and corresponding data is removed and we have 319 sessions (11 tasks with 29 participants) in total.

After assessments, we asked the assessors to organized *standard answers* for the 12 tasks. More specifically, the three assessors were instructed to search information about the tasks with the retrieval system that were used by the participants. Note that the assessors did not perform any search before assessments for individual search outcome to avoid potential biases, e.g., they may prefer the individual outcomes similar to the documents examined by them. Then, the assessors organized their own answers and summarized the *ideal search outcomes* based on both their own answers and the 29 participants'. In addition, all the recorded voices were converted to text, with discourse markers removed, which were regarded as users' individual search outcomes.

5.2   SSET vs. Human Assessment

With our proposed evaluation framework, Search Success Evaluation with Translation model (SSET), we attempt to evaluate what degree of success a searcher has achieved in a certain task. The normalized scores for three assessors are regarded as *Ground Truth* of the performance evaluation of search success evaluation model.

For each session (a user in a certain task), the input of SSET includes a *"reference"*, the ideal search outcome, and a *"translation"*, the individual search outcomes from each participants and the SSET outputs the degree of success in a value range.

We calculate the correlation of SSET1MM model and the Ground Truth. The SSET1MM model uses the *ideal search outcomes* organized by external assessors as *"references"* and use the *answers of questions* (individual search outcomes) as *"translations"*. The correlation on each task is shown in Table 2.

The results show that SSET1MM correlates with the human judgments on most of tasks. The Pearson's r is significant at 0.01 for 10 of 11 tasks, which makes this method as an automated understudy for search success evaluation when there is need for quick or frequent evaluations.

We notice that the performance of SSET1MM varies with the tasks. It may suggest that the SSET is task-sensitive, in other words, SSET1MM is not appropriate for all kinds of tasks. From the facet of search goal identified by Li et al. [20], we can classify the tasks into 2 categories: *specific* (well-defined and fully developed) and *amorphous* (ill-defined or unclear goals that may evolve along with the user's exploration). Thus, we find SSET performs better on the specific task (Task 5, 9, 4, 3, 6, 8, 7) rather than on the amorphous tasks (Task 2, 1, 12, 11, 7). For an amorphous task (e.g. find a ice breaker game), it is very difficult to construct a "perfect" search outcome including all possible answers. Therefore, SSET is more appropriate to evaluate the tasks which are well-defined and have restrained answers.

Table 2. Correlation between SSET1MM and the Ground Truth (*, **: correlation significant at 0.01, 0.001 level)

Tasks	Correlation with Ground Truth	
	Pearson's r	Kentall's τ-b
5	0.879**	0.363*
9	0.822**	0.600**
4	0.789**	0.670**
3	0.774**	0.551**
6	0.719**	0.524**
2	0.706**	0.378*
1	0.631**	0.295
12	0.630**	0.533**
8	0.629**	0.546**
11	0.552*	0.406*
7	0.537*	0.315

5.3   SSET vs. Offline/Online Metrics

SSET attempt to combine the advantages of offline and online evaluation methods. The tasks and ideal search outcomes organized by human experts offline can be reused easily and the individuals' search outcomes can be collected online efficiently and effectively. In this section, we investigate the relationship between SSET and offline/online metrics.

Previous work [13] reported that session cumulated gain (sCG) [12] correlated well with user satisfaction. We use sCG as a offline measure of the search outcome, which is the sum of each query's information gain. For each query, its gain is originally calculated by summing the gains across its results. In this work, we use the participants' subjective annotation ("how useful it is to solve the task?") as a proxy of the query's gain, e.g. $SearchOutcome = sCG = \sum_{i=1}^{n} gain(q_i)$.

The correlation between SSET1MM and sCG are shown in Table 3. There is weak correlation between SSET1MM and sCG. It is partly due to the difference in cognitive abilities between users. Consider the example in Sect. 1, two users search for "the side effects of red bulls", they issued similar queries, viewed similar SERPs and got quite close sCG scores. However, the information they gained for completing the task differed at quality and quantity. In other words, it means that the offline metric may lead to failure to evaluate in what degree the user has achieved success in complex tasks.

User satisfaction is a session/task level online evaluation metrics. In our experiment, we asked the users to rate their satisfaction for each task. The correlation of SSET1MM and user satisfaction is shown in Table 3.

Experiment shows less of a relationship between SSET1MM and user satisfaction. Jiang et al. [13] reported that the satisfaction was mainly affected by

Table 3. Correlation comparison between SSET and Offline/Online metrics (*, **: correlation significant at 0.01, 0.001 level)

Tasks	Correlation (Pearson's r)		
	SSET1MM vs. SAT	SSET1MM/#Queries vs. SAT	SSET1MM vs. sCG
7	−0.144	0.762**	0.186
2	0.027	0.574*	0.218
4	0.131	0.574*	0.208
6	0.232	0.568*	0.159
5	−0.001	0.554*	0.252
1	0.125	0.536*	0.285
3	−0.212	0.527*	0.140
8	0.257	0.432	0.196
9	−0.037	0.329	0.127
11	0.087	0.252	0.063
12	0.094	0.227	0.080

two factors, search outcome and effort. However, the search success evaluation mainly focuses on the search outcome of users. No matter the degree of success is assessed by external assessors or the SSET systems, they are not aware of the effort that the user has made to achieve the search outcome. This could be a plausible explanation for the closeness to uncorrelated between SSET and satisfaction. Jiang et al. proposed an assumption that the satisfaction is the value of search outcome compared with search effort. As our proposed SSET is also a measurement of search outcome, we investigate the correlation between SSET/Search Effort.

Search effort is the cost of collecting information with the search engine, e.g., formulating queries, examining snippets on SERPs, reading results, etc. We follow the economic model of search interaction proposed in [2]. For a particular search session, we can use Q (number of queries) as a proxy of search effort. Table 4 shows that there is strong correlation between SSET1MM/#queries and user's satisfaction for most of the tasks and our proposed SSET is able to act as an indicator of search outcome.

5.4   Performance of Automated Outcome Extraction

Development of search engine is based on ongoing updates. In order to validate the effect of a change to prevent its negative consequences, the developers compare various versions of the search engines frequently. This motivate us to improve SSET with automated methods for the organization of ideal/individual search outcomes.

In Table 4, we compared the correlation between 4 different SSET models and the Ground Truth (external assessments). SSET1MM is the model which

Table 4. Correlation (Pearson's r) comparison between different SSET models and the Ground Truth (*, **: correlation significant at 0.01, 0.001 level)

Tasks	Correlation (Pearson's r)			
	SSET1MM	SSET2MA	SSET3AM	SSET4AA
5	0.879**	0.907**	−0.063	−0.263
9	0.822**	0.808**	−0.193	−0.131
4	0.789**	0.724**	−0.243	−0.108
3	0.774**	0.769**	−0.107	−0.143
6	0.719**	0.625**	−0.165	−0.006
2	0.706**	0.691**	0.143	−0.222
1	0.631**	0.685**	−0.779	−0.032
12	0.630**	0.412	−0.035	0.313
8	0.629**	0.652**	−0.080	0.354
11	0.552*	0.385	0.144	0.268
7	0.537*	0.565*	0.132	0.146

use manually organized as ideal search outcome and users' answer for questions as individual search outcome. We use SSET1MM as a baseline to evaluate other SSET models.

SSET2MA performs almost as well as SSET1MM. It uses the same way to collect individual search outcomes (e.g. summarized by users) but constructs the ideal search outcomes automatically based on users' eye fixations on snippets. Thus, in practical environment, we can generate ideal search outcome based on group of users' behavior.

SSET3AM and SSET4AA correlates poorly with the Ground Truth. In these two models, we adopt the individual search outcome extraction method based on the user's eye fixations on SERPs. The individual search outcomes generated automatically differs a lot from their answers. The potential two reasons are: (1) the sparsity of user behavior makes it difficult to extract search outcome. (2) what the user has read is not equal to what he/she has perceived. Similar phenomenon has been observed by previous researches [21].

We also investigate the performance of SSET2MA based on different size of users' behaviors. We randomly split the all the participants into five groups, four groups has six participants while the remaining one has five. Then we construct multiple ideal search outcomes by sequentially adding group of users' fixations into the SSET2MA model. Then we compare the correlations between SSET2MA models and the Ground Truth.

The results are shown in Table 5, where $SSET2^k$ denotes the SSET2MA model based on the first k groups of users. As the size of users grows, the correlation between SSET2MA and the Ground Truth becomes stronger. The $SSET2MA^3$ almost performs as well as $SSET3AM^5$. In other words, in practice,

Table 5. Correlation (Pearson's r) comparison between SSET2MA models based on different size of users' behaviors and the Ground Truth (*, **: correlation significant at 0.01, 0.001 level)

Tasks	Correlation (Pearson's r) with the Ground Truth				
	SSET2^1	SSET2^2	SSET2^3	SSET2^4	SSET2^5
5	0.620**	0.792**	0.901**	0.907**	0.907**
9	0.508*	0.738**	0.800**	0.807**	0.808**
3	0.439	0.696**	0.769**	0.760**	0.769**
4	0.411	0.589*	0.701**	0.714**	0.724**
2	0.382	0.541*	0.660**	0.687**	0.691**
1	0.356	0.495*	0.629**	0.657**	0.685**
8	0.347	0.477	0.652**	0.654**	0.652**
6	0.298	0.433	0.589*	0.626**	0.625**
7	0.287	0.365	0.501*	0.561**	0.565*
12	0.101	0.276	0.327	0.414	0.412
11	0.220	0.287	0.342	0.383	0.385

we need about behavior data from about 15 people to construct a reliable ideal search outcome for SSET.

6   Conclusion and Futurework

Although previous offline/online evaluation frameworks have achieved significant success in the development of search engines, they are not necessarily effective in evaluate in what degree of success the search users have achieved. In this work, we put emphasis on the outcomes the users gained through multiple queries. We propose a Search Success Evaluation framework with Translation model (SSET). The search success evaluation is formalized as a machine translation evaluation problem. A MT evaluation algorithm called BLEU is adopted to evaluate the success of searchers. Experiments shows that evaluation methods based on our proposed framework correlates highly with human assessments for complex search tasks. We also propose a method for automatic generation of ideal search outcomes with the help of multiple users' search interaction behaviors. It proves effective compared with manually constructed ideal search outcomes. Our work can help to evaluate search success as an understudy of human assessments when there is need for quick or frequent evaluation. In the future work, we plan to adopt more MT evaluation methods in this framework and compare the performance in evaluate different types of tasks. Experiments with a relatively large scale of participants will be conducted based on crowdsourcing platforms.

References

1. Ageev, M., Guo, Q., Lagun, D., Agichtein, E.: Find it if you can: a game for modeling different types of web search success using interaction data
2. Azzopardi, L.: Modelling interaction with economic models of search. In: SIGIR 2014 (2014)
3. Berger, A., Lafferty, J.: Information retrieval as statistical translation. In: SIGIR 1999 (1999)
4. Cleverdon, J.M.C.W., Keen, M.: Factors determiningthe performance of indexing systems. Readings in Information Retrieval (1966)
5. Doddington, G.: Automatic evaluation of machine translation quality using N-gram cooccurrence statistics. In: HLT 2002 (2002)
6. Eisenberg, M.B.: Measuring relevance judgments. IPM **24**, 373–390 (1988)
7. Gao, J., He, X., Nie, J.-Y.: Clickthrough-based translation models for web search: from word models to phrase models. In: CIKM 2010 (2010)
8. Gescheider, G.A.: Psychophysics: the Fundamentals. Psychology Press, New York (2013)
9. Haghighi, A., Vanderwende, L.: Exploring content models for multi-document summarization. In: NAACL 2009 (2009)
10. Hassan, A., Jones, R., Klinkner, K.L.: Beyond DCG: user behavior as a predictor of asuccessful search. In: WSDM 2010 (2010)
11. Huffman, S.B., Hochster, M.: How well does result relevance predict session satisfaction? In: SIGIR 2007 (2007)
12. Järvelin, K., Price, S.L., Delcambre, L.M.L., Nielsen, M.L.: Discounted cumulated gain based evaluation of multiple-query IR sessions. In: Macdonald, C., Ounis, I., Plachouras, V., Ruthven, I., White, R.W. (eds.) ECIR 2008. LNCS, vol. 4956, pp. 4–15. Springer, Heidelberg (2008). doi:10.1007/978-3-540-78646-7_4
13. Jiang, J., Awadallah, A.H., Shi, X., White, R.W.: Understanding and predicting-graded search satisfaction. In: WSDM 2015 (2015)
14. Joachims, T.: Optimizing search engines using clickthrough data. In: KDD 2002 (2002)
15. Jungwha, C.: The interpretive theory of translation and its current applications (2003)
16. Kanoulas, E., Carterette, B., Clough, P.D., Sanderson, M.: Evaluating multi-query sessions. In: SIGIR 2011 (2011)
17. Kohavi, R., Longbotham, R., Sommerfield, D., Henne, R.M.: Controlled experiments on the web: survey and practical guide. Data Min. Knowl. Disc. **18**, 140–181 (2009)
18. Lease, M., Allan, J., Croft, W.B.: Regression rank: learning to meet the opportunity of descriptive queries. In: Boughanem, M., Berrut, C., Mothe, J., Soule-Dupuy, C. (eds.) ECIR 2009. LNCS, vol. 5478, pp. 90–101. Springer, Heidelberg (2009). doi:10.1007/978-3-642-00958-7_11
19. Lederer, M.: La traduction simultanée: expérience et théorie, vol. 3. Lettres modernes, Paris (1981)
20. Li, Y., Belkin, N.J.: A faceted approach to conceptualizing tasks in information seeking. In: IPM (2008)
21. Liu, Y., Wang, C., Zhou, K., Nie, J., Zhang, M., Ma, S.: From skimming to reading: a two-stage examination model for web search. In: CIKM 2014 (2014)
22. M. McGee.: Usability magnitude estimation. In: HFES 2003 (2003)

23. Over, P.: Trec-7 interactive track report. NIST SPECIAL PUBLICATION SP 1999)
24. Papineni, K., Roukos, S., Ward, T., Zhu, W.-J.: Bleu: a method for automatic evaluation of machine translation. In: ACL 2002 (2002)
25. Rayner, K.: Eye movements in reading, information processing: 20 years of research. Psychol. Bull. **124**, 372–422 (1998)
26. Sanderson, M., Paramita, M.L., Clough, P., Kanoulas, E.: Do user preferences and evaluation measures line up? In: SIGIR 2010 (2010)
27. Seleskovitch, D.: L'interprète dans les conférences internationales: problèmes de langage et de communication, vol. 1. Lettres modernes (1968)
28. Snover, M., Dorr, B., Schwartz, R., Micciulla, L., Makhoul, J.: A study of translation edit rate with targeted human annotation. In: AMTA 2006 (2006)
29. Stevens, S.S.: The direct estimation of sensory magnitudes: loudness. Am. J. Psychol. **69**, 1–25 (1956)
30. Wang, C., Liu, Y., Zhang, M., Ma, S., Zheng, M., Qian, J., Zhang, K.: Incorporating vertical results into search click models. In: SIGIR 2013 (2013)
31. Yang, Y., Lad, A.: Modeling expected utility of multi-session information distillation. In: ICTIR 2009 (2009)

Evaluating the Social Acceptability of Voice Based Smartwatch Search

Christos Efthymiou[✉] and Martin Halvey

Department of Computer and Information Sciences,
University of Strathclyde, Glasgow, UK
chefthimiou@hotmail.com, martin.halvey@strath.ac.uk

Abstract. There has been a recent increase in the number of wearable (e.g. smartwatch, interactive glasses, etc.) devices available. Coupled with this there has been a surge in the number of searches that occur on mobile devices. Given these trends it is inevitable that search will become a part of wearable interaction. Given the form factor and display capabilities of wearables this will probably require a different type of search interaction to what is currently used in mobile search. This paper presents the results of a user study focusing on users' perceptions of the use of smartwatches for search. We pay particular attention to social acceptability of different search scenarios, focussing on input method, device form and information need. Our findings indicate that audience and location heavily influence whether people will perform a voice based search. The results will help search system developers to support search on smartwatches.

Keywords: Smartwatch · Acceptability · Voice · Search · Information need

1 Introduction

Search using mobile devices is becoming more popular and this is set to continue as more devices with increased computing power become available. In addition, the range of devices people are using to access the web are increasing e.g. tablets, phones, smartwatches etc. As the range of devices increases the methods that people use to interact with these devices are also changing. In particular for mobile search this has resulted in the development of both voice based search systems e.g. Cortana, Siri etc. and proactive card based search systems [1], as opposed to traditional reactive search with ranked lists. A lot of effort and research has gone into developing both the hardware for these devices and also the intelligent software that allows voice based search for example. Much less research has looked into social factors surrounding these devices and their use i.e. how users will feel about using them for search on the move etc. Social acceptability issues have in part contributed to some high profile technology failures e.g. Google Glass. We want to address this lack of acknowledgment of social acceptability issue for smartwatches and in particular for search using smartwatches. Thus in this paper we investigate the social acceptability of using a smartwatch to search. In particular we focus on reactive search (i.e. search initiated by a user) and in particular on querying using a smartwatch where input is predominantly voice based. Specifically we focus on the following research questions:

© Springer International Publishing AG 2016
S. Ma et al. (Eds.): AIRS 2016, LNCS 9994, pp. 267–278, 2016.
DOI: 10.1007/978-3-319-48051-0_20

RQ1: Does information need determine when and where an individual would use a smartwatch for search?

RQ2: Does the form factor of the device determine when and where an individual would use a smartwatch for search?

RQ3: Does the input method determine when and where an individual would use a smartwatch for search?

RQ4: Does the expression of information need determine when and where an individual would use a smartwatch for search?

To address these research questions a lab based user study following a methodology proposed by Rico et al. [2] was conducted; videos depicting various interaction scenarios (in our case users searching on mobile devices i.e. phone and smartwatch) were presented and participant responses were elicited. The aim was to gather user perceptions of the social acceptability of the specific scenarios presented.

2 Related Work

2.1 Mobile Search Behaviour

Jones et al. [3] evaluated users' abilities on mobile phones, PDAs, and desktop interfaces for some of the earliest mobile search systems. They found that both the search speed and accuracy were worse on smaller screens. Church et al. [4] analysed almost 6 million individual search requests produced by over 260,000 individual mobile searchers over a 7 days period in 2006. At that time mobile search was only used frequently used by 8–10 % of mobile internet users. Church et al. also noted that users had a limited set of information needs, with a high number of transactional and navigational queries ([5]) and a high number of adult/pornographic queries. A number of researchers have looked at social setting and how that influences mobile information needs. According to Kassab and Yuan [10] mobile users are motivated to use mobile phones to seek information motivated from conversations with other people, to view their emails and download mobile applications. Church et al. [6] found that the majority of mobile users, use mobile search to seek information about trivia and pop culture things. Furthermore, they found that the social mobile search is more likely to take place in an unfamiliar locations. As a result Church et al. [7] developed the Social Search Browser (SSB), an interface that embodies social networking abilities with important mobile contexts. Location also plays an important part in social context. Ren et al. [8] investigated how mobile users use the web in large indoor spaces, specifically retail malls, to look for information. Church and Cramer [9] have recently looked at the requirements of place in local search in purpose to improve the location-based search in the future. Other researchers have looked at how search is changing as a result of device changes. Montanez et al. [10] have looked at the range of devices that people use to satisfy information needs, when particular devices are used and when people switch between these devices. As devices are evolving mobile search is beginning to move from being reactive into being more proactive. With this in mind some researchers have begun to look at card based retrieval and in particular how they are influenced by social situations [1]. Most wearable search systems operate a combination of reactive and proactive card based retrieval systems. This does not eliminate the need for reactive search and some sort of input e.g. systems like Siri and Cortana encourage voice input and a dialogue with the system.

2.2 Social Acceptability

Rico and Brewster [2] investigated the social acceptability of a set of gestures for interacting with mobile devices with respect to location and the audience. Within the area of wearables, Schaar and Ziefle [11] evaluated the acceptance of smart shirts in men and women and found that men were more accepting of the technology than women. Shinohara and Wobbrock [12] looked at the perception of a variety of users of assistive technology. Whilst this technology was extremely useful, some of participants felt stigmatised by their assistive technology. A study looking at the privacy behaviour of life loggers [13] found that for life loggers a combination of factors including time, location, and the objects and people appearing in a photo determines its 'sensitivity;' and that life loggers are concerned about the privacy of bystanders. Specifically for smartwatches, Pearson et al. [14] investigated the interaction and the process of attracting attention when glancing at someone else's watch. Bystanders not only could notice the information and content on someone else's smartwatch screen, but they were likely to interact with the wearer and tell them about information presented on their smartwatch screen. Moorthy and Vu [15] conducted an evaluation of the use of a Voice Activated Personal Assistant in the public space. Users were quite careful to reveal their private information in front of strangers and less careful with their non-private information. This behaviour difference was heightened in public locations and especially when obvious methods were used for information input.

3 User Evaluation

3.1 Information Need Survey

For our user evaluation we required a set of real mobile information needs, these were gathered through an online survey. For each information need we asked participants for a clear and precise single sentence that describes the information need (e.g. "What year Winston Churchill born?"). We then asked for a description of what constituted relevant information for this information need. We then asked participants to provide queries or keywords that they used to satisfy this information need. This follows the method used by Wakeling et al. for gathering information needs for their study of relevance of search results [16]. Finally, we asked what mobile device had been used to satisfy their information need (e.g. phone, tablet etc.) providing the make and the model of the mobile device. The survey had 83 respondents. 44 (53.7 %) were male, 38 (46.3 %) were female and 1 respondent declined to provide this information. Respondent's ages varied between 21–66 years old, the average age was 30.61. With respect to mobile search 11 (13.3 %) participants indicated they had low experience, 28 (33.7 %) had a medium level of experience and 44 (53 %) indicated that they have high experience. The mobile information needs provided were analysed. Similar to the information needs analysis by Church et al. [5] there were a large number of informational information needs (79.4 %), some transactional (17.8 %) and very little navigational (2.73 %). Initially the categories from Church et al. [5] were used for categorisation, but they did not appear to capture the information needs sufficiently well, so instead bottom up categorisation of information needs was created. One key difference between our work and that of Church et al. [5] is Church et al. used log analysis and our participants may be reluctant to outline some of their information needs e.g. Adult. This resulted in 11 categories, which were slightly different from those of Church et al., e.g. News and Weather featured in our categories, but Adult did not. Also News and Weather whilst common were not amongst the most popular categories. The three most popular categories were Directions, Entertainment and General Search (used interchangeably with Search with a capital S in parts of the paper), we use these in our user study. For reasons of space and as it is not the focus of the paper we do not go into detail about the categorisation. Our aim was to have representative information needs which could be used in our user study.

3.2   Procedure

We followed the procedure of Rico and Brewster [2], where participants watched a set of videos depicting interaction scenarios and were then asked to provide responses regarding the scenarios depicted in the video. A user evaluation in real locations was also considered, however that may require placing participants in potentially embarrassing and uncomfortable positions. In addition it would not allow us to consider as many locations and audiences as we can with this in lab study. In total there were 18 videos. Each video had an information depicted a person searching for information, each search had an information need (General Search, Directions, Entertainment) a query input (consisting of a device (Phone, Watch) and input method (Text, Voice) pair) and an expression type (Statement, Keyword). Information needs were based on the real information needs gathered in the online survey (see Sect. 3.1). Device allowed us to compare smartwatch and phone; mobile phones are the most common way for people to currently search whilst mobile. Input method allows comparison between the more common text entry and voice input. Three pairs were used, text entry on a smartwatch was omitted. Text entry on such a small screen is impractical and remains an area of open research [17]. Finally expression of the query could be a statement or in the form of keywords. For many voice based systems querying with statements is

Fig. 1. Screenshots from example videos.

Table 1. Example queries and conditions, not all are shown for reasons of space.

Query	Info need	Device	Input method	Expression
'What fireworks displays are being offered tonight?'	Entertainment	Phone	Voice	Statement
'I want to find data analytics articles'	Search	Phone	Text	Statement
'Glasgow Edinburgh bus times'	Directions	Watch	Voice	Keywords

common e.g. to view your steps on an Android device the statement is "okay Google, show me my steps". Whereas for text based search keywords are more prevalent. Keywords were the keywords provided by respondents and the statements were the single sentence that we requested. Before beginning participants were told the aim of the study was to assess the social acceptability of different mobile searches.

For each search scenario participants watched a video of the search being performed and answered multiple-choice questions. The videos lasted between four and ten seconds each and participants answered the associated questions after each video, they could request a video to be replayed. Each video portrayed a search being performed by a male actor sitting in front of a plain background. Where voice was used as input audio was provided, when text was an input the query ran along the bottom of the screen. The devices used in the video were a Nexus 5 phone and a Samsung Gear Live smartwatch. Figure 1 shows frames from two of the videos. Because participants were asked to imagine the locations and audiences where they might perform these searches the videos were designed to focus solely on the search scenario itself. The videos used in this study intentionally portrayed a plain scene without a defined context so that the setting would not distract viewers from evaluating the search input. After watching each video, participants were asked to select from a list of the locations where they would be willing to perform the given search. Users were then asked to select from a list of the types of audiences they would be willing to perform the search in front of. These audiences and locations are based on work by Rico and Brewster [2]. There were six locations (Driving, Home, Passenger, Pavement, Pub/Restaurant, Workplace) and six audiences (Alone, Colleagues, Family, Friends, Partner, Strangers). These responses intentionally asking participants to imagine themselves in these settings in a first person rather than second person view [18] in order to focus on one's personal actions rather than opinions of other's. The order of video presentation was randomised. Overall this study had a total of 4 independent variables; information need, device type, input type and expression type. Example videos and combinations of variables are presented in Table 1. For each of these independent variables, 2 dependent variables (audience acceptance and location acceptance) were analysed. As the data gathered was found to not be normally distributed, non-parametric statistical tests were used, those being Friedman Tests and Wilcoxon Sign Rank Tests. For each variable presented below we first look at the impact of the variable itself (e.g. information need), we then compare each location and audience between variables (e.g. Entertainment in front of Friends vs. Directions in front of Friends) and finally we look at the location and audience for each instance of each variable (e.g. compare all audiences for directions information needs) following Rico and Brewster [2].

3.3   Participants

There were 20 participants in our user study with age range from 23 to 32 and average age 25.15. Participants were mostly recruited from the University of Strathclyde. All of the participants lived in the UK and spoke English, but were from a range of countries. 5 (25 %) of the participants were female and 15 (75 %) were male.

4   Results

4.1   Information Need

A Friedman Test revealed significance for information needs have significant effect on location $(X^2(2) = 27.966, p < 0.001)$ (see Table 2). Pairwise comparisons with a Wilcoxon Sign Rank test (Bonferroni adjusted alpha $p = 0.0167$) revealed significant differences between Directions and Entertainment $(z = -3.495, p < 0.001)$ as well as Directions and Search $(z = -5.215, p < 0.001)$. With respect to audience (see Table 3) there was also a significant difference for information need $(X^2(2) = 78.072, p < 0.001)$. Pairwise comparisons revealed a significant difference between all combinations, Search-Entertainment $(z = -3.162, p = 0.002)$, Search-Directions $(z = -8.255, p < 0.001)$, and Entertainment-Directions $(z = -5.914, p < 0.001)$. Pairwise comparisons were made for each location and audience (Bonferroni adjusted alpha $p = 0.0014$). For location there were significant differences between Searching and Directions for Driving $(p = 0.001)$, Pavement $(p = 0.001)$ and Passenger $(p = 0.001)$. As can been seen in Table 2 these are locations where participants reported that they would be likely to look for Directions but not conduct a General Search. There were a higher number of significant differences in terms of audience. As with location the highest number were between Searching and Directions where there were significant differences for Family $(p = 0.001)$, Partner $(p = 0.001)$, Friends $(p = 0.001)$ and Strangers $(p = 0.001)$. Between Searching and Entertainment there were differences in audience for Partner $(p = 0.001)$ and Friends $(p = 0.001)$. Finally for Entertainment and Directions there was a difference for Strangers $(p = 0.001)$. Overall Directions was the most acceptable, followed by Entertainment and General Search being the least acceptable for all audiences.

Table 2. Average acceptance rates for different information needs by location.

	Home	Driv.	Pub	Pave.	Pass.	Work
Enter.	88 %	31 %	32 %	53 %	40 %	22 %
Search	95 %	18 %	16 %	36 %	38 %	41 %
Direc.	94 %	38 %	28 %	63 %	58 %	33 %

Table 3. Average acceptance rates for different information needs by audience.

	Alone	Fam.	Coll.	Part.	Frie.	Stran.
Enter.	93 %	73 %	23 %	81 %	88 %	19 %
Search	98 %	53 %	38 %	62 %	68 %	18 %
Direc.	99 %	84 %	38 %	90 %	93 %	42 %

Looking at each information need individually. For Entertainment, Home was the most acceptable search location (see Table 2) and is significantly more acceptable than any other location (p < 0.001 for all comparisons). Pavement is a more acceptable location than Driving (p = 0.001), Pub (p = 0.001) or the Workplace (p < 0.001). Workplace is also less acceptable than being a Passenger. With respect to Audience, searching for Entertainment related information is least acceptable in front of Strangers, which is significantly different to Alone (p < 0.001), Family (p < 0.001), Partner (p < 0.001) or Friends (p < 0.001). Searching Alone is also significantly different to in front of Colleagues (p < 0.001) or Family (p < 0.001), Alone is seen as the most acceptable audience (see Table 3). Colleagues as an audience is also significantly less acceptable than in front of Family (p < 0.001), Partners (p < 0.001) or Friends (p < 0.001). Family and Friends also has a significant difference (p < 0.001).

In terms of the Searching category there were significant differences between the acceptability of almost all locations. There was no significant difference for Workplace with Pavement (p = 0.446) and Passenger (p = 0.579). For Passenger there was no significant difference with Pavement (p = 0.670). All of these locations were mid-range in terms of acceptability compared to other locations (see Table 2). Pub and Driving were not significantly different (p = 0.732), both were seen as the least acceptable (see Table 2). Similarly for audience with respect to search most audiences had significantly different acceptance rates. Family had similar acceptance to Colleagues (p = 0.024) and Partner (p = 0.068). With Partner also having a similar acceptance rate to Friends (p = 0.131).

With respect to Directions, the majority of locations had significantly different acceptance rates. Workplace had a similar acceptance level to Driving (p = 0.327) and Pub (p = 0.355). Pub and Driving also had similar acceptance rates (p = 0.085). All had low acceptance rates in comparison to other locations. Passenger and Pavement also had non-significant differences in terms of acceptance rate (p = 0.257). In terms of audience, there were less significant differences for Directions, with relatively high acceptance rates. That being said the most acceptable audiences Alone was more acceptable than Family (p < 0.001), Colleagues (p < 0.001), Partner (p = 0.002) and Strangers (p < 0.001). Strangers the least acceptable audience was also significantly less acceptable than Family (p < 0.001), Partner (p < 0.001) and Friends (p < 0.001). Colleagues which was the second least acceptable audience and was significantly less acceptable than Partner (p < 0.001), Friends (p < 0.001) and Family (p < 0.001).

4.2 Device

The Device variable had a significant effect on location (z = −2.538, p = 0.011) and on audience (z = −2.121, p = 0.034), with the use of Phone being more acceptable on almost all cases, see Tables 4 and 5. Individual pair–wise comparisons between Phone and Watch for every location and audience were also conducted using a Wilcoxon Sign Rank test (Bonferroni adjusted alpha p = 0.004) and the only significant difference was for the Driving location (z = 4.808, p < 0.001), with using a Watch whilst Driving being more acceptable than a Phone. Although not significant the use of Watch rather than Phone for searching was more acceptable in front of Family (z = 2.252, p = 0.024). We also performed pairwise comparisons individually for both the Phone and Watch variables between every location and audience (Bonferroni adjusted alpha p = 0.003). For reasons of space we do not present all of the results but rather summarise them here. Looking at Phone first, there were significant differences between all locations with the exception of Driving-Pub (p = 0.088), Pavement-Passenger (p = 1.0) and Workplace-Pub (p = 0.083). In terms of audience, again for Phone almost all audiences are significantly different, the exceptions being Friends-Partner (p = 0.02) and Colleagues-Strangers (p = 0.107). Next looking at pairwise comparisons within the Watch variable we see that for location that Driving is as acceptable as on Pavement (p = 0.317) or as a Passenger (p = 0.121) and that Workplace is as acceptable as Pub (p = 0.303) or Passenger (p = 0.063). Searching in front of Friends (p = 0.059) and Partners (p = 0.819) is as acceptable as Family, with Friends and Partners also being similarly acceptable (p = 0.162). With searching in front of Strangers and Colleagues also being similarly acceptable (p = 0.04).

Table 4. Average acceptance rates for phone versus watch by location.

	Home	Driv.	Pub	Pave.	Pass.	Work
Phone	96 %	21 %	28 %	51 %	51 %	36 %
Watch	85 %	44 %	18 %	50 %	34 %	23 %

Table 5. Average acceptance rates for phone versus watch by audience.

	Alone	Fam.	Coll.	Part.	Frie.	Stran.
Phone	97 %	68 %	36 %	79 %	84 %	30 %
Watch	95 %	74 %	28 %	75 %	81 %	18 %

4.3   Input Method

In terms of Input Method there was no significant difference between Text and Voice for location (z = −1.472, p = 0.141), but there was for audience (z = −2.466, p = 0.014). Whilst overall location did not have an impact we see that for pairwise comparisons based on locations that there were significant differences (see Tables 6 and 8). With Text being more acceptable at Home, as a Passenger and in Work; whilst Voice input is deemed more acceptable whilst Driving. In terms of audience, using Text input is significantly more acceptable in front of Strangers and Colleagues (see Tables 7 and 8). Looking at pairwise comparisons for Text input per location (Bonferroni adjusted alpha p = 0.003), we found that Workplace is not significantly different to Pub (p = 0.01), Pavement (p = 0.325) and Passenger (p = 0.013). With Passenger and Pavement also not differing significantly (p = 0.048). With respect to audience we see no significant different between Partner and Friends (p = 0.007) and Family (p = 0.016), as well as no difference between Strangers and Colleagues (p = 0.752). Looking at Voice input, between locations there are no significant differences for Driving with Pavement (p = 0.03) and Passenger (0.514), and also between Pub and Workplace (p = 0.659). As for audience, Partner has no significant difference with Family (p = 0.024) and Friends (p = 0.101), and there is no significant differences between Strangers and Colleagues (p = 0.003).

Table 6. Average acceptance rates for voice versus text input by location.

	Home	Driv.	Pub	Pave.	Pass.	Work
Text	98 %	12 %	34 %	58 %	67 %	51 %
Voice	90 %	38 %	20 %	47 %	35 %	22 %

Table 7. Average acceptance rates for coice versus text input by audience.

	Alone	Fam.	Coll.	Part.	Frie.	Stran.
Text	98 %	68 %	48 %	77 %	84 %	47 %
Voice	96 %	71 %	26 %	78 %	83 %	16 %

Table 8. Pairwise comparison between text and voice for every location and audience. Significance value set using a Bonferroni correction at p = 0.004.

Location	p value	Audience	p value
Home	<0.001	Alone	0.102
Driving	<0.001	Family	0.302
Pub	0.123	Colleagues	<0.001
Pavement	0.101	Partner	0.435
Passenger	<0.001	Friends	0.262
Workplace	<0.001	Strangers	<0.001

4.4   Expression

For Expression of information need there was no significant effect on location ($z = -1.362$, $p = 0.173$), however, there was a significant difference for audience ($z = -3.500$, $p = 0.001$). Pairwise comparisons between all audiences revealed that the only significant difference between Statements and Keywords was for Partner ($z = -2.959$, $p = 0.003$). Looking at Keywords and Statements separately. For location and Statement no significant difference was found between Workplace and Driving ($p = 0.655$), Pub ($p = 0.485$) and Passenger ($p = 0.021$); Passenger also had no significant difference with Driving ($p = 0.081$) and Pavement ($p = 0.02$); Pub and Driving ($p = 0.258$) also had no significant difference. Audience seems to have more of an impact, with Partner having no significant difference to Family ($p = 0.006$) and Friends ($p = 0.016$); and no significant difference between Strangers and Colleagues ($p = 0.033$). Looking at location for Keywords we found a similar pattern to with Workplace not being significantly different to Driving ($p = 0.127$) or Pub ($p = 0.029$); Passenger and Pavement are not significantly different ($p = 0.763$); Driving and Pub are not significantly different ($p = 0.642$). In terms of audience, for Keywords, there is no significant difference for Partner with Family ($p = 0.071$) and Friends ($p = 0.201$); as well as no significant difference between Strangers and Colleagues ($p = 0.170$) (Tables 9 and 10).

Table 9. Average acceptance rates for statements versus keywords by location.

	Home	Driv.	Pub	Pave.	Pass.	Work
State.	91 %	31 %	26 %	50 %	41 %	29 %
Key.	93 %	27 %	24 %	51 %	50 %	34 %

Table 10. Average acceptance rates for statements versus keywords by audience.

	Alone	Fam.	Coll.	Part.	Frie.	Stran.
State.	96 %	64 %	34 %	73 %	81 %	25 %
Key.	97 %	76 %	33 %	82 %	86 %	28 %

5 Discussion

Information Need (RQ1): Analysis of the information needs provided in our online survey revealed that as Church found [5] there are a large number of informational mobile queries. The three most common categories from our analysis of mobile information needs were used for comparison in our in lab user study; those being Entertainment, Directions and General Search. There was a significant difference in terms of acceptability between all of the information needs. Directions, Entertainment and General Search were all viewed as being most to least acceptable in order. These information needs can be viewed as having a temporal aspect (Directions having an immediate temporal aspect). Also Directions and then Entertainment could be viewed as being easily displayed on the small screen of a smartwatch. In terms of audience questionnaire responses indicated that people are comfortable searching Alone. When they search in front of others they would rather do so in front of familiar audiences like Friends or Family, with searching for any information need in front of Strangers seen as being unacceptable. Some of the reasons for this are highlighted in the post evaluation interview, with issues of appearing strange and also privacy being highlighted as major concerns. With respect to location of performing a search again there are major differences, locations where a person might be alone or not actively engaged in other activities (i.e. Passenger and Pavement) are most acceptable for all information needs. The Workplace location has a different distribution to other locations, in that it is the only location where General Search is seen as being most acceptable in comparison to other locations.

Form Factor (RQ2): In our study device was seen to have a significant impact on the acceptability with respect to both audience and location. In general when directly comparing locations and audiences Phone was more acceptable than Watch, the exception being that search via smartwatch was significantly more acceptable when Driving. The trend and comparison when looking at pairwise comparisons for both Phone and Watch had similar distributions of acceptance. As a smartwatch is a relatively new technology this difference in acceptability may even out as smartwatches become more mainstream.

Input Method (RQ3): In our study location did not have a significant effect on acceptability. However pairwise comparisons revealed a difference between Home, Driving, Passenger and Workplace; with Text being more acceptable in all those locations except Driving. There was significant effect of audience in terms of acceptability. For Colleagues and Strangers Text was significantly more acceptable than Voice. One interesting thing to note was that for Partner, Family and Friend audience's Voice was slightly more acceptable than Text. The audiences that are familiar to the participants seem to be more acceptable for Voice, whereas those with unfamiliar people Voice becomes a more unacceptable input method.

Expression (RQ4): In general Keywords are more acceptable than Statements. However as with Input Method, location did not have a significant effect on the acceptability of using Statements or Keywords. In contrast Audience did have a significant effect. The only significant pairwise difference was for the Partner audience, with Keywords being more acceptable than Statements. Overall expression appears to have less of an impact than any of the other variables that we investigated.

6 Conclusion

Wearable technology and also voice based search are relatively new technologies for many people. These technologies open up new possibilities for search interaction whilst mobile. In comparison with more traditional desktop based search this also creates a range of new possibilities, but also factors that must be taken into account, for designing search interactions. In this paper we have focussed on social acceptability issues surrounding using voice based reactive search with smartwatches and also mobile phones. We conducted a lab study where we presented participants with various search scenarios and solicited responses to the acceptability of those scenarios. Overall these findings explain some of the reluctance for search to move beyond text input on mobile phones. This also validates some of the move towards card based proactive search, where the device displays "results" without a query. Our findings also demonstrate that there are some cases/locations/audiences where different ways of searching might be preferable. It is not yet possible to completely remove proactive search with user input, thus it is important to understand all of the factors that influence search interaction. This work is a first step in that direction and the results here provide some guidance on the types of information needs and scenarios that require proactive search and those that might be better served by reactive. A combination of both may ultimately provide the best user experience.

References

1. Shokouhi, M., Guo, Q.: From queries to cards: re-ranking proactive card recommendations based on reactive search history. In: 38th International ACM SIGIR Conference on Research and Development in Information Retrieval, pp. 695–704 (2015)

 2. Rico, J., Brewster, S.: Usable gestures for mobile interfaces: evaluating social acceptability. In: 28th ACM Conference Conference on Human Factors in Computing Systems, pp. 887–896 (2010)
 3. Jones, M., Buchanan, G., Thimbleby, H.: Improving web search on small screen devices. Interact. Comput. **15**, 479–495 (2003)
 4. Church, K., Smyth, B., Bradley, K., Cotter, P.: A large scale study of European mobile search behaviour. In: 10th International Conference on Human Computer Interaction with Mobile Devices and Services, pp. 13–22 (2008)
 5. Broder, A.: A taxonomy of web search. ACM SIGIR Forum **36**, 3–10 (2002). ACM
 6. Church, K., Cousin, A., Oliver, N.: I wanted to settle a bet!: understanding why and how people use mobile search in social settings. In: 14th International Conference on Human-Computer Interaction with Mobile Devices and Services, pp. 393–402 (2012)
 7. Church, K., Neumann, J., Cherubini, M., Oliver, N.: SocialSearchBrowser: a novel mobile search and information discovery tool. In: Proceedings of the 15th International Conference on Intelligent User Interfaces, pp. 101–110 (2012)
 8. Ren, Y., Tomko, M., Ong, K., Sanderson, M.: How people use the web in large indoor spaces. In: 23rd ACM International Conference on Conference on Information and Knowledge Management, pp. 1879–1882 (2014)
 9. Church, K., Cramer, H.: Understanding requirements of place in local search. In: 33rd Annual ACM Conference Extended Abstracts on Human Factors in Computing Systems, pp. 1857–1862 (2015)
10. Montañez, G.D., White, R.W., Huang, X.: Cross-device search. In: 23rd ACM Conference on Conference on Information and Knowledge Management, pp. 1669–1678 (2015)
11. Schaar, A.K., Ziefle, M.: Smart clothing: perceived benefits vs. perceived fears. In: 5th IEEE Pervasive Computing Technologies for Healthcare, pp. 601–608 (2011)
12. Shinohara, K., Wobbrock, J.O.: In the shadow of misperception: assistive technology use and social interactions. In: 29th ACM Conference Conference on Human Factors in Computing Systems, pp. 705–714 (2011)
13. Hoyle, R., Templeman, R., Armes, S., Anthony, D., Crandall, D., Kapadia, A.: Privacy behaviors of lifeloggers using wearable cameras. In: ACM International Joint Conference on Pervasive and Ubiquitous Computing, pp. 571–582 (2014)
14. Pearson, J., Robinson, S., Jones, M.: It's about time: smartwatches as public displays. In: 33rd ACM Conference Conference on Human Factors in Computing Systems, pp. 1257–1266 (2015)
15. Easwara Moorthy, A., Vu, K.-P.L.: Privacy concerns for use of voice activated personal assistant in the public space. Int. J. Hum.-Comput. Interact. **31**, 307–335 (2015)
16. Wakeling, S., Halvey, M., Villa, R., Hasler, L.: A comparison of primary and secondary relevance judgements for real-life topics. In: ACM SIGIR Conference on Human Information Interaction and Retrieval, pp. 173–182 (2016)
17. Leiva, L.A., Sahami, A., Catalá, A., Henze, N., Schmidt, A.: Text entry on tiny QWERTY soft keyboards. In: 33rd Annual ACM Conference on Human Factors in Computing Systems, pp. 669–678 (2015)
18. Montero, C.S., Alexander, J., Marshall, M.T., Subramanian, S.: Would you do that?: understanding social acceptance of gestural interfaces. In: 12th International Conference on Human Computer Interaction with Mobile Devices and Services, pp. 275–278 (2010)

How Precise Does Document Scoring Need to Be?

Ziying Yang, Alistair Moffat$^{(\boxtimes)}$, and Andrew Turpin

Department of Computing and Information Systems,
The University of Melbourne, Melbourne, Australia
`ziyingy@student.unimelb.edu.au`, {`ammoffat,aturpin`}`@unimelb.edu.au`

Abstract. We explore the implications of tied scores arising in the document similarity scoring regimes that are used when queries are processed in a retrieval engine. Our investigation has two parts: first, we evaluate past TREC runs to determine the prevalence and impact of tied scores, to understand the alternative treatments that might be used to handle them; and second, we explore the implications of what might be thought of as "deliberate" tied scores, in order to allow for faster search. In the first part of our investigation we show that while tied scores had the potential to be disruptive to TREC evaluations, in practice their effect was relatively minor. The second part of our exploration helps understand why that was so, and shows that quite marked levels of score rounding can be tolerated, without greatly affecting the ability to compare between systems. The latter finding offers the potential for approximate scoring regimes that provide faster query processing with little or no loss of effectiveness.

1 Introduction

Batch evaluation techniques are widely used in information retrieval system measurement. Each system that is to be compared generates a ranking, or *run*, for each of a set of topics, with documents included in the run and also ordered within the run on the basis of some computed textual *similarity score* relative to the given query. Possible similarity computations include the Okapi BM25 mechanism of Robertson et al. [10] and the language modeling techniques of Ponte and Croft [9]. Static score components such as Pagerank or other assessments of document quality can also be included. Those runs are then mapped to numeric *effectiveness values* using a set of relevance judgments and an *effectiveness metric*, which generates a single number as an assessment of the quality, or utility, of that run in the eyes of the user that is presumed to have inspected it. Finally, the effectiveness values are aggregated in some way across topics to get an overall performance measure which is often used, with a suitable statistical test, as a basis for answering the question "is System A demonstrably better than System B?".

In this work we consider the consequences of allowing *tied similarity scores* (or just *ties*) in the ranking. The obvious issue is that ties admit a level of ambiguity in the effectiveness metric values, and hence (potentially) in the outcome

© Springer International Publishing AG 2016
S. Ma et al. (Eds.): AIRS 2016, LNCS 9994, pp. 279–291, 2016.
DOI: 10.1007/978-3-319-48051-0_21

of a system versus system comparison, since a group of documents that all share the same computed similarity score could be presented to the user in any permutation that is consistent with the scores being non-increasing. Our first goal is thus to quantify the extent to which past Text Retrieval Conference (TREC) evaluation exercises have been affected by tied similarity scores, and determine whether the presence of ties may have caused ambiguity to flow through into system scores. In this part of the project we make use of a range of tie-breaking regimes, including the rules embedded in the well-known trec_eval program, and conclude that while ties have had the potential to be significantly disruptive, in practice they did not influence the outcomes of the measurements that were undertaken.

A second related goal is to ask whether the deliberate introduction of ties might be useful in some way. For example, a range of approaches in which similarity scoring might be approximated or otherwise quantized have been suggested over the years including, for example the quantized document weights of Moffat et al. [8], or the impact-ordered indexes of Anh and Moffat [1]. If we allow that the retrieval system might gain tangible efficiency benefits from assigning scores with low precision to documents, then we may end up with large numbers of ties in the runs that the system generates, and being able to estimate the extent to which ties can be tolerated before there is risk of degraded system retrieval effectiveness is a key component of the approximation. In experiments using submitted TREC runs, we show that quite marked levels of approximation can be tolerated before system scores change significantly, and hence that relatively low-precision scoring can be employed if it boosts efficiency.

2   Ties, and Methods for Dealing with Them

Terminology. We suppose that the similarity scores generated for a query partition the document ranking – the *run* – into *groups* within which the documents have the same score. Let b_g be the rank in the run at which the gth equi-score group commences, with, by definition, $b_1 = 1$; and let e_g be the rank of the last document in that group, with $b_{g+1} = e_g + 1$. That is, the gth group of tied documents spans the items $[b_g \ldots e_g]$, and contains $s_g = b_{g+1} - b_g$ documents. We further define G_g to be the multiset of gain values associated with the documents in the gth group, $G_g = \{r_k \mid b_g \leq k \leq e_g\}$, with $r_k \in \{0,1\}$ the gain associated with the document at rank k; and define t_g to be the total gain associated with the gth group, $t_g = \sum\{r_k \mid b_g \leq k \leq e_g\}$. For example, consider the ten-item ranking shown in Fig. 1, with each document given a single letter label for convenience, and with five different computed similarity scores. The second row shows a presumed relevance value for each corresponding document ("0" and "1"); and the third row lists the similarity scores that are presumed to have led to that ranking.

If the scores are ignored and only the list of relevance values is employed, computation of (for example) the metric precision at depth $k = 5$ (P@5) yields a score of $2/5 = 0.4$, because there are two "1"s among the first five gain values.

Similarly, the ranking shown has a reciprocal rank (RR) score of $1/3 = 0.333$, since the first relevant document appears at rank $k = 3$. Other metrics such as average precision (AP), rank-biased precision (RBP) [7], and normalized discounted cumulative gain (NDCG) [5], can also be computed, based solely on that third "gain" row, without consideration of the document labels in the first row, or their scores in the second row.

When scores are included, the situation changes. Now documents M and S can be seen to have the same similarity score, and are part of a tied group. That means that P@5 might be either $2/5$ or $3/5$, depending on the tie-breaking rule employed to order them. Similarly, RR might be $1/2$ or $1/3$, because of the tie involving documents H and A and C (but note that there is no possible arrangement in which RR can be $1/4$).

rank, k	1	2	3	4	5	6	7	8	9	10
document, d_k	D	H	A	C	M	S	W	B	E	J
gain, r_k	0	0	1	1	0	1	1	0	0	1
score	9.8	9.3	9.3	9.3	8.4	8.4	8.2	8.0	8.0	8.0
groups	$b_1{=}1$	$b_2{=}2$			$b_3{=}5$		$b_4{=}7$	$b_5{=}8$		

Fig. 1. Example run showing five equi-score groups.

Run Order. A range of mechanisms have evolved to deal with tied scores. The first and most obvious option is to do as has already been suggested in connection with the example shown in Fig. 1, and that is to ignore the document scores and process the run in the order in which the documents are presented – in effect, pushing the responsibility for tie-breaking back to the retrieval system, whether or not it accepts it. This approach presumes that the system has employed more information than is captured in the final score, perhaps via further precision in the internal computation above and beyond what is passed to the evaluation regime, or perhaps via a secondary-key ordering process that is not part of the scores at all. However the system's ordering arises, respecting the sequential presentation of documents is a plausible default way of handling tied scores.

External Tie-Break Rule. A second option is to make use of some external fixed ordering criterion and use it to reorder the documents within each tied group, thereby obtaining a canonical representation for the run. For example, the documents in each group might be sorted according to their document identifier, or according to their length, or according to their URL or filename. As one specific example of this type of approach, the widely-used `trec_eval` program (see http://trec.nist.gov/trec_eval/) sorts tied groups into decreasing order of document identifier before performing its various effectiveness metric computations.

Optimistic and Pessimistic Limits. A third way of handling runs with ties is to compute the best and worst scores that might arise, and then present a score range rather than a score value. The advantage of this approach is that it makes clear when scores contain potential ambiguity, in a way that mirrors the residuals of Moffat and Zobel [7], which provide guidance as to the metric weight assigned to unjudged documents. To compute an optimistic upper score bound, the t_g relevant documents within the gth group are assumed to appear in the first rank positions, that is, $[b_g \ldots b_g + t_g - 1]$, and the metric score then computed in the usual way. Similarly, to get a pessimistic lower score bound, the t_g relevant documents in the group are assumed to appear as a block as deep in the run as is possible, at ranks $[e_g - t_g + 1 \ldots e_g]$. In the example shown in Fig. 1, the ordering "H then A then C" (and similarly in the other groups) is used to derive a lower bound on the score, and the ordering "A then C then H" (and so on) is used to obtain an upper bound. If a document is unjudged, then for many metrics (but notably, not for AP or NDCG) it should be assumed to be non-relevant for the purposes of establishing the lower bound, and assumed to be relevant for the purposes of establishing the upper bound.

Averaging Across Permutations. While the worst-case bounds can be informative, they are also somewhat pessimistic, and computing the average, or expected, value of the metric across all possible permutations of documents within each of the tied score groups provides a useful balance. If every permutation of documents in each group is equally likely, then computing the expectation is simply the process of computing the metric for each permutation and taking their average. For a small number of small groups, this $O(\prod_g(s_g!))$ brute-force approach is computationally feasible. But if there are many blocks, or if there are any large blocks, it is expensive. Fortunately, the summation over all permutations telescopes for most metrics, leading to a tractable computation. McSherry and Najork [6] describe this process in detail, and present an incremental formulation for average precision that computes the expected score across all possible permutations of documents in each group. A similar computation can be used to compute an expected (across permutations within groups) RR score.

For weighted-precision metrics such as RBP, a similar process can be adopted. The set of gain values associated with each group is summed and averaged, and then that average gain applied at each rank position, and weighted according to the decay function. For the example shown in Fig. 1, and an RBP parameter $p = 0.5$, the expected RBP0.5 score is computed as

$$0.5 \times \frac{0}{1} + (0.25 + 0.125 + 0.0625) \times \frac{2}{3} + (0.0313 + 0.0156) \times \frac{1}{2} + \cdots$$

We use these formulations for expected AP, expected RR (not to be confused with the metric ERR), and expected RBP in the experiments described in the next two sections.

3   Ties in TREC Experimentation

TREC Resources. In this section we examine the role that ties may have had on past TREC evaluations. The primary resource we make use of are the 103 runs submitted as part of the 1998 TREC7 Ad-Hoc experimentation round [13], see `trec.nist.gov`, and Harman [4] for a broad overview. Each run is a list of (up to) 1,000 responses from that system for each of 50 topics, with each row in the run file including fields for *docnum*, *rank*, and *score*. There are thus three possible ways that each run could be interpreted:

- by the line number ordering implicit in the presentation of the run;
- by (increasing, or at least, non-decreasing) values in the *rank* field;
- by (decreasing, or at least, non-increasing) values in the *score* field.

Line numbers are unique within each system-topic combination, and do not admit ties, but both ranks and scores might provide ties in runs. To explore the prevalence of ties, the TREC7 Ad-Hoc runs were analyzed. Somewhat surprisingly, we discovered that there were 254 instances in the archived runs where scores were increasing rather than non-increasing in terms of the line ordering, and that five systems were affected by this inconsistency. The primary reason appears to be incorrect sorting of scores when exponential formatting is being used. For example, in the run `bbn1`, for topic 355, the second-to-last score in the run is -1.37; and final score is $-7.763e-05$. In fact, that last document's correct position is some 700 locations higher, at rank 304, the rank that row was labeled with. When rank ordering was similarly checked the situation was even more confused, and 7.3 % of the documents in the archived runs (358,631 entries in total) were mis-ordered according to their stated ranks. That is, the supplied document ordering in the runs corresponds to neither increasing rank nor to non-decreasing score.

To resolve this apparent mislabeling, we re-sorted all of the TREC7 submissions, taking care to treat the exponential formats correctly. We used decreasing numeric score as the primary key, and then increasing rank as a secondary key. This is guaranteed to give rise to runs in which there are no score-based out-of-order items. We then counted the occurrences of score ties at the document, topic, and system level; and the occurrence of rank contradictions, where a "contradiction" is a pair of adjacent documents that when sorted by score have ranks that indicate the opposite ordering. Table 1 shows the results of this processing.

Table 1. Ties occurring in 103 TREC7 Ad-Hoc runs after score-based re-sorting: the percentage of systems, system/topic combinations, and documents that include tied scores; and the corresponding percentages of score-rank contradictions. There are 103 systems, 103×50 system-topic combinations, and 4,900,042 documents. Note that not all runs contain 1,000 documents.

	Percentage affected		
	Systems	System-topics	Documents
Tied scores	95.2	91.0	14.0
Rank/score contradictions	6.8	4.2	1.4

As can be seen, 14 % of the documents in the runs have the same score as their predecessor document in that run, a fact that provides the motivation for our work here; and, of equal concern, a further 1.4 % of the documents cannot be placed in a manner that is consistent with both their assigned score and their assigned rank, with seven of the 103 systems affected. We can only assume that the cause of the latter issue was programming errors at the time the runs were created by the corresponding research groups. There were no ties on rank in any of the TREC7 runs.

To ensure that the results in the remainder of the paper were not affected by programming mistakes and other experimental misunderstandings on the part of the 1998 TREC7 participants, we then took the top 80 systems, as ordered by average AP score over the 50 topics, discarding the other 23 systems from further evaluation. Similar restrictions have also been employed by other authors.

Ties in TREC7. The primary evaluation metric used in TREC7 was average precision, as implemented in the program trec_eval (version 9.0). Working with the 80 score/rank-sorted runs, we next sought to examine the effect that the score-ties had on AP scores for systems. Figure 2 plots those systems. The horizontal axis is the trec_eval score for that system, expressed as a mean AP value over the 50 topics. By inspecting the trec_eval source code we were able to confirm that it (a) ignored line ordering in the input runs; (b) used exponential number formats correctly when performing its sorting-by-score step; and (c) resolved score ties by reverse sorting on document number, paying no attention to the supplied *rank* field. The scale on the vertical axis in Fig. 2 is the AP score range measured by taking the difference between the pessimal and optimal topic scores, and then averaging across topics to get a system range. The higher up the axis a system is plotted, the greater the uncertainty in its score.

Each system is plotted as a segment. The right and left ends of the segment reflect the scores that would be generated by the optimistic and pessimistic orderings for each of the tied groups; the trec_eval score is shown as a circle; and the "average across permutations" score as a triangle. The color of each point reflects the number of document ties for that system, in terms of Table 1. The vertical axis is truncated at 10^{-6}, and the points plotted along that line have a score difference of 10^{-6} or below. At the top of the graph, many tied scores lead to wide score ranges, with the trec_eval ordering being just one of them, usually not too far from the average overall. But for some systems the optimal-to-pessimal spread is wide, and as can be seen in the overlapping vertical extents, ties may have affected the relative ordering of the top few systems ($AP \geq 0.30$). At the bottom of the graph, only a tiny minority of systems have no tied scores at all; but for most evaluations the ties that do exist do not result in any appreciable score range, with optimal-to-pessimal ranges less than 10^{-4} when averaged across topics.

Ties in Other Years. We carried out the same analysis on several other TREC rounds, and found similar rates of tied scores in general (Table 1), and instances of systems with wide potential score ranges. However we found no further years

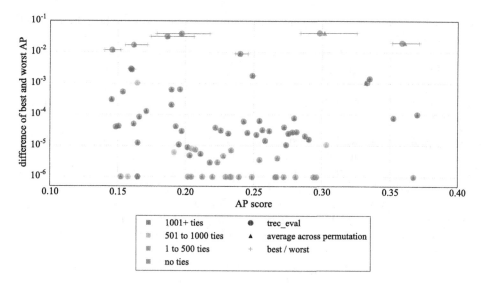

Fig. 2. Imprecision in AP scores caused by ties in a set of 80 TREC7 runs.

in which the ordering of the top few systems might have been affected by the tie-breaking rule employed.

4   Deliberate Score Grouping

We now consider whether the deliberate use of tied scores has a discernible effect on retrieval effectiveness.

Score Approximation. Scoring documents using modern similarity computations involves non-trivial amounts of arithmetic, especially if phrase components or term proximity components are being used. Regimes such as WAND [3] seek to minimize the number of documents scored, while still giving rise to exactly the same ranking for the top-k documents, an approach that meets the requirements for being *rank-safe to depth* k. That is, the WAND process ensures that all of the documents in the first k places of the ranking are in their right positions, but makes no guarantee for documents beyond depth k. This is a relatively stringent requirement, and other computation-pruning techniques might also be considered that provide more flexible trade-offs.

In particular, we now consider the following weaker requirement: that each document must be scored in a manner that guarantees that it is in the correct *band* of the ranking, where the bands are defined geometrically based on a parameter $\rho > 1$. More precisely, let $b_1 = 1$, and thereafter let $b_{g+1} = \lceil \rho \cdot b_g \rceil$. The gth band, for $g \geq 1$, spans the ranks from b_g to $e_g = b_{g+1} - 1$ inclusive. For example, if $\rho = 2$, then the bands are $[1 \ldots 1]$, $[2 \ldots 3]$, $[4 \ldots 7]$, and so on; and if (say) $\rho = 1.62$ (the golden ratio) the bands are $[1 \ldots 1]$, $[2 \ldots 3]$, $[4 \ldots 6]$, $[7 \ldots 11]$, and so on, with widths given by the Fibonacci sequence. The smaller the value of ρ, the smaller the band is that spans any given position in the ranking, and the nearer the approximate ranking is to the "true" and exact ranking. In the limit, as ρ approaches 1, the retrieval system is obliged to place each document at its final "correct" position; that is, $\rho = 1$ corresponds to a "full" computation in which all document relationships are finalized. But when $\rho > 1$, we allow the retrieval system to economize on its computational costs and return groups of documents $[b_g \ldots e_g]$, with equal scores assumed within each band.

Worst-Case Bounds. It is straightforward to show that when $\rho > 1$ the first group containing more than one document starts at rank $v = b_v = 1 + \lfloor 1/(\rho - 1) \rfloor$. That fact implies that the approximate scoring mechanism is rank-safe to depth $v - 1$, and more generally, allows bounds on the imprecision in scores to be computed. For example, consider the metric reciprocal rank (RR). With the vth group the first one with multiple documents in it, the loss of score that can arise when permutation-based averaging is applied is given by

$$\Delta\mathrm{RR} = \frac{1}{b_v} - \frac{1}{e_v - b_v + 1} \sum_{k=b_v}^{e_v} \frac{1}{k},$$

where the bound arises because the worst situation is when the original run has its first relevant document at rank b_v, and no other document in that group is relevant. Table 2 gives some $\Delta\mathrm{RR}$ values; when $\rho \leq 2$, all are less than 0.1.

It is also possible to compute worst-case differences for rank-biased precision (RBP, see Moffat and Zobel [7]). In the case of RBP, the maximum difference score difference arises when the run has a sequence of relevant documents at the start of each of its groups, followed by non-relevant documents for the rest of each group. The exact number $1 \leq t_g \leq (e_g + b_g)/2$ of relevant documents required in the initial run for the gth group varies according to both p (the RBP parameter) and ρ, and is chosen independently in each group to maximize the difference

$$\left(\sum_{k=b_g}^{b_g+t_g-1} (1-p)p^{k-1} \right) - \left(\frac{t_g \cdot w_g}{e_g - b_g + 1} \right),$$

where $w_g = \sum_{k=b_g}^{e_g}(1-p)p^{k-1}$ is the sum of the RBP weights associated with that gth group. The overall bound on the difference, $\Delta\mathrm{RBP}$, is the sum of the group maximum differences. Table 2 includes $\Delta\mathrm{RBP}$ differences for two values of the RBP parameter p. Recall-based metrics such as average precision (AP) cannot be analyzed as readily, because assuming additional documents to be relevant might decrease rather than increase the score. Experimental results showing that practice that AP has less divergence of scores than does RBP are presented in the next subsection.

Effectiveness Score Differences in Practice. Given these worst-case bounds, the next question we ask is this: to what extent does an allowance for rank-based score imprecision affect effectiveness scores in practice? To respond to this question, we again make use of the 1998 TREC7 resources, taking the same

Table 2. Worst-case metric score differences associated with geometric grouping of documents in runs, controlled by parameter ρ. It is not possible to derive equivalent bounds for AP.

ρ	Metric		
	RR	RBP0.5	RBP0.85
1.1	0.0038	0.0002	0.0087
1.2	0.0119	0.0052	0.0231
1.4	0.0417	0.0429	0.0482
1.7	0.0833	0.0945	0.0777
2.0	0.0833	0.1016	0.0971

system runs as were already examined in Sect. 3, and for each run, mapping it to a set of equivalent banded runs based on a set of ρ values, with the documents ranked in band g in each of those runs assigned a synthetic score of $1/g$. The original system scores that were part of the TREC7 data were ignored as the grouping operation was being carried out, and original file order was used as the reference point for each run. As already detailed in Sect. 3, 23 low-scoring systems were removed as part of the experimental methodology.

Figure 3 shows the results of this experimentation, plotted as a sequence of box-whisker elements using four different effectiveness metrics and a single representative value of $\rho = 1.4$. In all cases the score difference calculated is the across-permutations computation that was illustrated in Sect. 2 when applied to the deliberately-tied rankings, subtracted from the score the same metric achieved on the original submitted ranking for that same topic. We followed standard protocols and assumed that unjudged documents were not relevant for the purposes of scoring the runs.

Figure 3 shows that the average score variation arising from the banding process is small, and that there are nearly as many system-topic combinations that gain from the approximation process as there are that lose from it. Most RR values are unaffected (both quartiles are zero, for all of the ρ values tested), and the two deep metrics (RBP0.85 and AP) also have small inter-quartile ranges on the computed score differences. The average original metric scores across all system-topic combinations for RR, RBP0.5, RBP0.85, and AP are, respectively, 0.6939, 0.5556, 0.4677, and 0.2311; and hence the smaller AP score differences are in part a matter of relative scale. The shallow metric RBP0.5 suffers the most from the score grouping process; even so, it is only when $\rho > 1.5$, the first value for which ranks 2 and 3 are placed in the same group, that the differences are large. When $\rho \leq 2$ the first group always contains a single document.

Table 3 explores whether the small score differences identified in Fig. 3 can be regarded as being significant. To generate the table, each of the 80 systems was scored for the 50 topics using the original runs, and then re-scored using the grouped runs. The set of original topic scores was then multiplied by 0.99, and compared to the grouped scores, using a one-tail paired t-test. If a p value less

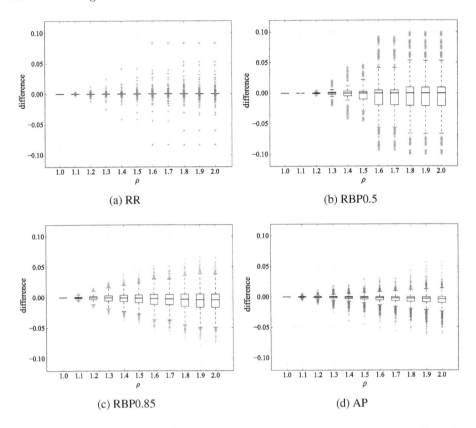

Fig. 3. Variation in metric effectiveness score across a set of 80 runs and 50 topics (that is, 50×80 points are plotted in each column), as a function of ρ, for four different retrieval effectiveness metrics. The whiskers indicate the last outlier still within 1.5 times of the inter-quartile range from the corresponding quartile (the limits of the boxes).

than or equal to 0.05 was generated by that test, that system was counted as being one for which the grouping process degraded the system score by 1 % or less. The closer the count of such systems is to 80, the greater the confidence we can have that the grouping process will not give notably inferior system scores overall, where "notably inferior" is defined (at first) as being a 1 % degradation in measured score. Those values are shown in the left half of Table 3, and the corresponding counts when "notably inferior" is defined as being a 3 % degradation are shown in the right half. The relationship between ρ and score fidelity is reflected by the decreasing numbers down each column of the table, and as ρ increases, the possible implications of changes in score also increase. When the "tolerable degradation limit" was further reduced to 95 %, all 16 entries for metric and ρ were 80.

Table 3. Number of systems (maximum 80) for which a t-test across 50 topics yields confidence at the $p \leq 0.05$ level that the grouped runs yield a metric score greater than or equal to 99 % (left) and 97 % (right) of the original run score.

ρ	Relative to 99 % of original score				Relative to 97 % of original score			
	RR	RBP0.5	RBP0.85	AP	RR	RBP0.5	RBP0.85	AP
1.1	80	80	80	80	80	80	80	80
1.2	80	80	80	80	80	80	80	80
1.4	77	44	65	44	80	80	80	80
1.7	37	11	14	0	80	67	80	77
2.0	38	10	3	0	80	61	71	20

System Comparison Sensitivity. Effectiveness measurements are also used to compare systems in a pairwise manner. In a final experiment, we explore the implications that score rounding has on the ability of metrics to differentiate between systems. The normal approach to comparing systems is to take their computed scores across a set of topics, and perform a paired t-test to explore the null hypothesis that the two systems are in fact the same. The process of carrying out the t-test generates a p value; the smaller the p value, the smaller the chance that the two systems being compared are giving the same performance on the data used. To establish significance, a threshold value α is employed, often $\alpha = 0.05$, with $p \leq \alpha$ being regarded as a significant outcome.

To measure the effect that score rounding has on system comparisons, we took the 50 topics of the TREC7 collection and the 80 runs associated with it that we have been using, and computed, for each of eleven different values of ρ, the set of p values generated for the $80 \times 79/2$ distinct system pairs. In all cases when $\rho > 1$, the averaging processes described in Sect. 2 were used; when $\rho = 1$, each run was processed in sorted-by-score order, and then the scores were discarded.

Figure 4 shows that score grouping has almost no effect at all on the ability to distinguish between systems using a statistical test (the *discrimination ratio* of the metric, see Sakai [11]), across the four metrics used in our experiments. For example, the plot in the lower-right for AP shows when $\rho = 1.0$ that 62.2 % of the system pairs yield "significant at $p = 0.05$" comparison outcomes; at $\rho = 1.4$, that fraction is 62.1 %, with only 0.1 % of false positives, and 0.2 % false negatives. The situation is similar for the other metrics, with the discrimination ratios (down to 45 % for RR) determined primrily by the effective evaluation depth, and only a small fraction of false positives and negatives.

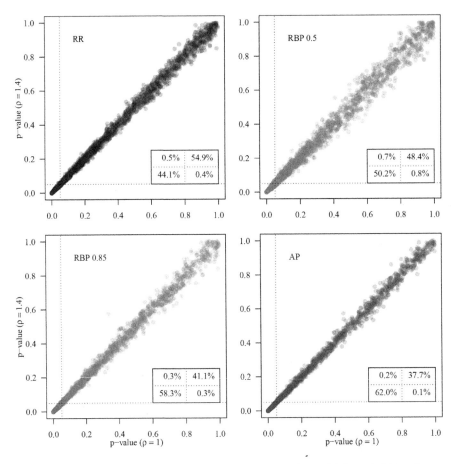

Fig. 4. Correlation of p values for all pairs of systems ($80 \times 79/2 = 3,160$ points per pane), with the p value from a paired t-test using the original system scores across 50 topics plotted on the horizontal axis, and the p value for the corresponding system pair with grouped runs ($\rho = 1.4$) on the vertical axis. The dotted lines at are $p = 0.05$, with the grid showing the percentage of data points in each quadrant.

5   Conclusion and Future Work

We have explored the impact of score ties on the evaluation of retrieval system effectiveness, as measured using binary relevance judgments and three established effectiveness metrics. Ties have the potential to affect system comparisons, and using TREC data, we showed that a small number of systems did indeed generate runs with very ambiguous score outcomes, but that – fortunately – the overall conclusions from those rounds of experimentation were unlikely to have been compromised. We further demonstrated that allowing a controlled grouping of scores in runs – in a sense, permitting the deliberate introduction of ties – resulted in only small changes in the ability to compare systems. This approach represents a novel direction in which retrieval efficiency improvements might be achieved. We have not yet addressed the question of how those efficiency gains might be achieved, and a clear direction for future work is to reexamine the computation embedded in standard similarity scoring regimes and existing dynamic pruning heuristics, to identify and measure ways in which processing economies might accrue through the use of inexact scoring.

Another area for future work is in the space of test collection construction. Previous investigations [2,12,14] have explored the reliability and quality of the collected judgments; it may be that the pooled documents can be stratified according to the groups they appear in, and less emphasis placed on judgment quality for deeper pools, relying instead on averaging effects to preserve overall evaluation quality.

References

1. Anh, V.N., Moffat, A.: Pruned query evaluation using pre-computed impacts. In: Proceedings of SIGIR, pp. 372–379 (2006)
2. Bailey, P., Craswell, N., Soboroff, I., Thomas, P., de Vries, A.P., Yilmaz, E.: Relevance assessment: are judges exchangeable and does it matter. In: Proceedings of SIGIR, pp. 667–674 (2008)
3. Broder, A.Z., Carmel, D., Herscovici, M., Soffer, A., Zien, J.: Efficient query evaluation using a two-level retrieval process. In: Proceedings of CIKM, pp. 426–434 (2003)
4. Harman, D.K.: The TREC test collections (Chap. 2). In: Voorhees, E.M., Harman, D.K. (eds.) TREC: Experiment and Evaluation in Information Retrieval, pp. 21–52. MIT Press, Cambridge (2005)
5. Järvelin, K., Kekäläinen, J.: Cumulated gain-based evaluation of IR techniques. ACM Trans. Inf. Sys. **20**(4), 422–446 (2002)
6. McSherry, F., Najork, M.: Computing information retrieval performance measures efficiently in the presence of tied scores. In: Macdonald, C., Ounis, I., Plachouras, V., Ruthven, I., White, R.W. (eds.) ECIR 2008. LNCS, vol. 4956, pp. 414–421. Springer, Heidelberg (2008)
7. Moffat, A., Zobel, J.: Rank-biased precision for measurement of retrieval effectiveness. ACM Trans. Inf. Syst. **27**(1), 2.1–2.27 (2008)
8. Moffat, A., Zobel, J., Sacks-Davis, R.: Memory efficient ranking. Inf. Process. Manag. **30**(6), 733–744 (1994)
9. Ponte, J.M., Croft, W.B.: A language modeling approach to information retrieval. In: Proceedings of SIGIR, pp. 275–281 (1998)
10. Robertson, S.E., Walker, S., Jones, S., Hancock-Beaulieu, M., Gatford, M.: Okapi at TREC-3. In: Proceedings of TREC, pp. 109–126 (1994)
11. Sakai, T.: Alternatives to BPref. In: Proceedings of SIGIR, pp. 71–78 (2007)
12. Scholer, F., Turpin, A., Sanderson, M.: Quantifying test collection quality based on the consistency of relevance judgements. In: Proceedings of SIGIR, pp. 1063–1072 (2011)
13. Voorhees, E.M., Harman, D.K.: Overview of the seventh text retrieval conference (TREC-7). In: Proceedings of TREC, pp. 1–23. NIST Special Publication 500-242 (1998)
14. Voorhees, E.M.: Variations in relevance judgements and the measurement of retrieval effectiveness. Inf. Process. Manag. **36**(5), 697–716 (2000)

Short Paper

Noise Correction in Pairwise Document Preferences for Learning to Rank

Harsh Trivedi[✉] and Prasenjit Majumder

Dhirubhai Ambani Institute of Information and Communication Technology,
Gandhinagar, India
harshjtrivedi94@gmail.com, prasenjit.majumder@gmail.com

Abstract. This paper proposes a way of correcting noise in the training data for Learning to Rank. It is natural to assume that some level of noise might seep in during the process of producing query-document relevance labels by human evaluators. These relevance labels, which act as gold standard training data for Learning to Rank can adversely affect the efficiency of learning algorithm if they contain errors. Hence, an automated way of reducing noise can be of great advantage. The focus in this paper is on noise correction for pairwise document preferences which are used for pairwise Learning to Rank algorithms. The approach relies on representing pairwise document preferences in an intermediate feature space on which ensemble learning based approach is applied to identify and correct the errors. Up to 90 % errors in the pairwise preferences could be corrected at statistically significant levels by using this approach, which is robust enough to even operate at high levels of noise.

1 Introduction

Learning to rank is an approach to automatically build a ranking model, based on the training data using machine learning technologies [6]. The training data for learning to rank when used for document retrieval usually consists of queries, the associated documents, and relevance labels for query-document pairs which are assigned by human judges. Several previous works have shown that human judges may not agree with each others in the task of assigning relevance labels to query-document pairs [1,9,10]. Since human annotation is costly, especially in web-search which requires large amount of training data, one can usually not afford to have several annotators to make multiple judgments. As a result, such relevance judgements are prone to be biased, unreliable and noisy. Xu et al. have shown that errors in training data can significantly degrade the performance of ranking functions trained by learning to rank algorithms [11]. So, automatic error correction for training data of learning to rank can be of great advantage.

Primarily, there are 3 types of learning to rank algorithms: pointwise, pairwise and listwise [5]. In this paper, the focus is on training data of pairwise Learning to Rank algorithms which take pairwise preferences of documents for each query as the learning instances. Using the proposed method, noise present in the pairwise preferences can be considerably reduced. To test it's efficiency

S. Ma et al. (Eds.): AIRS 2016, LNCS 9994, pp. 295–301, 2016.
DOI: 10.1007/978-3-319-48051-0_22

different levels of artificial noise are injected in the data. On this noisy data, noise reduction process is applied and the output is compared to the original human generated data, which is assumed to be correct for the sake of the evaluation. Since the effectiveness is tested on a wide range of injected noise, it also checks the robustness of the proposed process to initial noise present in the data.

There have been few attempts on improving the quality of training data for Learning to Rank. Geng et al. proposed a way of computing training data quality for Learning to Rank with a concept of "Pairwise Preference Consistency" (PPC). They have shown a way to select the most optimal subset of the initial training data which maximizes the PPC score [3]. However, because of selection of a subset there is a possibility of loosing some important examples which are discarded in this process. Hence, in this attempt an error correction, rather than error elimination approach is targeted. Xu et al. proposed a method of error correction, by leveraging the information from click-through data [11]. However, it is not natural to assume the availability of such data in all cases. To the best of our knowledge, there hasn't been any work yet, that deals with improving the quality of training data for learning to rank by error correction rather than error elimination solely on the based on training data itself.

In contrast to ranking, there has been good amount of work on improving the quality of training data for classification [2]. Ensemble learners are often used for this purpose in classification data. For example, many classifiers are learnt from different samples of training data and used to classify the data. If there is a good amount of agreement among the classifiers then only that instance is kept, otherwise discarded. A similar approach is used here to correct the highly probable error-some instances and there by reducing noise in data. However, instead of elimination, correction of the highly suspicious preference pairings is performed. Hence unlike noise elimination, there is no risk of loosing important training instances in process of noise correction.

The remaining paper is organized as follows: Sect. 2 describes the proposed approach, Sect. 3 elaborates on the experimental setup, Sect. 4 discusses the results and Sect. 5 concludes and discusses future scope of this project.

2 Approach

Learning to Rank training data contains queries, the associated documents, set of features extracted from each query-document pair and the relevance label of documents for the corresponding query. Formally, given query q, there is a set of documents $D = \{d_1, d_2..d_n\}$ and for each query-document pair (q, d_i) there exists a feature vector $\bar{f}(q, d_i)$ and relevance label $rel(q, d_i)$. First of all, transformation of this representation to pairwise preference sets is performed as following:

2.1 Pairwise Preferences Sets

We define a **partial pairwise preference set** as:

$$\{[\bar{F}(q : d_i > d_j), 1] : rel(q, d_i) > rel(q, d_j) \text{ and } d_i, d_j \in D\} \tag{1}$$

and **full pairwise preference set** as:

$$\{[\bar{F}(q : d_i > d_j), 1] \cup [\bar{F}(q : d_j > d_i), 0] : rel(q, d_i) > rel(q, d_j) \text{ and } d_i, d_j \in D\} \tag{2}$$

where,
$\bar{F}(q : d_i > d_j)$ is document preference pair vector representation, which is taken as:

$$[\bar{F}(q : d_i > d_j)] = [\bar{f}(q, d_i) - \bar{f}(q, d_j)] \tag{3}$$

The preference pair $(q : d_i > d_j)$ is represented with feature vector $[\bar{f}(q, d_i) - \bar{f}(q, d_j)]$. Its class label is 1 if $rel(q, d_i) > rel(q, d_j)$ and 0 if otherwise. This means that for a given query q, if A is set of relevant documents and B is set of irrelevant documents, then there is a set $\{(q : a > b)|a \in A, b \in B\}$ for which class label is 1. Also, at the same time, there is a set $\{(q : b > a)|a \in A, b \in B\}$ for which the class label is 0. Hence, in all there are $2 \times |A| \times |B|$ number of pairwise instances, half of which are tagged positive and other half negative. Partial and Full pairwise preference set are easily inter-convertible from each others.

2.2 Noise Injection

Once partial pairwise preference set of original noiseless data is performed, different levels of noise are injected in it. For noise level p, each pairwise document preference is reversed ($\equiv$ class label is flipped) with probability p and kept the same with probability $p - 1$. The partial preference set is then converted to full preference representation.

2.3 Two Phase Process

For each query, a 2-phase process on the full pairwise preference set is performed.

Phase 1. x-fold cross validation on the full pairwise preference set with classifier a is performed. For each of the x parts of the data, classifier a is trained on remaining $x - 1$ parts and used to label the remaining part. The preference pair is identified as faulty (error) if the predicted label doesn't match the actual label. This process is repeated for $x \in \{3, 5, 7, 10\}$ and $a \in \{MultilayeredPerceptron\}$[1]. Intersection of all preference pairs are made which are identified as faulty by

---

[1] Weka - machine learning software was used for classification [4] .

any combination of x and a. It is worth noting that taking such intersection highly improves the precision of fault identification. Once, these suspected faulty preference pairs are extracted, they are removed from the full pairwise preference set. A separate set is made from them, basically decomposing the initial data in 2 parts: purer and noisier sub-sample.

The choices of x and a were empirically found to be working efficiently. We do not claim that this is the best choice, but it is at least a good choice for performing this task. Also, Multilayer Perceptron classifier with default parameters was found to be giving far better results for this task than any other classifier available in weka software.

Phase 2. The purer sub-sample of full preference set is used to train the classifier b. The trained model is then used for detecting the faulty preferences in noisier sub-sample. Here $b \in$ {Multilayered Perceptron, Random Forest }. Finally a union of these faults (errors) predicted by each classifier b is taken and they are considered as the final pairwise preference faults which need to be flipped.

2.4 Noise Measurement

Once the appropriate flips are made, measurement of the noise of updated paired representation is done. It is computed as the number of incorrect document preference pairs to the total number of preference pairs. The idea of computing Document Pair noise is taken from [7] in which it is referred to as pNoise. From the study, they have concluded that document pair noise captures the true noise of ranking algorithms, and can well explain the performance degradation of ranking algorithms. Hence, it has been used to evaluate the effectiveness of the noise-correction process by the reduction in document pair noise achieved by the method[2].

3   Experimental Setup

Experiments are performed on 3 standard Learning to Rank LETOR 3.0 datasets [8]: OHSUMED, TREC-TD-2003, TREC-TD-2004. Noise levels of {0.05, 0.1, 0.15, 0.2, 0.25, 0.3, 0.35, 0.4, 0.45, 0.5} are injected in each of these datasets and checked to what extent noise can be corrected depending on the initial noise present. The process is performed thrice and the average results are reported.

OHSUMED contains 106 queries with approximately 150 documents associated with each query. TREC-TD-2003 contains 50 queries with approximately 1000 documents associated with each query and TREC-TD2004 contains 75 queries with approximately 1000 documents for each query. OHSUMED represents query-documents pair by a set of 45 features, while TREC-TD 2003, 2004 use 44 features each. OHSUMED has 3 relevance levels {2, 1, 0} while TD2003 and TD2004 have {1, 0}.

---

[2] document pair noise will be referred to as noise henceforth.

Table 1. Noise correction on OHSUMED

Injected noise	Post correction noise	Percentage noise reduction	Queries improved	Queries worsened
0.05	0.029	**42.00 %**[*]	100	6
0.1	0.050	**50.00 %**[*]	96	10
0.15	0.085	**43.33 %**[*]	105	1
0.2	0.091	**54.50 %**[*]	105	1
0.25	0.132	**47.20 %**[*]	103	3
0.3	0.147	**51.00 %**[*]	105	1
0.35	0.204	**41.71 %**[*]	103	3
0.4	0.269	**32.75 %**[*]	100	6
0.45	0.419	**6.80 %**[*]	90	16
0.5	0.493	**1.40 %**	51	55

Table 2. Noise correction on TREC-TD-2003

Injected noise	Post correction noise	Percentage noise reduction	Queries improved	Queries worsened
0.05	0.002	**96.00 %**[*]	50	0
0.1	0.006	**94.00 %**[*]	50	0
0.15	0.019	**87.33 %**[*]	50	0
0.2	0.013	**93.49 %**[*]	50	0
0.25	0.023	**90.80 %**[*]	50	0
0.3	0.030	**90.00 %**[*]	50	0
0.35	0.064	**81.71 %**[*]	50	0
0.4	0.108	**73.00 %**[*]	50	0
0.45	0.393	**12.66 %**[*]	39	11
0.5	0.483	**3.40 %**	23	27

Table 3. Noise correction on TREC-TD-2004

Injected noise	Post correction noise	Percentage noise reduction	Queries improved	Queries worsened
0.05	0.002	**96.00 %**[*]	75	0
0.1	0.004	**96.00 %**[*]	75	0
0.15	0.009	**94.00 %**[*]	75	0
0.2	0.011	**94.50 %**[*]	75	0
0.25	0.027	**89.20 %**[*]	75	0
0.3	0.032	**89.33 %**[*]	75	0
0.35	0.093	**73.42 %**[*]	75	0
0.4	0.109	**72.75 %**[*]	75	0
0.45	0.392	**12.88 %**[*]	64	11
0.5	0.510	**−2.00 %**	37	38

4   Results

Tables 1, 2 and 3 show computed noise before and after applying the noise-correction process across different levels of injected noise. They also show the number of queries for which the noise decreased and the number of queries for which the noise increased after the process. To check if this reduction in noise was statistically significant, t-tests were performed using noise levels before and after the process across all the queries. Improvements marked by (*) symbol denote statistical significance with p-value < 0.05.

The 2-phase process reduces significant amount of noise up to noise level of 0.4. After this, curve takes a very steep turn and almost fails to reduce noise at statistically significant levels around noise level of 0.5. However, the process has been proved robust enough to correct errors even at high noise level of 0.45 in each of the 3 datasets.

The difference in noise reduction between OHSUMED and TREC-TD datasets is due to an inherent characteristic of the datasets. OHSUMED has 3 relevance labels $\{2, 1, 0\}$ and so it's preference set contains 3 kinds of document pairs: $(d_2, d_1), (d_1, d_0)$ & (d_2, d_0) from which anomalies are to be found. Whereas, TREC-TD datasets contain only 2 relevance labels $\{1, 0\}$ and so have only 1 kind of document pair (d_1, d_0). So the noise reduction is efficient in case of TREC-TD compared to OHSUMED in which there are mixed document pairs because of which error detection is difficult.

5   Conclusion and Future Scopes

This paper proposes a simple yet very efficient approach to correct the errors in pairwise preferences for learning to rank. The proposed approach was able to reduce up to 90 % of induced noise at statistically significant levels depending on the initial noise injected in it. The robustness of this process has also been checked by inducing different noise levels. On response to this, the process was able to correct errors at statistically significantly even at high noise level of 0.45. The proposed model has been checked on three different Learning to Rank data-sets and shown to work efficiently on each of them.

In reality some documents are difficult to assign relevance than others. All mistakes are not equally probable. So, a more realistic method for noise injection which considers this can help to better evaluate this approach. Apart of that, reduction in noise of pairwise document preferences should have direct positive impact on efficiency of pairwise learning to rank algorithms. Different Learning to Rank algorithms have different levels of robustness against noise [7]. Hence, as a future work, it would also be interesting to analyse the effect of noise correction of training data on efficiency of various pairwise learning to rank algorithms.

References

1. Bailey, P., et al.: Relevance assessment: are judges exchangeable and does it matter. In: Proceedings of the 31st Annual International ACM SIGIR Conference on Research and Development in Information Retrieval, pp. 667–674. ACM (2008)
2. Brodley, C.E., Friedl, M.A.: Identifying mislabeled training data. J. Artif. Intell. Res. **11**, 131–167 (1999)
3. Geng, X., et al.: Selecting optimal training data for learning to rank. Inf. Process. Manag. **47**(5), 730–741 (2011)
4. Hall, M., et al.: The WEKA data mining software: an update. ACM SIGKDD Explor. Newslett. **11**(1), 10–18 (2009)
5. Hang, L.I.: A short introduction to learning to rank. IEICE Trans. Inf. Syst. **94**(10), 1854–1862 (2011)

6. Liu, T.-Y.: Learning to rank for information retrieval. Found. Trends Inf. Retrieval **3**(3), 225–331 (2009)

7. Niu, S., et al.: Which noise affects algorithm robustness for learning to rank. Inf. Retrieval J. **18**(3), 215–245 (2015)

8. Qin, T., et al.: LETOR: a benchmark collection for research on learning to rank for information retrieval. Inf. Retrieval **13**(4), 346–374 (2010)

9. Voorhees, E.M.: Variations in relevance judgments, the measurement of retrieval effectiveness. Inf. Process. Manag. **36**(5), 697–716 (2000)

10. Voorhees, E., Harman, D.: Overview of the fifth text retrieval conference (TREC-5). In: NIST Special Publication SP, pp. 1–28 (1997)

11. Jingfang, X., et al.: Improving quality of training data for learning to rank using click-through data. In: Proceedings of the Third ACM International Conference on Web Search and Data Mining, pp. 171–180. ACM (2010)

Table Topic Models for Hidden Unit Estimation

Minoru Yoshida[✉], Kazuyuki Matsumoto, and Kenji Kita

Institute of Technology and Science, University of Tokushima,
2-1, Minami-josanjima, Tokushima 770-8506, Japan
{mino,matumoto,kita}@is.tokushima-u.ac.jp

Abstract. We propose a method to estimate hidden units of numbers written in tables. We focus on Wikipedia tables and propose an algorithm to estimate which units are appropriate for a given cell that has a number but no unit words. We try to estimate such hidden units using surrounding contexts such as a cell in the first row. To improve the performance, we propose the table topic model that can model tables and surrounding sentences simultaneously.

1   Introduction

Numbers in text documents can be a good source of knowledge. In comparison to the number of reports on the mining of numeric data stored in databases, those focusing on numbers written in text documents have been few.

Tables are a standard medium for expressing numeric data in documents. However, the problem associated with numbers written in tables is that some numbers are provided with no units when they are obvious from the contexts. (Imagine a table of ranking in some sports event. Typically, in such tables, the "ranks" of the competitors are provided without any unit word such as "rank".) We consider the problem of recovering units omitted from numbers in tables.

The estimation of such hidden units greatly contribute to the application on tables. For example, the number of cells that match the query about numbers in tables in the form of "number + unit" will increase and the search or mining results will be greatly improved.

The contributions of this study are two-fold. First, we propose a new task of estimating hidden units of numbers, especially focusing on tables. Second, we propose a *table topic model*, that can naturally model the tables and its surrounding sentences. We tested our model using regularized logistic regressions on the hidden-unit estimation task.

2   Related Work

To the best of our knowledge, this is the first work that proposes a task of estimating hidden units in tables. However, there are some works considering related problems.

Although there have been few research efforts in analyzing numbers written in text, some researchers have attempted the task. Yoshida et al. [10] proposed

© Springer International Publishing AG 2016
S. Ma et al. (Eds.): AIRS 2016, LNCS 9994, pp. 302–307, 2016.
DOI: 10.1007/978-3-319-48051-0_23

to find frequent adjacent strings to number range queries. Although this research can be used to estimate units of numbers in free text, numbers in tables show different behaviors, and a different approach is needed. Narisawa et al. [5] proposed collecting numerical expressions to provide some "common sense" about numbers. Takamura and Tsujii [8] proposed to obtain numerical attributes of physical objects from the numerical expressions found in texts. Although their research is more application-oriented, our unit estimation algorithm improves the coverage of such systems by providing clues for semantics of numbers.

There are many researches on using HTML tables as a source of knowledge. For example, [3] provided a system that can extract knowledge about attributes of objects from a large number of Web tables. Recently, Sarawagi et al. [7] proposed a system that can answer numerical queries on Web tables. Their method assumes that units are explicitly shown in Web tables, and our unit estimation algorithm can be used for preprocessing for such systems. Govindaraju et al. [4] proposed to merge knowledge extracted from texts and tables using standard NLP toolkits. Wang et al. [9] proposed a method to classify the sentences around a table into table-related ones or not. To the best of our knowledge, the current paper is the first work that applies topic models to analyze the table semantics, especially to find relations between tables and the sentences around them. Many previous researches have tried to find attribute positions in tables [11]. However, we can not avoid estimation errors by using such methods as preprocessing. Rather, we decided not to estimate attribute positions, and consider both positions (i.e., the first row and the first column) as features instead.

3   Problem Setting and Data

We used the Japanese Wikipedia articles downloaded in 2013 as our corpus.[1] A large portion of the number cells (i.e., the cells that contain numbers) in Wikipedia tables omit units.[2] Randomly-selected 482 documents (all of which contain one or more tables) are used in our experiments. Among them, we randomly selected 297 number cells that have no unit and annotated them with appropriate units by hand. We call them *hidden units*.

The training data for our system is a list of (x_i, y_i) pairs where each x_i, or *data*, represents each cell, consisting of a list of the number in the cell and context words related to the cell.[3] y_i, or *label*, is a unit in the cell. In our problem setting, the task is to estimate hidden *label* (unit) given each *data*.

As features for classification, we use the number itself (described in later sections)[4] and the context words. Context words are the words in the positions that are likely to be related to the cell, such as the top cell (i.e., the cell in the

---

[1] The total number of tables founded in the corpus was 255,039.

[2] We observed that 39.1 % cells out of randomly selected number-cells were number-only cells (i.e., cells without any unit).

[3] Note that in this paper the x_i is assumed to be a vector whose value is 1 in the i-th dimension where i is the ID of every context word for the cell.

[4] If we see the number "1987", we think of it as a number that indicates a year.

first row) in the same column used for estimation of hidden units.[5] We use the
following ones.

Same Column and Row: the words and numbers found in the cells related
to the current cell. We take *all* cells in the same column (same as the current
cell), and the *leftmost* cell (i.e., cell in the first column) in the same row
(same as the current cell).[6]

Headings: Subtitles in the documents are extracted using Wiki rules and the
subtitle nearest to the current table is used as a context.

4   Hidden Unit Estimation Using Table Topic Models

We use (regularized) logistic regression [1] as a standard algorithm for multi-
class classification. In addition to the contexts described in the previous section,
we also use topic IDs related to the target cell as features for logistic regression.
For topic modeling, we used a *variant* of Latent Dirichlet Allocation (LDA) [2].

LDA assumes for each word in documents the following generative processes:

1. Select parameter θ_d according to the Dirichlet distribution: $\theta \sim Dir(\alpha)$
2. For each word,
   (a) Select topic z according to the multinomial distribution: $z \sim Multi(\theta_d)$
   (b) Select word w according to the word distribution of topic z: $p(w|z)$

where d is the id of the current document. Topic z is typically estimated by
Gibbs sampling after θ is integrated out.

On the other hand, our table topic model is a variant of LDA which assigns
one topic to each *column* of tables based on the assumption that all the cells
in the same column in tables belong to the same topic. It is almost always true
when we look at the row-wise tables (i.e., tables in which attributes are lined in
one row). We observed that most of the tables in our data set were row-wise[7],
so the above assumption is valid to some extent for Wikipedia tables. Topics are
shared between tables and sentences outside the tables, which enables us to take
advantage of the sentences for better estimation of table topics.

We regard each column of tables as a list of words belonging to the same
topic, resulting in one topic for each column. Therefore, we model each column
(denoted by the word list $C = < c_1, c_2, \cdots, c_{|C|} >$) by Mixture of Unigrams
(Multinomial Mixtures) assuming the following generative processes:

1. Select topic z according to the multinomial distribution: $z \sim Multi(\theta_d)$
2. For each word, select word c_i according to the word distribution of topic z:
   $p(c_i|z)$.

---

[5] For example, if the unit word is "yen", the surrounding words are likely to contain
the word "price".

[6] We observed that using *all* "same row" cells worsen the accuracy in preliminary
experiments, so we do not use those cells.

[7] In our data set, 266 (93.7%) out of 284 tables (which is the tables that contains one
or more hand-annotated cells) were row-wise.

Gibbs sampling for this model is similar to that for LDA. For each column, we calculate the probabilities when we assign each topic to the column, and sample the topic according to the calculated probabilities. Note that the topic sampling for the words outside of the tables is the same as LDA.

4.1 Models for Numbers

We propose a new model for numbers that is different from Naive ones (e.g., replacing them with zeros). Our idea is to use staged models[8] for modeling a whole number by using the idea of "significant digits" that are popular in scientific measurements. Here, each number d is converted into a list of digits e_i (we call them "codes") which consists of the position of the most significant digits followed by the most N significant digits.[9,10]

For example, the number string "95300"[11] can be converted into the list <4, 9, 5, 3> because its significant-digit expression is 9.53×10^4. Note that taking some first digits of the list corresponds to the abstraction of source number d. Because currently we set $N = 2$, we use the list <4, 9, 5> for representing "95300", ignoring the final digit.

In the table topic model, each digit is assumed to be generated from the multinomial distribution, and the multinomial distribution is assumed to be generated from the Dirichlet distribution.

1. Each digit e_i is drawn from the distribution $H_{e_1...e_{i-1}}$: $e_i \sim H_{e_1...e_{i-1}}$.
2. Each distribution H_i is drawn from the Dirichlet distribution $Dir(H)$.

where $H_{e_1...e_{i-1}}$ is the multinomial distribution defined for each sequence of digits $e_1 \ldots e_{i-1}$. We simply assume the Dirichlet parameter H is a uniform (i.e., the value is the same for all digits.)

For logistic regression, we used the abstracted number expressions themselves as features. For example, the number "95300" is converted into the code <4,9,5>, resulting in the feature N495. We call this expression "quantization". This scheme is equivalent to using the "significant digit" expressions directly. We also consider another expression which we call "history", which include all the history before reaching to the whole expression, e.g., N4, N49, N495 for the above case for the purpose of comparison with the generative model mentioned above.

5 Experiments

We conducted the experiments on our hand-annotated corpus described in Sect. 3. We consider only units that appeared two or more times in the corpus (41 unit types in total). This resulted in reducing the size of test set from

---

[8] It is inspired by the Polya-tree models for modeling of continuous values.

[9] We also use some additional digits such as for signs, but omit them here for the sake of simplicity.

[10] We set $N = 2$ currently.

[11] We use some rules to parse the number string, so different expressions like "95,300" are also available.

297 cells into 270 cells. Logistic regression was performed using Classias [6]. We used the L1-regularization with the parameter $C = 0.1$, which performed the best in our preliminary experiments. We performed the 5-fold cross validation by dividing each corpus into five subsets.[12]

We estimated the topics for every word in the sentences and every column in the tables. We tested several settings for the number of topics, and selected 10 which performed the best in our preliminary experiments. We ran Gibbs sampling 5 times and all the accuracy values were averaged.[13]

We compared our model with the baselines that does not use topic IDs ("no topic"). We used two types of topics: one was estimated using both tables and their surrounding sentences ("topic2") and the other was estimated using tables only ("topic1"). We also tested two baseline number expressions: "nonum" which uses no number expressions as features, and "raw", which uses number expression as is (as mere string).

Table 1 shows the results. We observed that using the estimated topics improved performance except for the case of the "history" expression. It suggests that the estimated topics effectively modeled the words outside the tables, which contributed to improving the classification performance for cells inside the tables, as well as clustering effects which assemble the columns with the same topic ID, which increased effective features used to estimate the hidden units.

We observed the best performance when we used the "raw" expressions and topic IDs. We think this is mainly because our Table Topic Model uses coded expressions of numbers, which is the expressions different from the "raw" expression, thus topic IDs and "raw" numbers worked as complementary features to each other.

The result for topic2 was slightly better than topic1, which suggests that modeling the tables and surrounding sentences simultaneously have a good effect on modeling the table topics.

Table 1. Estimation accuracies for various settings

Number experssion	No topic	Topic1	Topic2
Nonum	69.63	74.30	74.67
History	72.96	72.96	73.18
Quantization	71.11	74.96	75.71
Raw	69.63	76.00	76.60

---

[12] We divided the corpus in such a way that the cells from the same table are not included in the same subset. The accuracy is calculated by summing up the correct/incorrect of predictions on each cell, i.e., the accuracy is micro-averaged one.

[13] Each Gibbs sampling performed 500 iterations. The distribution of the sampled topic IDs in the final 200 iterations were used as the input features for the logistic regression (i.e., we added each topic ID observed for the column of each cell in the test data with their relative frequency as a weight.).

6   Conclusion and Future Work

We proposed a method to estimate "hidden units" of number cells in tables, which uses the new topic model for tables. Experiments showed that table topics contributed to improving accuracies. Future work includes further investigation of modeling the texts around the tables that considers linguistic features such as dependency relations.

Acknowledgement. This work was supported by JSPS KAKENHI Grant Numbers JP15K00309, JP15K00425, JP15K16077.

References

1. Andrew, G., Gao, J.: Scalable training of l1-regularized log-linear models. In: Proceedings of ICML 2007, pp. 33–40 (2007)
2. Blei, D.M., Ng, A.Y., Jordan, M.I.: Latent Dirichlet allocation. J. Mach. Learn. Res. **3**, 993–1022 (2003)
3. Cafarella, M.J., Halevy, A.Y., Wang, D.Z., Wu, E., Zhang, Y.: Webtables: exploring the power of tables on the web. Proc. VLDB Endow. **1**(1), 538–549 (2008)
4. Govindaraju, V., Zhang, C., Re, C.: Understanding tables in context using standard NLP toolkits. In: Proceedings of ACL2013 (2013)
5. Narisawa, K., Watanabe, Y., Mizuno, J., Okazaki, N., Inui, K.: Is a 204 cm man tall or small? Acquisition of numerical common sense from the web. In: Proceedings of the ACL, vol. 1, pp. 382–391 (2013)
6. Okazaki, N.: Classias: a collection of machine-learning algorithms for classification. http://www.chokkan.org/software/classias/
7. Sarawagi, S., Chakrabarti, S.: Open-domain quantity queries on web tables: annotation, response, and consensus models. In: Proceedings of KDD, pp. 711–720 (2014)
8. Takamura, H., Tsujii, J.: Estimating numerical attributes by bringing together fragmentary clues. In: Proceedings of NAACL-HLT2015 (2015)
9. Wang, H., Liu, A., Wang, J., Ziebart, B.D., Yu, C.T., Shen, W.: Context retrieval for web tables. In: Proceedings of ICTIR 2015, pp. 251–260 (2015)
10. Yoshida, M., Sato, I., Nakagawa, H., Terada, A.: Mining numbers in text using suffix arrays and clustering based on Dirichlet process mixture models. In: Zaki, M.J., Yu, J.X., Ravindran, B., Pudi, V. (eds.) PAKDD 2010. LNCS, vol. 6119, pp. 230–237. Springer, Heidelberg (2010)
11. Zanibbi, R., Blostein, D., Cordy, J.R.: A survey of table recognition. Int. J. Doc. Anal. Recogn. **7**(1), 1–16 (2004)

Query Subtopic Mining Exploiting Word Embedding for Search Result Diversification

Md Zia Ullah[(✉)], Md Shajalal, Abu Nowshed Chy, and Masaki Aono

Department of Computer Science and Engineering,
Toyohashi University of Technology, Toyohashi, Aichi, Japan
{arif,shajalal,nowshed}@kde.cs.tut.ac.jp, aono@tut.jp

Abstract. Understanding the users' search intents through mining query subtopic is a challenging task and a prerequisite step for search diversification. This paper proposes mining query subtopic by exploiting the word embedding and short-text similarity measure. We extract candidate subtopic from multiple sources and introduce a new way of ranking based on a new novelty estimation that faithfully represents the possible search intents of the query. To estimate the subtopic relevance, we introduce new semantic features based on word embedding and bipartite graph based ranking. To estimate the novelty of a subtopic, we propose a method by combining the contextual and categorical similarities. Experimental results on NTCIR subtopic mining datasets turn out that our proposed approach outperforms the baselines, known previous methods, and the official participants of the subtopic mining tasks.

Keywords: Subtopic mining · Word embedding · Diversification · Novelty

1 Introduction

According to user search behavior analysis, query is usually unclear, ambiguous, or board [10]. Issuing the same query, different users may have different search intents, which correspond to different subtopic [8]. For example, with an ambiguous query such as "eclipse," users may seek different interpretations, including "eclipse IDE," "eclipse lunar," and "eclipse movie." With a broad query such as "programming languages," users may be interested in different subtopic, including "programming languages java," "programming languages python," and "programming languages tutorial." However, it is not clear which subtopic of a broad query is actually desirable for a user [11]. Search engine often fails to capture the diversified search intents of a user if the issued query is ambiguous or broad and results in a list of redundant documents. As these documents may cover a few subtopic or interpretations, the user is usually unsatisfied.

In this paper, we address the problem of *query subtopic mining*, which is defined as: "given a query, list up its possible subtopic which specialises or disambiguates the search intent of the original query." In this regard, our contributions are threefold: (1) some new features based on word embedding,

© Springer International Publishing AG 2016
S. Ma et al. (Eds.): AIRS 2016, LNCS 9994, pp. 308–314, 2016.
DOI: 10.1007/978-3-319-48051-0_24

(2) a bipartite graph based ranking for estimating the relevance of the subtopic, and (3) estimating the novelty of the subtopic by combining a mutual information based similarity and categorical similarity.

The rest of the paper is organized as follows. Section 2 introduces our proposed subtopic mining approach. Section 3 discusses the overall experiments and results that we obtained. Finally, concluding remarks and some future directions of our work are described in Sect 4.

2   Mining Query Subtopic

In this section, we describe our approach to query subtopic mining, which is composed of subtopic extraction, features extraction, and ranking. Given a query, first we extract candidate subtopics from multiple resources. Second, we extract multiple semantic and content-aware features to estimate the relevance of the candidate subtopics, followed by a supervised feature selection and a bipartite graph based ranking. Third, to cover the possible search intents, we introduce a novelty measure to diversify the subtopic.

2.1   Subtopic Extraction

Inspired by the work of Santos [9], our hypothesis is that *suggested queries* in across search engines hold some intents of the query. For a query, we utilize the suggested queries, provided by the search engines. If a query is matched with the title of a Wikipedia disambiguation page, we extract the different meanings from that page. Then, we aggregate the subtopic by filtering out the candidates, which is the part of the query or exactly similar.

2.2   Features Extraction

Let $q \in \mathcal{Q}$ represents a query and $\mathcal{S} = \{s_1, s_2, ..., s_k\}$ represents a set of candidate subtopics extracted in Sect. 2.1. We extract multiple local and global features, which are broadly organized as word embedding and content-aware features. We propose two semantic features based on locally trained word embedding and make use of *word2vec*[1] model [6].

In order to capture of the semantic matching of a query with a subtopic, we first propose a new feature, the maximum word similarity (MWS) as follows:

$$f_{MWS}(q, s) = \frac{1}{|q|} \sum_{t \in q} sem(t, s)$$

$$sem(t, s) = \max_{w \in s} f_{sem}(\boldsymbol{t}, \boldsymbol{w})$$

(1)

where $\boldsymbol{t}$ and $\boldsymbol{w}$ are the word vector representations from *word2vec* model, corresponding to two words t and w, respectively. The function f_{sem} returns the cosine similarity between two word vectors.

---

[1] *word2vec* (https://code.google.com/p/word2vec/).

To estimate the global importance of a query with a subtopic, we propose our second feature, the mean vector similarity (MVS) as follows:

$$f_{MVS}(q,s) = f_{sem}\left(\frac{1}{|q|}\sum_{t\in q} t, \frac{1}{|s|}\sum_{w\in s} w\right) \tag{2}$$

Among content-aware features, we extract features based on term frequency, including *DPH* [9], *PL2* [9], and *BM25* [9]; language modeling, including Kullback-Leibler (*KL*) [9], Query Likelihood with Jelinek-Mercer (*QLM-JM*) [9], Query Likelihood with Dirichlet smoothing (*QLM-DS*) [9], and Term-dependency Markov random field (*MRF*) [9]; lexical, including edit distance, sub-string match (*SSM*) [5], term overlap [5], and term synonym overlap (*TSO*) [5]; web hit-count, including normalized hit count (*NHC*), point-wise mutual information (*PMI*), and word co-occurrence (*WC*); and some query independent features, including average term length (*ATL*), topic cohesiveness (*TC*) [9], and subtopic length (*SL*).

2.3   Subtopic Ranking

To remove noisy and redundant features, we normalize the features using *Min-Max* and employ elastic-net regularized regression.

Bipartite Graph Based Ranking. Many real applications can be modeled as a bipartite graph, such as Entities and Co-List [1] in a Web page. We hypothesize that a relevant subtopic should be ranked at the higher position by multiple effective features and intuitively, an effective feature should be ranked at higher position by multiple relevant subtopics. On this intuition, we represent a set of features $\mathcal{F} = \{f_1, f_2, \cdots, f_m\}$ and a set of candidate subtopics $\mathcal{S} = \{s_1, s_2, \cdots, s_n\}$ as a bipartite graph, $\mathcal{G} = (\mathcal{F} \cup \mathcal{S}, \mathcal{E})$, and introduce weight propagations from both sides. The weight $w_{i,j} = 1/\sqrt{\log_2(rank(L_{f_i}, s_j) + 2.0)}$ of an edge between a feature f_i and a subtopic s_j, where $rank(L_{f_i}, s_j)$ returns the position of the subtopic s_j in the ranked list L_{f_i} for the feature f_i.

Let $\mathcal{M}$ be a bi-adjacency matrix of $\mathcal{G}$, $\mathcal{W}_1 = \mathcal{D}_{\mathcal{F}}^{-1}\mathcal{M}$, and $\mathcal{W}_2 = \mathcal{D}_{\mathcal{S}}^{-1}\mathcal{M}^T$, where $\mathcal{D}_{\mathcal{F}}$ and $\mathcal{D}_{\mathcal{S}}$ are the row-diagonal and the column-diagonal matrices of $\mathcal{M}$.

The weight propagations from the set $\mathcal{S}$ to the set $\mathcal{F}$ and vice versa are represented as follows:

$$\begin{aligned} F_{k+1} &= \lambda_1 \mathcal{W}_1 S_k + (1 - \lambda_1)\ F_0 \\ S_{k+1} &= \lambda_2 \mathcal{W}_2 F_{k+1} + (1 - \lambda_2)\ S_0 \end{aligned} \tag{3}$$

where $0 < \lambda_1, \lambda_2 < 1$, F_0 and S_0 are the initial weight vectors, F_k and S_k denotes the weight vectors after the k-th iterations.

From the iterative solution of the Eq. (3), we have

$$S^* = (I - \lambda_1\lambda_2\mathcal{W}_2\mathcal{W}_1)^{-1}\ [(1 - \lambda_1)\lambda_2\mathcal{W}_2 F_0 + (1 - \lambda_2)\ S_0] \tag{4}$$

Given λ_1, λ_2, $\mathcal{W}_1$, $\mathcal{W}_2$, F_0, and S_0, we estimate the scores S^* directly by applying Eq. (4). These scores S^* are considered as the relevance scores, rel(q, $\mathcal{S}$), which is utilized in diversification.

Subtopic Diversification. To select the maximum relevant and the minimum redundant subtopic, we diversify the subtopic using the MMR [2] framework, which can be defined as follows:

$$s_i^* = \underset{s_i \in R \backslash C_i}{\operatorname{argmax}} \ \gamma \ rel(q, s_i) + (1 - \gamma) \ novelty(s_i, C_i) \tag{5}$$

where $\gamma \in [0, 1]$, $rel(q, s_i)$ is the relevance score, and $novelty(s_i, C_i)$ is the novelty score of the subtopic s_i. R is the ranked list of subtopic retrieved by Eq. (4). C_i is the collection of subtopic that have already been selected at the i-th iteration and initially empty.

Since subtopics are short in length and they might not be lexically similar. We hypothesize that if two subtopics represent the similar meaning, they may belong to the similar categories and retrieve similar kinds of documents from a search engine. Therefore, we propose to estimate the novelty of a subtopic by combining the contextual and categorical similarities as follows:

$$novelty \ (s_i, C_i) = - \max_{s' \in C_i} \left(1.0 - \sqrt{JSD(s_i, s')} + \sum_{x \in X} \frac{[(s_i, s') \in x]}{|x|} \right) \tag{6}$$

where $JSD(s_i, s')$ is estimated through the Jensen-Shannon divergence of the word probability distributions of the top-k documents refer to the subtopics s_i and s'. X is the set of clusters obtained by applying the frequent phrase based soft clustering on the candidate subtopics, $|x|$ is the number of subtopics belong to the cluster x, and $[(s_i, s') \in x] = 1$ if true, zero, otherwise.

3 Experiments and Evaluations

In this section, we evaluate our proposed method (*W2V-BGR-Nov*) and compare the performance with previous methods, including [3,4,7], the diversification methods MMR [2], XQuAD [9], and the baseline, MergeBGY, merging of query completions from Bing, Google, and Yahoo. For relevance estimation, linear ranking is used in the MergeBGY, whereas Eq. (4) is used for MMR and XQuAD. Moreover, the cluster label of the frequent phrase based soft clustering of candidates is considered as the sub-topics for XQuAD. For estimating novelty, cosine similarity is utilized for MergeBGY, MMR, and XQuAD. We estimate evaluation metrics, including I-rec@10, D-nDCG@10, and D#-nDCG@10; and use the two-tailed paired t-test for statistical significance testing (p < 0.05).

3.1 Dataset

The INTENT-2 and IMINE-2 [12] test collections include 50 and 100 topics, respectively. As resources, query completions from Bing, Google, and Yahoo were collected and included in the datasets. To estimate the features, including Eqs. (1) and (2), we retrieved the top-1000 documents from the clueweb12-b13 corpus based on language model for each topic and locally trained *word2vec*. The parameters in the *word2vec* tool are Skip-gram architecture, window width of 10, dimensionality of 200, and the sampling threshold of 10^{-3}.

3.2   Important Features and Parameter Tuning

We trained elastic-net on INTENT-2 dataset, however, we employed on IMINE-2 dataset, and vice versa. We extracted in total 27 features and the selected features were as follows: *MWS, MVS, DPH, QLM-JM, MRF, SSM, TSO, NHC, WC,* and *ATL*. It turned out that our proposed features *MWS* and *MVS* are important and were chosen during feature selection. Through empirical evaluation, we found the optimal insensitive range of values of λ_1 and λ_2 in Eq. (4) as [0.6 – 0.8] and [0.4 – 0.6], respectively. We found the optimal value of γ in Eq. (5) as 0.85, which reflects that *MMR* rewards relevance than diversity in mining subtopic.

3.3   Experimental Results

The comparative performances are reported in Tables 1 and 2 for INTENT-2 and IMINE-2 topics. The results show that overall *W2V-BGR-NOV* is the best. In terms of diversity (i.e. I-rec@10), *W2V-BGR-NOV* significantly outperforms all baselines except [3] for INTENT-2 topics. Though previous methods utilize multiple resources which often cause noisy subtopics, however, our proposed estimation of subtopic novelty in Eq. (6) eliminates redundant subtopics and benefits more diverse subtopics. In terms of relevance (i.e. D-nDCG@10), *W2V-BGR-NOV* outperforms all baselines except HULTECH-Q-E-1Q for IMINE-2 topics. Our proposed word embedding based features, followed by bipartite graph based ranking capture better semantics to estimate the relevance of the subtopics.

Table 1. Comparative performance of our proposed W2V-BGR-NOV with previous methods for INTENT-2 topics. The best result is in bold. † indicates statistically significant difference and ◇ indicates statistically indistinguishable from the best

	Method	I-rec@10	D-nDCG@10	D#-nDCG@10
Our proposed	*W2V-BGR-NOV*	**0.4774**†	**0.5401**†	**0.5069**†
Baseline	MergeBGY	0.3365	0.3181	0.3273
Previous methods	Kim and Lee [4]	0.4457	0.4401	0.4429
	Moreno et al. [7]	0.4249	0.4221	0.4225
	Damien et al. [3]	0.4587◇	0.3625	0.4106
Diversification methods	XQuAD [9]	0.3637	0.5055	0.4346
	MMR [2]	0.3945	0.4079	0.4048

Table 2. Comparative performance of our proposed W2V-BGR-NOV with the known previous methods for IMINE-2 topics. Notation conventions are the same as in Table 1.

	Method	I-rec@10	D-nDCG@10	D#-nDCG@10
Our proposed	*W2V-BGR-NOV*	**0.8349**†	**0.6836**†	**0.7602**†
IMINE-2 participants	KDEIM-Q-E-1S	0.7557	0.6644	0.7101
	HULTECH-Q-E-1Q	0.7280	0.6787◇	0.7033
	RUCIR-Q-E-4Q	0.7601	0.5097	0.6349
Diversification methods	XQuAD [9]	0.6422	0.6571	0.6510
	MMR [2]	0.7572	0.6112	0.6908

In terms of D#-nDCG@10, which is a combination of I-rec@10 (0.5) and D-nDCG@10 (0.5), *W2V-BGR-NOV* significantly outperforms all the baselines. The overall result demonstrates that our proposed *W2V-BGR-NOV* is effective in query subtopic diversification.

4 Conclusion

In this paper, we proposed mining and ranking query subtopic by exploiting word embedding and short-text similarity measure. We introduced new features based on word embedding and bipartite graph based ranking to estimate the relevance of the subtopic. To diversify the subtopic covering multiple intents, we proposed to estimate the novelty of a subtopic by combining the contextual and categorical similarities. Experimental results demonstrate that our proposed approach outperforms the baseline and known previous methods. In the future, we will evaluate the effectiveness of the mined subtopics by employing search diversification.

Acknowledgement. This research was supported by JSPS Grant-in-Aid for Scientific Research (B) 26280038.

References

1. Cao, L., Guo, J., Cheng, X.: Bipartite graph based entity ranking for related entity finding. In: 2011 IEEE/WIC/ACM International Conference on Web Intelligence and Intelligent Agent Technology (WI-IAT), pp. 130–137. IEEE (2011)
2. Carbonell, J., Goldstein, J.: The use of MMR, diversity-based reranking for reordering documents and producing summaries. In: SIGIR, pp. 335–336. ACM (1998)
3. Damien, A., Zhang, M., Liu, Y., Ma, S.: Improve web search diversification with intent subtopic mining. In: Zhou, G., Li, J., Zhao, D., Feng, Y. (eds.) NLPCC 2013. CCIS, vol. 400, pp. 322–333. Springer, Heidelberg (2013)
4. Kim, S.J., Lee, J.H.: Subtopic mining using simple patterns and hierarchical structure of subtopic candidates from web documents. Inf. Process. Manag. **51**(6), 773–785 (2015)
5. Metzler, D., Kanungo, T.: Machine learned sentence selection strategies for query-biased summarization. In: SIGIR Learning to Rank Workshop, pp. 40–47 (2008)
6. Mikolov, T., Sutskever, I., Chen, K., Corrado, G.S., Dean, J.: Distributed representations of words and phrases and their compositionality. In: NIPS, pp. 3111–3119 (2013)
7. Moreno, J.G., Dias, G., Cleuziou, G.: Query log driven web search results clustering. In: SIGIR, pp. 777–786. ACM (2014)
8. Ren, P., Chen, Z., Ma, J., Wang, S., Zhang, Z., Ren, Z.: Mining and ranking users intents behind queries. Inf. Retr. J. **18**(6), 504–529 (2015)
9. Santos, R.L.T.: Explicit web search result diversification. Ph.D. thesis, University of Glasgow (2013)
10. Song, R., Luo, Z., Nie, J.Y., Yu, Y., Hon, H.W.: Identification of ambiguous queries in web search. Inf. Process. Manag. **45**(2), 216–229 (2009)

11. Wang, C.J., Lin, Y.W., Tsai, M.F., Chen, H.H.: Mining subtopics from different aspects for diversifying search results. Inf. Retr. **16**(4), 452–483 (2013)
12. Yamamoto, T., Liu, Y., Zhang, M., Dou, Z., Zhou, K., Markov, I., Kato, M.P., Ohshima, H., Fujita, S.: Overview of the NTCIR-12 IMine-2 task. In: NTCIR (2016)

Assessing the Authors of Online Books in Digital Libraries Using Users Affinity

B. de La Robertie[✉]

Université de Toulouse, IRIT UMR5505, 31071 Toulouse, France
baptiste.delarobertie@irit.fr

Abstract. Information quality generated by crowd-sourcing platforms is a major concern. Incomplete or inaccurate user-generated data prevent truly comprehensive analysis and might lead to inaccurate reports and forecasts. In this paper, we address the problem of assessing the authors of users generated published books in digital libraries. We propose to model the platform using an heterogeneous graph representation and to exploit both the users' interests and the natural inter-users affinities to infer the authors of unlabelled books. We formalize the task as an optimization problem and integrate in the objective a prior of consistency associated to the networked users in order to capture the neighboors' interests. Experiments conducted over the *Babellio* platform (http:// babelio.com/), a French crowd-sourcing website for book lovers, achieved successful results and confirm the interest of considering an affinity-based regularization term.

Keywords: User-generated-content · Labels propagation · Classification

1 Introduction

Over the past decade, crowd-sourcing platforms have entered mainstream usage and rapidly become valuable organizational resources, offering rich heterogeneous and relational data. However, to properly exploit the user-generated data and to produce comprehensive analysis, associated digital business must face several issues of quality and consistency. Even by clamping down signups, meta-data associated to users generated contents can be doubtfull or incomplete, justifying the needs of quality and consistency assessment tools.

In this work, the challenge of assessing the authors of unlabelled books in digital libraries is addressed. An heterogeneous graph is used to represent the platform and the relations between the different entities and a classification problem is formulated to predict the authors of unlabelled nodes. The *homophily patterns* lying between the interests of the users and their friends are first empirically demonstrated. Based on this observation suggesting that close friends tend to have similar favorite readings, an affinity-based regularization term in integrated in a dedicated objective function in order to smooth latent representations of the users.

© Springer International Publishing AG 2016
S. Ma et al. (Eds.): AIRS 2016, LNCS 9994, pp. 315–321, 2016.
DOI: 10.1007/978-3-319-48051-0_25

The paper is organized as follow. Section 2 introduces previous research closely related to our problem. Section 3 motivates the general ideal of our work. Section 4 describes the proposal. Finally Sects. 5 and 6 provide experimental setup, evaluations and conclusions.

2   Related Work

Several research has empirically demonstrated [1,5,6,8] or exploited [2,4,10] many types of correlations between the structural properties of a graph and the associated users properties. Cook et al. [5] show that people's affinity networks are highly correlated with several behavioral and sociodemographic characteristics, exploring geography, family ties, education, social class and others. In [8], the social structure of the Facebook affinity network of several American institutions in studied. The authors has examined the *homophily patterns* using assortativity coefficients based on observed ties between nodes, considering both microscopic and macroscopic properties. They show different realizations of networks and, for example, observe that women are more likely to have friends within their common residence while this characteristic for male-only networks exhibit a larger variation. Backstrom et al. [1] have studied the ways in which communities grow over time, and more importantly, how different groups come together or attract new members. By taking the case of the *LiveJournal* platform, they have shown how the affinity graph structure of a member impacts his propensity to join new communities. Similar results have been suggested over the collaboration networks of scientists. For example, in [3], authors suggest that two researchers are more likely to collaborate if both have already collaborated with a third common scientist. As in [4,10], we suppose that two nodes connected in a network will tend to have similar latent representations. Thus, we propose to capture *homophily patterns* using an *affinity-based* regularization term.

3   Motivations

In this section, we make use of the affinity graph of the members of the *Babelio* platform to demonstrate that linked users tend to have similar favorite books.
Let consider the affinity relation V such that $(i,j) \in V$ iff user i and user j are friends on the platform. Let $\boldsymbol{f}_i^k$ be a characteristics vector such that $f_{i,j}^k$ is the number of books written by author j for which user i has given k stars (from 1 to 5). From the averaged distance function S^k formalized in Eq. (1), we define the *inter-relation* and *extra-relation* distances metric as follow:

$$S^k = \frac{1}{N} \sum_i \sum_{j \in \mathcal{N}_i} \frac{||\boldsymbol{f}_i^k - \boldsymbol{f}_j^k||_2}{|\mathcal{N}_i|} \tag{1}$$

For a neighborhood $\mathcal{N}_i = \{j : (i,j) \in V\}$, i.e., the friends of the user associated to node i, the *inter-relation* metric captures the averaged distance between all nodes and their neighboors. For $\mathcal{N}_i = \{j : (i,j) \notin V\}$, i.e., users who are not

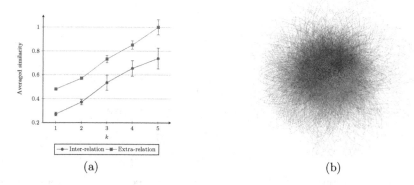

Fig. 1. (a) Normalized averaged *Intra-relations* and *extra-relations* measures for ratings $k \in \{1, 2, 3, 4, 5\}$ over the experimental graphs. (b) An affinity graph extracted from the *Babelio* platform colored by users' favorite authors. (Color figure online)

friends with node i, we define the *extra-relation* metric. In practice, the latter is defined over a random subset of $\mathcal{N}_i$ such that the neighboroods' size of both metrics are equal. Figure 1(a) reports the normalized evolution of both distances metrics in function of k over the four graphs used for the experimentations and described in Sect. 5.

Firstly, we observe that the *inter-relation* distance (in blue) is globally lower than the *extra-relation* one. In other words, connected nodes are more likely to read books of similar authors than non connected ones. This first observation constitutes the core of our proposal and justifies the regularization term proposed in Sect. 2 that constraints users to have similar latent representations. Secondly, from Fig. 1(b), which shows the main component of the affinity graph G_1 used in the experiments, we observe two distinct patterns. A color is associated to each author, and nodes are colored according to their favorite ones. Areas of uniform colors clearly reflect homophily patterns showing that users tend to naturally create communities sharing similar reading.

4   Model

Notations. Let $\mathcal{U} = \{u_i\}_{1 \leq i \leq n}$ be the set of users, $\mathcal{B} = \{b_j\}_{1 \leq j \leq m}$ the set of books and $\mathcal{A} = \{a_l\}_{1 \leq l \leq p}$ the set of authors, with $|\mathcal{U}| = n$, $|\mathcal{B}| = m$ and $|\mathcal{A}| = p$. Let $G_{pref} = (U_{pref}, V_{pref})$, with $U_{pref} = \mathcal{U} \cup \mathcal{B}$ and $V_{pref} = \{(u_i, b_j, v_{ij})\}_{i \leq n, j \leq m}$, be a bi-partite graph associating the interest $v_{ij} \in \mathbb{R}$ of user $u_i \in \mathcal{U}$ to book $b_j \in \mathcal{B}$. In addition, let $G_{friends} = (U_{friends}, V_{friends})$ with $U_{friends} = \mathcal{U}$ be an affinity graph: users u_i and u_j are friends iff $(u_i, u_j) \in V_{friends}$. Let $\boldsymbol{\alpha}_i \in \mathbb{R}^k$, $\boldsymbol{\beta}_j \in \mathbb{R}^k$ and $\boldsymbol{\gamma}_l \in \mathbb{R}^k$ be the latent representations of the users, books and authors respectively, with k being the dimension of the common latent space. Finally, let $\boldsymbol{y}_j \in \mathbb{R}^p$ be the labels vector associated to book

b_j. In particular, $y_{j,l} = 1$ if a_l is the author of book b_j, -1 otherwise. The goal is to reconstruct the labels vectors $\boldsymbol{y}_j$ for each unlabelled book.

Formulation. Predicting books' author is viewed as a classification task where the variable $y_{j,l} \in \{-1, +1\}$ has to be explained. In this work, we assume a set of linear classifiers per books, where the prediction $\tilde{y}_{j,l}$ for a pair $(b_j, a_l) \in \mathcal{B} \times \mathcal{A}$ is given by the linear model $f_l(b_j) = \langle \boldsymbol{\gamma}_l; \boldsymbol{\beta}_j \rangle$. Given a particular loss function $\Delta : \mathbb{R}^2 \to \mathbb{R}$, we propose to optimize the following objective:

$$\mathcal{L} = \sum_{(b_j, a_l) \in \mathcal{B} \times \mathcal{A}} \Delta(y_{j,l}, f_l(b_j)) + \sum_{(u_i, b_j) \in V_{pref}} d(\boldsymbol{\alpha}_i, \boldsymbol{\beta}_j) + \sum_{(u_i, u_j) \in V_{friends}} d(\boldsymbol{\alpha}_i, \boldsymbol{\alpha}_j) \quad (2)$$

The first term computes the classification error related to the authors' predictions associated to each book. A Hinge loss function $\Delta(y_{j,l}, f_l(b_j)) = \max(0, 1 - y_{j,l} f_l(b_j))$, which is suitable for classification problems, was used in our experiments. The last two terms are aimed to smooth and propagate the decision variables through the different relations and capture the proposed intuition. The regularization d is done using the L_2 norm. Therefore, close friends and related favorite books tend to have similar representations in $\mathbb{R}^k$. We call the last term the *affinity regularization term*. Finding the representations of the users, books and authors such that $\mathcal{L}$ is minimized is equivalent to solve:

$$(\boldsymbol{\alpha}^*, \boldsymbol{\beta}^*, \boldsymbol{\gamma}^*) = \arg\min_{\alpha, \beta, \gamma} \mathcal{L} \quad (3)$$

Since the Hinge loss is a convex function, standard approaches based on gradient descent can be used. In particular, we have:

$$\frac{\partial \mathcal{L}}{\partial \boldsymbol{\alpha}_i} = \sum_{(u_i, b_j) \in V_{pref}} 2(\boldsymbol{\alpha}_i - \boldsymbol{\beta}_j) + \sum_{(u_i, u_j) \in V_{friends}} 2(\boldsymbol{\alpha}_i - \boldsymbol{\alpha}_j) \quad (4)$$

$$\frac{\partial \mathcal{L}}{\partial \boldsymbol{\beta}_j} = \sum_{(u_i, b_j) \in V_{pref}} 2(\boldsymbol{\beta}_j - \boldsymbol{\alpha}_i) - \sum_{1 \leq l \leq p} y_{j,l} \boldsymbol{\gamma}_l \quad (5)$$

$$\frac{\partial \mathcal{L}}{\partial \boldsymbol{\gamma}_l} = \sum_{b_j \in \mathcal{B}} -y_{j,l} \boldsymbol{\beta}_j \quad (6)$$

In practice, we solved Eq. (3) using *L-BFGS* [7], a quasi-Newton method for nonlinear optimizations. The parameters $\boldsymbol{\alpha}$, $\boldsymbol{\beta}$, $\boldsymbol{\gamma}$ exhibit kn, km and kp decision variables respectively. Thus, our model parameters $\boldsymbol{\theta} = (\boldsymbol{\alpha}, \boldsymbol{\beta}, \boldsymbol{\gamma})$ define a metric space in $\mathbb{R}^{k(n+m+p)}$.

5   Experiments

Dataset. For the experiments, several subsets of the *Babelio*[1] platform were used. Founded in 2007, *Babelio* is an emerging French crowd-sourcing portal

---

[1] http://babelio.com/.

Table 1. Four graphs used for the experiments.

	G_1	G_2	G_3	G_4
Authors	5	10	50	100
Books	525	937	5 615	11 462
Users	5 425	7 178	19 659	25 297
Preferences	25 470	28 852	156 022	251 477
Affinities	60 067	93 548	259 360	312 792

for book lovers, where internauts can share their favorite readings. Members can critic books by leaving textual comments and assigning from 1 to 5 stars. These engagement signals are made public by the platform, allowing members to network each others using a friendship functionality. Table 1 summarized the four graphs used for the experiments.

Evaluation Metric. For every book j, let σ^j be the permutation over the authors induced by the predicted scores $\tilde{\boldsymbol{y}}_j$. The ranking induced by σ^j is evaluated using the *Normalized Discounted Cumulative Gain* [9], computed as follow:

$$NDCG(\sigma^j, k) = \frac{DCG(\sigma^j, k)}{DCG(\sigma^{j,*}, k)} \text{ with } DCG(\sigma^j, k) = \sum_{i=1}^{k} \frac{2^{y_{j,\sigma^j(i)}} - 1}{\log(1+i)}$$

where $\sigma^{j,*}$ is the optimal ranking for book j, consisting in placing the real authors of a book in first positions. Thus, we capture how far the prediction is from the optimal rank. The average of the NDCG values over all the books is reported.

Protocol. For each graph, two optimizations, with identical initial values, are performed:

- **Prefs. + Aff.** The proposed objective as formalized in Equation (2).
- **Prefs. only.** The proposed objective without considering the *affinity regularization term*.

Since the initialization may affect the solution, only the best runs according to the introduced evaluation metric are reported. For each run, the dataset is randomly splitted into a train and a test datasets as follow: for each author, $x\%$ of his books are used for training, and the rest for testing. Results over the test dataset are reported.

Results. Several values of k have been tested and only the best runs are reported. Results are summarized in Table 2. Proposed solution globally improves the baseline in generalization by roughly 3 %, confirming our intuition and the interest of smoothing the users representations.

Table 2. Evaluation of the solutions using the NDCG metric over the test datasets.

| | Training | | | | | | | |
| | 10 % | | | | 50 % | | | |
Graph	G_1	G_2	G_3	G_4	G_1	G_2	G_3	G_4
Prefs.	64.79	56.18	59.82	58.61	68.72	60.20	63.81	62.06
Prefs. + Aff.	65.52	59.04	61.12	60.76	69.59	64.91	67.57	64.97
Inprovement	+1.11 %	+4.84 %	+2.12 %	+3.53 %	+1.25 %	+7.25 %	+5.56 %	+4.47 %

6   Conclusions

We address the problem of assessing the authors of unlabelled books in digital libraries. To this end, the *homophily patterns* lying between the interests of the users and their friends are empirically demonstrated and incorporated as a regularization term in a dedicated objective function. By postulating that friends are more likely to share favorite readings, we force connected node to have similar representations. Experiments demonstrate significant quality improvement compared to the baseline that does not consider inter-users relationship. As future work, we will pursue our study by integrating the numerical votes in the system and new members caracteristics.

References

1. Backstrom, L., Huttenlocher, D., Kleinberg, J., Lan, X.: Group formation in large social networks: membership, growth, and evolution. In: Proceedings of the 12th ACM SIGKDD International Conference on Knowledge Discovery and Data Mining, KDD 2006, pp. 44–54. ACM, New York (2006)
2. He, Y., Wang, C., Jiang, C.: Discovering canonical correlations between topical and topological information in document networks. In: Proceedings of the 24th ACM International on Conference on Information and Knowledge Management, CIKM 2015, pp. 1281–1290. ACM, New York (2015)
3. Hou, H., Kretschmer, H., Liu, Z.: The structure of scientific collaboration networks in scientometrics. Scientometrics **75**(2), 189–202 (2008)
4. Jacob, Y., Denoyer, L., Gallinari, P.: Learning latent representations of nodes for classifying in heterogeneous social networks. In: Proceedings of the 7th ACM International Conference on Web Search and Data Mining, WSDM 2014, pp. 373–382. ACM, New York (2014)
5. Cook, J.M., McPherson, M., Smith-Lovin, L.: Birds of a feather: homophily in social networks. Ann. Rev. Sociol. **27**, 415–444 (2001)
6. Newman, M.: Networks: An Introduction. Oxford University Press Inc., New York (2010)
7. Nocedal, J.: Updating quasi-Newton matrices with limited storage. Math. Comput. **35**(151), 773–782 (1980)
8. Traud, A.L., Mucha, P.J., Porter, M.A.: Social structure of Facebook networks. CoRR, abs/1102.2166 (2011)

9. Yining, W., Liwei, W., Yuanzhi, L., Di, H., Wei, C., Tie-Yan, L.: A theoretical analysis of NDCG ranking measures. In: Proceedings of the 26th Annual Conference on Learning Theory (2013)
10. Zhou, D., Bousquet, O., Lal, T.N., Weston, J., Schlkopf, B.: Learning with local and global consistency. In: Advances in Neural Information Processing Systems 16, pp. 321–328. MIT Press (2004)

Reformulate or Quit: Predicting User Abandonment in Ideal Sessions

Mustafa Zengin[(⊠)] and Ben Carterette

University of Delaware, Newark, USA
zengin@udel.edu

Abstract. We present a comparison of different types of features for predicting session abandonment. We show that under ideal conditions for identifying topical sessions, the best features are those related to user actions and document relevance, while features related to query/document similarity actually hurt prediction abandonment.

1 Introduction

Detecting when a user is close to abandoning a search session is an important problem in search for complex information needs in order to take precautions and respond to users' information needs. Session abandonment and its types has been studied extensively. Li et al. [1] introduced and made the first distinction between good and bad abandonment. They defined good abandonment as an abandoned query for which the searcher's information need was successfully satisfied, without needing to clickthrough to additional pages. Chuklin and Serdyukov examined query extensions [2] and editorial and click metrics [3] for their relationship to good or bad abandonment. In [4] they developed machine learned models to predict good abandonment by using topical, linguistic, and historic features. Diriye [5] studied abandonment rationales and developed a model to predict abandonment rationale. Beyond the SERP (search engine result page), White and Dumais [6] studied aspects of search engine switching behavior, and develop and evaluate predictive models of switching behavior using features of the active query, the current session, and user search history.

In this paper, we present a comparison of different feature sets for detecting session abandonment. In Sect. 2, we describe the TREC Session track data that we use for our experiments–since this data consists of clearly-segmented sessions on a single topic, it is low-noise and thus ideal for analyzing relative feature effectiveness. In Sect. 3 we describe our feature sets. In Sect. 4 we describe our experiments and results, and we conclude in Sect. 5.

2 Data

We use the 2014 TREC Session track data [7], which consists of a large amount of logged user actions in the course of full search sessions for pre-defined topical

© Springer International Publishing AG 2016
S. Ma et al. (Eds.): AIRS 2016, LNCS 9994, pp. 322–328, 2016.
DOI: 10.1007/978-3-319-48051-0_26

information needs, for training and testing our models. As ensured by the track protocol, sessions in this data have clearly marked start and end points, and are entirely related to a given topic. Thus this data is much "cleaner" than standard search engine log data. From this data we used 1063 full sessions, each of which is made up of a sequence of "interactions". Each interaction consists of one query, up to 10 ranked results for that query from a search engine, and user clicks and dwell times on those results.

3   Methods

Our main assumption for a proposed session abandonment prediction model is that users abandonment decision base primarily on their positive or negative search experience through the search session according to their information need. We formulate the search experience through the features of document titles, URLs, and snippets on the current search result page, the queries submitted to the search engine in the history of the users' current session, the relevancy of seen documents on the current search result page and through out the current search session, dwell time on relevant, non-relevant and on whole clicked documents, total duration of the interaction and the session. Additionally, knowing that other users have seen and/or clicked on the same snippets in their own sessions may influence whether a user feel fulfilled about the information gathered in his/her own session.

The first set of features is derived from those that have been used in LETOR [8] and other datasets [9,10] for learning to rank: in particular, retrieval model scores between query and document URL/title/snippet; statistics about query term frequencies (normalized and non-normalized) in document URL/title/snippet; URL length and depth; number of inlinks and outlinks; spam score as computed by Waterloo's model for web spam [11]; and the Alexa ranking for the domain are all features that are either directly in LETOR data or are similar to other LETOR features that we cannot compute in our own data. Most of the textual features can be computed using our Indri index of ClueWeb12; our index also stores spam scores and inlink counts. The Alexa ranking is available through Alexa's API. Table 1 summaries all LETOR features.

The second set of features (Table 2) includes features based on document relevance, durations of actions, similarity between queries and topic statement, and similarities between a user's actions and those of other users working on the same topic. We describe these in more detail below.

TREC Session track users are provided with a topic description to use to guide their interactions with the search engine. We extract keyphrases from the topic description using the Alchemy API extraction tool. We also combined all queries that were submitted by other users working on the same topic. For each interaction, by using the previously submitted queries in prior interactions by the user, we calculate an approximation of submitted user queries to newly formed Alchemy topic keyphrases and other users' combined query.

Table 1. List of LETOR features. tf stands for term frequency, sl stands for stream length. Type-1 features are calculated by using query and URL, title, snippet. Type-2 features are calculated by using URL, title and snippet. Type-3 features are calculated by using query and snippet. Type-4 features are calculated by using only URL and type-5 features are calculated by using only the web document.

Feature	Type	Feature	Type
covered query term cuont/ratio	1	stream length of the field	2
sum of tf of the field	1	min/max/mean/var of tf of the field	1
sum of sl normalized tf	1	min/max/mean/var of sl-normalized tf	1
sum of tf*idf	1	min/max/mean/var of tf*idf	1
boolean model	1	vector space model	1
BM25 (Okapi) score	3	LM with Dirichlet smoothing	1
LM with Jelinek-Mercer smoothing	1	LM with two-stage smoothing	1
number of slashes in URL	4	length of URL	4
Waterloo spam score	5	number of inlinks to the web page	5
alexa ranking of web page	5	-	-

The approximation of query Q_i to Q_j is computed as follows:

$$Approximation(Q_i, Q_j) = |Q_i \cap Q_j|/|Q_j| \qquad (1)$$

We de-case all letters, replace all white space with one space, remove all leading and trailing white spaces, replace punctuation, remove duplicate terms, and remove stop words. Thus each query is represented as a bag of non-stopwords.

In order to calculate $|Q_i \cap Q_j|$, we consider two terms equivalent if any one of the following four criteria are met:

1. The two terms match exactly.
2. The Levenshtein edit distance between the two terms is less than two.
3. The stemmed roots of the two terms match exactly.
4. The WordNet Wu and Palmer measure is greater than 0.5.

For relevance features, we used graded relevance judgments of retrieved documents to the topic description as provided. High positive scores of relevancy indicates that the document is more relevant to the topic. Features we used include relevancy grade of each document in the interaction, total relevant and non-relevant document counts in the interaction, and total relevant and non-relevant document counts appeared in the session up to the current interaction.

Duration-based features extracted from the session data include start time of each interaction, time spent in each interaction, time spent in each clicked document, total time spent in clicked relevant documents which have relevance judgement of 1, total time spent in highly relevant clicked documents which have relevance judgement of 2 or higher, total time spent by user on scanning and reading documents including the last interaction.

Other users' sessions on the same topic are also used to extract features. Their queries, clicked documents, documents appeared on sessions were collected.

Table 2. List of session and interaction features.

Feature	Notes
int_order_num	Interaction number in a session
int_start_time	Interaction start time
int_duration	Interaction duration
int_duration_nr	Dwell time on NR documents in interaction
int_duration_r	Dwell time on R documents in interaction
int_duration_hr	Dwell time on HR documents in interaction
search_duration	Time spent on current and previous interactions
doc_rank_n_rl	Relevancy of document at rank n
doc_duration_n	Time spent on document at rank n
topic_alchemy_app	Query approximation to alchemy keywords
other_queries_app	Query approximation to other users' queries
non_rel_count_ses	Number of NR documents appeared in session
rel_count_ses	Number of R and HR documents appeared in session
non_rel_count_int	Number of NR documents appeared in interaction
rel_count_in	Number of R documents appeared in interaction
seen_clicked_docs_count_ses	Number of seen clicked documents from other sessions
seen_clicked_docs_ratio_ses	Ratio of seen clicked documents from other sessions
seen_clicked_docs_count_int	Number of seen clicked documents in the interaction from other sessions
seen_clicked_docs_ratio_int	Ratio of seen clicked documents from other sessions
seen_doc_ratio	Ratio of seen documents from other sessions through out the current session

The features related with other user sessions are, total number of other users' clicked documents appearance in current and previous interactions, its ratio, total number of other users' sessions' documents appearance in current and previous interactions and its ratio.

Finally, interaction order number is the number of the interaction within a session. Actions including the first query and any clicks on results retrieved for that query are associated with interaction order number 1; actions starting from the second query up to just before the third query are associated with interaction order number 2; and so on.

4   Experiments

We compare our session abandonment feature sets by their effectiveness at predicting actual user abandonments, using standard classification evaluation measures like precision, recall, AUC, and classification accuracy. We trained and tested random forest models using all sessions from the TREC 2014 Session track. For each interaction that appears in the Session track data, we have one instance for training/testing: the 0/1 label indicating whether the session was abandoned immediately after that interaction or not.

In total there are 2400 instances in the data (2400 interactions across 1063 sessions), of which 1763 are non-abandonment and 637 are abandonment. Since the class distribution is so skewed, we re-balanced the training data with SMOTE, which creates artificial data for the under-represented class [12]. We trained and tested using four-fold cross-validation and report micro-averaged

Table 3. Precision/recall/AUC/accuracy of our four models

Method	Precision	Recall	AUC	Accuracy
Baseline	0.438	0.320	0.718	0.710
LETOR	0.273	0.267	0.529	0.617
Session+Interaction	0.536	0.752	0.841	0.762
LETOR+Session+Interaction	0.452	0.457	0.681	0.708

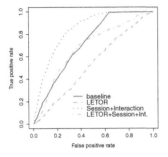

Fig. 1. ROC curves for different session abandonment prediction methods

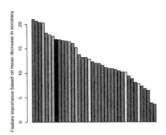

Fig. 2. Feature importance based on mean decrease in accuracy. Key: blue: duration features; green: features extracted using other sessions with same topic; black: interaction id; purple: query approximation features; red: relevance features; yellow: duration and relevance features (Color figure online)

evaluation measures aggregated across all four testing splits. Note that only training data, not testing data, in each fold is rebalanced with SMOTE.

We tested 4 models. The first model, a simple baseline, uses only the interaction order number. The second model uses the LETOR features listed in Table 1 for every document appeared in the interaction. The third model uses the session and interaction features listed in Table 2, and the fourth model uses all features. Table 3 and Fig. 1 summarizes the performance of the four models. The baseline using only interaction order number for training performs better than model 2 (which only uses LETOR features) in all measures, *and* model 4 (which uses all features) in AUC and accuracy (though that model scores slightly higher in precision and substantially higher recall). The best achieving model is the model that uses only session and interaction features. It improves precision over the baseline by 22 %, recall by 135 %, AUC by 17 % and accuracy by 7 %.

Figure 2 shows the feature importance of the random forest that is trained with session and interaction based features. Mean importance of the features that are extracted by using the other sessions with the same topic is 17.46,

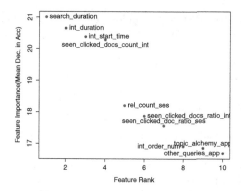

Fig. 3. Top 10 features based on mean decrease in accuracy

only interaction id is 16.91, query approximation features is 16.77, relevancy based features is 12.89, duration based features is 12.89 and lastly duration and relevancy based features is 10.29.

Figure 3 shows the top 10 features based on the mean decrease in accuracy. The first three features are unsurprisingly based on session and interaction duration. The fourth feature is the number of clicked documents appeared in the current interaction. The fifth is the number of relevant documents seen in session including the current interaction. The last two features are query approximations.

5   Conclusions

We have compared two different sets of features for predicting abandonment: one set is more like LETOR, including many features related to query/document similarity; the other includes a collection of features derived from the session and topic. We tested them in a setting with low noise: fully-segmented sessions on a single topic, with the topic description available as well as sessions by other users for the same topic. We found that the second set is far superior to the first set, which actually degrades effectiveness at predicting abandonment. We also found that the inclusion of durability-based features in the second set is not the sole reason they perform better.

References

1. Li, J., Huffman, S., Tokuda, A.: Good abandonment in mobile and PC internet search. In Proceedings of the 32nd International ACM SIGIR Conference on Research and Development in Information Retrieval, pp. 43–50. ACM, July 2009
2. Chuklin, A., Serdyukov, P.: How query extensions reflect search result abandonments. In: Proceedings of the 35th International ACM SIGIR Conference on Research and Development in Information Retrieval, pp. 1087–1088. ACM, August 2012

3. Chuklin, A., Serdyukov, P.: Good abandonments in factoid queries. In: Proceedings of the 21st International Conference on World Wide Web, pp. 483–484. ACM, April 2012

4. Chuklin, A., Serdyukov, P.: Potential good abandonment prediction. In: Proceedings of the 21st International Conference on World Wide Web, pp. 485–486. ACM, April 2012

5. Diriye, A., White, R., Buscher, G., Dumais, S.: Leaving so soon?: understanding and predicting web search abandonment rationales. In: Proceedings of the 21st ACM International Conference on Information and Knowledge Management, pp. 1025–1034. ACM, October 2012

6. White, R.W., Dumais, S.T.: Charactering and predicting search engine switching behavior. In: CIKM, pp. 87–96 (2009)

7. Carterette, B., Kanoulas, E., Clough, P.D., Hall, M.: Overview of the TREC 2014 Session track. In: Proceedings of TREC (2014)

8. Liu, T.Y., Xu, J., Qin, T., Xiong, W., Li, H.: Letor: benchmark dataset for research on learning to rank for information retrieval. In: Proceedings of SIGIR Workshop on Learning to Rank for Information Retrieval, pp. 3–10, July 2007

9. Chapelle, O., Chang, Y.: Yahoo! learning to rank challenge overview. In: Yahoo! Learning to Rank Challenge, pp. 1–24, Chicago (2011)

10. Serdyukov, P., Dupret, G., Craswell, N.: WSCD: workshop on web search click data. In: Proceedings of the Sixth ACM International Conference on Web Search and Data Mining, pp. 787–788. ACM, February 2013

11. Cormack, G.V., Smucker, M.D., Clarke, C.L.: Efficient and effective spam filtering and re-ranking for large web datasets. Inf. Retrieval **14**(5), 441–465 (2011)

12. Chawla, N.V., Bowyer, K.W., Hall, L.O., Kegelmeyer, W.P.: SMOTE: synthetic minority over-sampling technique. J. Artif. Intell. Res., 321–357 (2002)

Learning to Rank with Likelihood Loss Functions

Yuan Lin[1], Liang Yang[2], Bo Xu[2], Hongfei Lin[2(✉)], and Kan Xu[2]

[1] WISE Lab, Dalian University of Technology, Dalian, China
zhlin@dlut.edu.cn
[2] School of Computer Science and Technology,
Dalian University of Technology, Dalian, China
{hflin,xukan}@dlut.edu.cn, {yangliang,xubo2011}@mail.dlut.edu.cn

Abstract. According to a given query in training set, the documents can be grouped based on their relevance judgments. If the group with higher relevance labels is in front of the one with lower relevance judgments, the ranking performance of ranking model could be perfect. Inspired by this idea, we propose a novel machine learning framework for ranking, which depends on two new samples. The first sample is one-group constructed of one document with higher relevance judgment and a group of documents with lower relevance judgment; the second sample is group-group constructed of a group of documents with higher relevance judgment and a group of documents with lower relevance judgment. We also develop a novel preference-weighted loss function for multiple relevance judgment data sets. Finally, we optimize the group ranking approaches by optimizing initial ranking list for likelihood loss function. Experimental results show that our approaches are effective in improving ranking performance.

Keywords: Information retrieval · Learning to rank · Group ranking

1 Introduction and Motivation

Ranking documents with respect to given queries is the central issue for designing effective web search engines, as the ranking model decide relevance of search results. Many approaches are proposed for ranking. The purpose of ranking is to permute the relevance documents on the top positions of the result list. Learning to rank [1] is an effective machine learning approach to improve the performance of information retrieval. However, the previous approaches always take the every document as a single object respectively, although the documents may be labeled by same relevance judgment. It may increase the training time and computational complexity. In addition, if we take the documents with the same relevance judgment as a group, the ranking task is reduced from ranking the multiple documents to ranking several groups. Moreover, the permutation in a single group is also effective to the final performance. And the different combinations of group sample can also affect the results of ranking. In this paper we try to improve classic approaches of learning to rank by presenting a new framework, named group wise framework. The loss function of our framework is based on novel samples one-group sample and group-group sample. In order to acquire higher ranking accuracies, we also explore optimal

S. Ma et al. (Eds.): AIRS 2016, LNCS 9994, pp. 329–334, 2016.
DOI: 10.1007/978-3-319-48051-0_27

selection of the initial ranking list for group samples. For the data set with more than two relevance judgments, we develop a new loss function by weighting their group samples.

The contributions of this paper are as follows: (1) we investigate the problems of the existing likelihood loss based approaches; (2) we propose a relevance preference based approach to optimize the group ranking loss function with more than two relevance judgments; (3) we investigate the influence of initial ranking permutation of documents and develop a permutation selection to improve its likelihood loss construction.

2   Related Work

Recently, there has been a great deal of research on learning to rank. They are usually catego-rized into three approaches: the pointwise, pairwise and listwise approaches, which are based on different input samples for learning. Most approaches of learning to rank are based on the three frameworks. However, there are a few studies that have improved the learning to rank methods using other input samples. The roles of ties, which are document pairs with same relevance are, investigated in the learning process [2], which shows that it is feasible to improve the performance of learning ranking functions by introducing the new samples.

ListMLE [3] is an effective ranking approach of listwise that formalizes learning to rank as a problem of minimizing the likelihood function of a probability model. Xia et al. [4] presented a new framework, which is used to optimize the top-k ranking. They demonstrate it is effective to improve the ranking performance of top-k and also derive sufficient conditions for a listwise ranking method to be consistent with the top-k true loss. In our previous work [5], we conducted some preliminary experiments to expand their work. In this paper, we divide the list of documents with respect to a query into several groups based on their relevance judgments. There are two patterns in the frame-work to construct samples: One-group and Group-group. We also define the new loss function of group sample based on likelihood loss to examine whether it is effective to improve the ranking performance.

3   Group Ranking Framework

In this section, we briefly introduce the framework of group ranking [5], which is used to improve the ranking accuracies. It is developed from the framework of listwise and the framework of top-k ranking. We analyze likelihood based approaches and point out the existing problems. In order to improve the two frameworks, we proposed the frame-work of group ranking framework.

Group ranking framework is similar to the top-k ranking framework. However, the samples of the two methods are different, and the meaning of k is different. We denote the ranking function based on the neural network model w as f. Given a group feature vector $f(x)$ assigns a score to x. For the loss function of group ranking, we employs gradient decent to perform the optimization. When we use likelihood loss function as metric, the group loss function becomes:

$$L(f;x^g, y^g) = \sum_{s=1}^{r} (-f(x^g_{y^g(s)}) + \ln(\sum_{i=s}^{n} \exp(f(x^g_{y^g(s)})))) \tag{1}$$

Where r is decided by the number of documents with the higher label in the group ranking samples: for one-group sample r is set to 1, but for the group-group sample r is set to the length of the group with the higher label. X^g is a group sample and y^g is a optimum permutation for x^g.

Relevance Preference Based Loss Function. Although the group-wise approaches can get an appropriate level of ranking performance, there are still several problems with the construction of loss function.

First, for multiple level relevance labeled data set, the main issue is that the group samples with different preferences are considered as the equivalent ones in the training process, which may neglect the difference of original relevant labels for the group of documents. In this paper, we propose a relevance preference based approach to solve the problem. Based on this approach, our method can lead to more significant improvements than original group ranking approaches in retrieval effectiveness.

The framework of group ranking improves likelihood loss based algorithm. Especially for the data set with binary relevance labels, it can get a more significant performance than the multiple relevance labeled data set. The reason for that may cause the different relevant judgment labels for the two types of data set. For the group sample construction of binary label data set, the importance of each sample is no differences. However, there are often multiple relevance judgments, such as 2 (definitely relevant), 1 (possibly relevant), 0 (irrelevant). Thus the group sample constructed by the documents with labels 2 and 1 is indeed different from that constructed by the documents with labels 2 and 0. So it is imperfect that the group ranking framework takes all the samples as equivalent. In this paper we propose the preference weighted loss function to deal with this problem. In the case of learning to rank web documents, preference data are given in the form that one document is more relevant than another with respect to a given query. For each group sample, there is a preference for two group of documents, which reflects the relevance difference between the two labels. We introduce the preference to improve the group ranking loss function, which is based on Likelihood. We define the loss function for each sample as follows:

$$L(f;x^{g_i}, y^{g_i}, p) = weight(g_i, p) * L(f;x^{g_i}, y^{g_i}) \tag{2}$$

Where p is a parameter depended on the preference, weight (g_i, p) is a weight function that depends on the group and preference, which can introduce the relevance of the documents into the learning process. We defined the weight function for the sample as follows:

$$weight(g_i, p) = \frac{label_h(g_i) - label_l(g_i)}{\sum_{j=1}^{m} (label_h(g_j) - label_l(g_j))} \tag{3}$$

g_i is the *i-th* of group samples, $label_h(g_i)$ and $label_l(g_i)$ are the two types of relevant labels in the group sample. In this way, we can define the weight by the preference, the larger the difference between the two labels, the bigger the weight. We can introduce the relevant labels and preference into the loss function to improve the original group ranking loss functions. We name this approach W-GroupMLE for short.

Initial Ranking Permutation Optimization. In the training set, as matter of fact that the relevance labels of document 1 and 2 are equal, the descent loss seems to be not necessary. However, it is still a key clue to improve our approaches. For the likelihood loss of ListMLE, y is a randomly selected optimum permutation which satisfies the condition for any two documents x_i, x_j, if $label(x_i) > label(x_j)$, the x_i is ranked before x_j in y. However, the optimal ranking list for y is not unique, different orders of documents for the selected y may result in different loss functions. For the group-group ranking, there are still a group documents with higher label in the list, so it is similar to ListMLE, where different permutations may generate different loss function. It is an important issue to distinguish which function could get the best performance. Furthermore, for the documents 1 and 2, if the basis of the initial ranking decides that document 1 should be permuted before document 2, it may be more effective to select the ranking model f_2. In this paper for the initial ranking list y, instead of selecting optimum permutation randomly, we choose the ranking features with the best ranking performance as the basis of optimum permutation in training set.

4 Experiments

We evaluate our methods on the Letor3.0 data set released by Microsoft Research Asia. This data set contains two collections: the OHSUMED collection and the.Gov collection. The collections we use to evaluate our experiments are the OHSUMED and TD2003, TD2004. As evaluation measure, we adopt NDCG@N and MAP to evaluate the performance of the learned ranking function.

Effectiveness of Preference Weighted Likelihood Loss. Varying from TD2003 and Td2004 data set, there are three level labels: 2 (definitely relevance), 1 (possibly relevant) and 0 (irrelevant) in the OHSUMED data set. It is not correct to take all the group ranking samples as equal, because there are samples with different preferences. According to the Sect. 4, we use the weighted likelihood loss to optimize the training process of OHSUMED data set. In this section, we examine the effectiveness of group ranking methods. The results of group ranking methods on OHSUMED collection are shown in Table 1. From the Table 1, we can see that the W-GroupMLE can significantly boost the ranking accuracies based on one-group sample (W-GroupMLE1) and group-group sample (W-Group MLE2). In addition, preference weighted method gets best performance, which clearly validates our argument that it can improve the ranking performance by introducing the preference to loss function based on relevant label. The group sample whose preference is big should be set a large weight for training. It is effective to optimize the loss function.

Table 1. Ranking accuracies on OHSUMED

Methods	MAP	N@1	N@3	N@10
Top10 ListMLE	0.4441	0.5156	0.4772	0.4360
One-group	0.4414	0.4942	0.4599	0.4219
Group-group	0.4517	0.5589	0.4888	0.4572
W-GroupMLE1	0.4451	0.5123	0.4677	0.4442
W-GroupMLE2	0.4521	0.5621	0.4912	0.4583

Effectiveness of Optimum Permutation Selection. First, we list the feature number with the best and the worst performance (short for max and min) evaluated by MAP on the Table 2. We can see different features can gets different performances. However, on the Letor3.0 data set, although the features with the best performance are different, the worst is the same one in the same data set for 5 folds. The Letor3.0 ranks the documents with respect to one query by the feature which is from the scores of BM25 for the whole document. The above experiments are based on these ranks for the likelihood loss functions. However, we find that the BM25 feature do not get the best performance on the data set.

Table 2. Ranking feature performance for Letor data set (MAX/MIN)

Data set	Fold1	Fold2	Fold3	Fold4	Fold5
OHSUMED	8/24	11/24	11/24	10/24	10/24
TD2003	46/59	41/59	41/59	43/59	46/59
TD2004	46/59	41/59	41/59	41/59	41/59

Second, we examine the performance of the best and the worst features compared with the BM25 scores. The results evaluated by MAP are listed on Table 3. For the evaluations, the features with the best performance all achieve better ranking accuracies than the other features used as initial ranking list. Overall, as a good information retrieval method, BM25 also gets a better performance than the worst performance features, but it cannot outperform the best performance features which are learned from training data of single fold. Moreover, the group-group ranking methods almost all outperform the one-group ranking method by introducing the information of initial ranking list to the loss function on the training process.

Table 3. Ranking performance of different group ranking methods evaluated by MAP

Data set	MAX	MIN	BM25	One-group
OHSUMED	0.4529	0.4284	0.4521	0.4451
TD2003	0.2811	0.2808	0.2811	0.2511
TD2004	0.2467	0.2325	0.2386	0.2183

Finally, in this section, we also examine the performance of our methods compared with the existing methods. We select several representative ranking algorithms compared with our group ranking. We choose OHSUMED and .Gov collections as the test collections, because they have different type relevant judgments. The performance of the ranking methods is shown in Fig. 1.

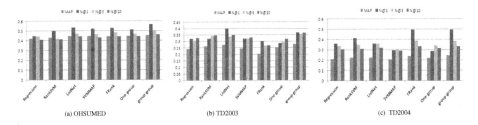

(a) OHSUMED   (b) TD2003   (c) TD2004

Fig. 1. Ranking accuracies of ranking methods

The results on the test collections show that the group-group ranking method, based on preference weighted likelihood loss and optimum permutation selection, attains a notable improvement over existing ranking method by learning a suitable model based on group loss function and group samples. These findings indicate that our method in comparison to the collection with multiple relevant labels is more effective.

5   Conclusion

In this paper, we develop a preference weighted likelihood loss function to boost ranking accuracies. In addition, the experimental results show that the optimization of initial ranking list for likelihood loss function is also effective to improve the ranking perform-ance of group ranking approaches. As future work, we will continue to study the theory basis of the group ranking framework and examine whether the ranking performance can be improved further; and we will also exploit the odds to using other existing methods to implement the group ranking methods.

Acknowledgement. This work is partially supported by grant from the Natural Science Foundation of China (No. 61402075, 61572102, 61572098), and the Fundamental Research Funds for the Central Universities.

References

1. Liu, T.Y.: Learning to rank for information retrieval. Found. Trends Inf. Retrieval **3**(3), 225–331 (2009)
2. Zhou, K., Xue, G. R., Zha, H.Y., Yu, Y.: Learning to rank with ties. In: Proceedings of SIGIR, pp. 275–282 (2008)
3. Xia, F., Liu, T.Y., Wang, Z.J., Li, H.: Listwise approach to learning to rank - theory and algorithm. In: ICML, pp. 192–199 (2008)
4. Xia, F., Liu, T.Y., Li, H.: Statistical consistency of top-k ranking. In: Proceedings of NIPS, pp. 2098–2106 (2009)
5. Lin, Y., Lin, H.F., Ye, Z., Jin, S., Sun, X.L.: Learning to rank with groups. In: Proceedings of CIKM, pp. 1589–1592 (2010)

Learning to Improve Affinity Ranking
for Diversity Search

Yue Wu[1], Jingfei Li[1], Peng Zhang[1], and Dawei Song[1,2(✉)]

[1] School of Computer Science and Technology, Tianjin University, Tianjin, China
{yuewuscd,jingfl}@foxmail.com, {pzhang,dwsong}@tju.edu.cn
[2] The Computing Department, The Open University, Milton Keynes, UK

Abstract. Search diversification plays an important role in modern search engine, especially when user-issued queries are ambiguous and the top ranked results are redundant. Some diversity search approaches have been proposed for reducing the information redundancy of the retrieved results, while do not consider the topic coverage maximization. To solve this problem, the Affinity ranking model has been developed aiming at maximizing the topic coverage meanwhile reducing the information redundancy. However, the original model does not involve a learning algorithm for parameter tuning, thus limits the performance optimization. In order to further improve the diversity performance of Affinity ranking model, inspired by its ranking principle, we propose a learning approach based on the learning-to-rank framework. Our learning model not only considers the topic coverage maximization and redundancy reduction by formalizing a series of features, but also optimizes the diversity metric by extending a well-known learning-to-rank algorithm LambdaMART. Comparative experiments have been conducted on TREC diversity tracks, which show the effectiveness of our model.

Keywords: Search diversification · Affinity ranking · Learning-to-rank

1 Introduction

Search diversification plays an important role in modern search engine, especially when user-issued queries are ambiguous and the top ranked results are redundant. Some diversity search approaches have been proposed (e.g., Maximal Marginal Relevance (MMR) [1] and its numerous variants [5,7,9]) for reducing the information redundancy of the retrieved results, while do not consider the topic coverage maximization.

In order to address the aforementioned drawbacks of traditional implicit diversity approaches, Zhang et al. [8] proposed an innovative method named Affinity Ranking (AR) model which pursues the query subtopics coverage maximization and information redundancy reduction simultaneously. Specifically, AR applies a content-based document graph to compute the information coverage score for each document and imposes a penalty score to the information coverage score in order to reduce the information redundancy, then ranks documents

© Springer International Publishing AG 2016
S. Ma et al. (Eds.): AIRS 2016, LNCS 9994, pp. 335–341, 2016.
DOI: 10.1007/978-3-319-48051-0_28

according to the final document score which linearly combines the query relevance information score and the diversity information (i.e., topic coverage information and redundancy reduction information) score of the document. However, the original Affinity ranking model uses a predefined heuristic ranking function which can only integrate limited features and has many free parameters to be tuned manually. A direct idea to solve this problem is to borrow machine learning methods to train the Affinity ranking model. Intuitively, the Affinity ranking model is similar to the traditional retrieval ranking model (e.g., query likelihood Language Model) which ranks documents in descending order according to document scores. Therefore, improving the Affinity ranking model with learning-to-rank technique is reasonable and feasible. To do this, in this paper, we addressed three pivotal problems, i.e., (i) how to redefine the ranking function which can incorporate both relevance information and diversity information within an unified framework; (ii) how to learn the ranking model by optimizing the diversity evaluation metric directly; (iii) how to extract diversity features (i.e., topic coverage features and redundancy reduction features) inspired by the Affinity ranking model. Particularly, we propose a learning based Affinity ranking model by extending a well-known Learning-to-Rank method (i.e., LambdaMART). Extensive comparative experiments are conducted on diversity tracks of TREC 2009-2011, which show the effectiveness of our method.

2   Model Construction

2.1   Overview of Affinity Ranking Method

This subsection gives a brief description of the Affinity Ranking model [8] which maximizes the topic coverage and reduces the information redundancy. At first, they introduce a directed link graph named Affinity Graph to compute the information richness score which represents how many the query subtopics have been covered for each document. Similar to the PageRank, the information richness score for each document is obtained through running the random walk algorithm. The documents with largest information richness score (subtopic coverage information) will be returned to users. Meanwhile, in order to reduce the information redundancy, they compute the Affinity ranking score by deducting a diversity penalty score for each document as described in the Algorithm 1. However, improving the diversity may bring harm to the relevance quality. In order to balance the diversity ranking and relevance ranking, their final ranking function linearly combines both original relevance score and Affinity ranking score, and then they sorts the documents in descending order according to the final combination score.

In order to obtain a good diversity performance (in term of the diversity evaluation measures) and incorporate more features, we propose a learning approach which is illustrated in the following parts.

Algorithm 1. The greedy algorithm for diversity penalty.

Input: $InfoRich(d_i)$: information richness score, D: candidate document set, $\hat{M}_{ji}$: the weight of
link in the graph, $\hat{M}_{ji}InfoRich(d_i)$: penalty score.
Output: Affinity ranking score $AR(d_i)$ for every document d_i
   for $d_i \in D$ **do**
      $AR(d_i) = InfoRich(d_i)$
   end for
   while $D \neq empty$ **do**
      $d_i = \arg\max_{d_i \in D}(AR(d_i))$
      $D \leftarrow D - d_i$
      for $d_j \in D$ **do**
         $AR(d_j) = AR(d_j) - \hat{M}_{ji}InfoRich(d_i)$
      end for
   end while

2.2 Learning Diversity Ranking Method

We build a learning based Affinity ranking model with the help of learning-
to-rank technique (the LambdaMART [6] algorithm) to improve the diversity
ability of Affinity ranking model. In following parts, we will redefine the ranking
function, label, the objective function of learning algorithm, and then describe
the features to build our learning model.

Learning Algorithm for Diversity Search. For the original LambdaMART,
the ranking score of each document can be computed by ranking function $f(x) =
w^T x$. The x is the document feature vector which only considers the relevance.
However, for diversity task, we need to incorporate both relevance, redundancy
reduction and topic coverage maximization. Inspired by the ranking function of
the Affinity Ranking model, we can extend the ranking function as described in
the Eq. 1:

$$f(w_1, w_2, w_3, x, y, z) = w_1^T x + w_2^T y + w_3^T z \qquad (1)$$

where the w_1, w_2 and w_3 encode the model parameters, the x, y is topic coverage
maximization and redundancy reduction feature vector respectively while the z
is relevance feature vector. Even if we have the reasonable ranking function, it
is still a big challenge to redefine the objective function for using the diversity
metric to guide the training process.

Unlike others, the LambdaMART algorithm defines the derivative of objective function with the respect to document score rather than deriving them from
the objective function. For the document pair $< i, j >$ (the document i is more
relevant than document j), the derivative λ_{ij} is:

$$\lambda_{ij} = \frac{-\sigma}{1 + e^{\sigma(s_i - s_j)}}|\Delta Z_{ij}| \qquad (2)$$

where σ is the shape parameter of the sigmoid function, s_i is the model score
of document i and the $|\Delta Z_{ij}|$ is the change value of evaluation metric when
swapping the rank positions of document i and j. We know that LambdaMART
can be extended to optimize any IR metric by simply replacing $|\Delta Z_{ij}|$ in Eq. 2.

However, the evaluation metric needs to satisfy the property that if irrelevant document ranks before the relevant document after swapping (that is, wrong swapping), the metric should decrease (i.e., $\Delta Z_{ij} < 0$). So if we extend derivative λ_{ij} by using the current diversity metric (e.g., α-$NDCG$ [3] or ERR-IA [2]), some adjustments should be made. The relevance label of a document is one value in original LambdaMART to decide the relevant-irrelevant document pair used in the Eq. 2, while our label should a multiple values (in which each value represents whether the document is relevant to the each query subtopic) in order to compute the change value of diversity metric. So we assume that the document covering at least one query subtopic is more relevant than document covering no any query subtopics. Thus the label of the document covering at least one query subtopic is bigger than the document covering no any query subtopics. Therefore, the document label used in the training procedure contains two part. And then after defining the relevant-irrelevant document pair, we should show diversity metrics satisfy the above property. We choose the α-$NDCG$ as the representative because ERR-IA is same in rewarding the relevant document ranking before the irrelevant document. In the top k results of a return list for query q, for example, there are m documents which covers at least one query subtopic where four documents among the m is relevant to the query subtopic t. We denote the ranking positions of the four documents as p_1, p_2, p_3, p_4 where $0 < p_1 < p_2 < p_3 < p_4 < k$. If one relevant document (we use d_{p_2} in the following proof case, which means the document at the position p_2) swaps with another irrelevant document which ranking position is beyond k, we have proved that $\Delta Z < 0$. Let Z is the α-$NDCG@k$ before the swapping while $\tilde{Z}$ is the α-$NDCG@k$ after the swapping (the value of α is between 0 and 1). When only considering the query subtopic t, we have $\Delta Z_t = \frac{\tilde{Z}_t - Z_t}{ideaDCG@k} = \frac{1}{ideaDCG@k} \sum_{j=1}^{k} \frac{G_t[j] - G_t[j]}{\log_2(1+j)} = \frac{1}{ideaDCG@k}\left(\left(\frac{(1-\alpha)}{\log_2(1+P_1)} - \frac{(1-\alpha)}{\log_2(1+P_1)}\right) + \left(0 - \frac{(1-\alpha)^2}{\log_2(1+P_2)}\right) + \left(\frac{(1-\alpha)^2}{\log_2(1+P_3)} - \frac{(1-\alpha)^3}{\log_2(1+P_3)}\right) + \left(\frac{(1-\alpha)^3}{\log_2(1+P_4)} - \frac{(1-\alpha)^4}{\log_2(1+P_4)}\right)\right) < \left(\frac{(1-\alpha)^2}{\log_2(1+P_3)} - \frac{(1-\alpha)^2}{\log_2(1+P_2)}\right) + \left(\frac{(1-\alpha)^3}{\log_2(1+P_4)} - \frac{(1-\alpha)^3}{\log_2(1+P_3)}\right) - \frac{(1-\alpha)^4}{\log(1+P_4)} < 0$. The same is true for every subtopics. Thus, the diversity metric satisfies the above property. Through above adjustments, it is suitable for using the diversity metric as part of objective function to guide the training process.

Feature Extraction. At first, we formalize the diversity features (which aim at maximizing the topic coverage and reducing the information redundancy) inspired by the Affinity ranking. For topic coverage maximization features, we use the information richness score used in the Affinity ranking model. For information redundancy reduction features, we formalize it according to Algorithm 1:

$$f(q, d_i, D_q) = t(q, d_i) - \sum_{d_j \in D_q} p(d_j, d_i) \tag{3}$$

where document set D_q is the already selected document set, $t(q, d_i)$ measures how many query topics has been covered by the document d_i, $p(d_j, d_i)$ denotes the penalty score that the document d_j deploy to d_i for the information

Table 1. Diversity and Relevance features for learning on ClueWeb09-B collection

Feature	Description
TopicCovFea0	information richness score in the [8]
RedReduceFea1	$t(q, d_i)$ is information richness score, $p(d_j, d_i)$ is penalty score in the [8]
RedReduceFea2	$t(q, d_i)$ is TF-IDF score, $p(d_j, d_i) = \sqrt{t(q, d_i)}\sqrt{t(q, d_j)}f(d_i, d_j)$
RedReduceFea3	$t(q, d_i)$ is BM25 score, $p(d_j, d_i)$ is same to RedReduceFea2
RedReduceFea4	$t(q, d_i)$ is LMIR with ABS smoothing, $p(d_j, d_i)$ is same to RedReduceFea2
RedReduceFea5	$t(q, d_i)$ is LMIR with DIR smoothing, $p(d_j, d_i)$ is same to RedReduceFea2
RedReduceFea6	$t(q, d_i)$ is LMIR with JM smoothing, $p(d_j, d_i)$ is same to RedReduceFea2
RelFea7	sum of query term frequency for every document
RelFea8	length for the every document
RelFea9-13	sum,min,max,mean,variance of document term frequency in collection
RelFea14-18	sum,min,max,mean,variance of document tfidf in collection
RelFea19-23	tfidf score, BM25 score, LMIR score with ABS, DIR, JM smoothing

redundancy. In this paper, we can use different form of $t(q, d_i)$ and $p(d_j, d_i)$ to produce diversity features for capturing redundancy reduction information. Then, for relevance features, we use some common features which consider the query-dependent features, document-dependent features and query-document features. Detailed features are shown in Table 1, where $f(d_i, d_j)$ denotes cosine similarity between documents d_i and d_j represented with TF-IDF vectors.

3   Experiments and Results

3.1   Experimental Setting

We evaluate our method using the diversity task of the TREC Web Track from 2009-2011, which contains 148 queries. We use the ClueWeb09 category-B as the document collection and the official evaluation metrics of diversity task (α-$NDCG$ [3] where α is 0.5 and ERR-IA [2]). All approaches are tested by re-ranking the original top 1000 documents retrieved by the Indri search engine (implemented with the query likelihood Language Model abbreviated with LM) for each query. For all approaches with free parameters, 5-fold cross validation is conducted. We tested 5 baseline approaches including the original query likelihood Language Model (LM), MMR [1], quantum probability ranking principle (QPRP)[9], RankScoreDiff [4] and Affinity Ranking model (AR) [8].

3.2   Result and Analysis

In this section, we report and analyze the experiment results to investigate the effectiveness of the proposed diversity model. If our model uses α-$NDCG$ in objective function, it is denoted as LAR(α-$NDCG$) while it is denoted as LAR(ERR-IA) for using ERR-IA. At first, we compare AR model with other

Table 2. Diversification performance of the models

Metric	LM	MMR	QPRP	RankScoreDiff	AR	LAR $(\alpha - NDCG)$	LAR $(ERR - IA)$
α-NDCG	0.2695	0.2681↓	0.1663↓	0.2705↑	0.2711↑	0.3442↑	0.3560↑
ERR-IA	0.1751	0.1715↓	0.1266↓	0.1767↑	0.1765↑	0.2536↑	0.2655↑

baselines to show the diversity ability of Affinity ranking model. From Table 2, we find AR model has better performance than other baselines. Moreover, we find that the result list does not achieve good diversity ability in term of diversity evaluation α-NDCG and ERR-IA for two approaches [1,9] which only reduce the redundancy. The experiment result shows that a group of document with low redundancy can not achieve large subtopic coverage. For RankScoreDiff approach [4], one only considers the subtopic coverage maximization, also outperform MMR and QPRP [1,9]. The experiment results illustrate that query subtopic coverage maximization is more important than low information redundancy for diversity search. Secondly, we compare our model with AR model to prove that our model (both LAR(α-NDCG) and LAR(ERR-IA) model) improve the diversity ability of AR model significantly. For using α-NDCG as evaluation metric, the improvement percentages of our model compared with the AR model is 26.96 % for LAR(α-NDCG) and 31.31 % for LAR(ERR-IA) respectively. When using ERR-IA as evaluation metric, the improvement percentage is 43.68 % for LAR(α-NDCG) and 50.42 % for LAR(ERR-IA) respectively.

4   Conclusions and Future Work

In this paper, we build a learning diversity model within the framework of learning-to-rank to improve the diversity ability of Affinity Ranking model. Our motivation comes from that the Affinity Ranking model can reduce the redundancy and make topic coverage maximization. Beyond that, the ranking principle of Affinity Ranking model makes it possible to build learning model with help of learning-to-rank approach. The final comparative experiments have shown that our approach is effective. In the future, we will propose better topic coverage representation technique to formalize the better diversity features.

Acknowledgements. This work is supported in part by the Chinese National Program on Key Basic Research Project (973 Program, grant No. 2013CB329304, 2014CB744604), the Chinese 863 Program (grant No. 2015AA015403), the Natural Science Foundation of China (grant No. 61272265, 61402324), and the Tianjin Research Program of Application Foundation and Advanced Technology (grant no. 15JCQNJC41700).

References

1. Carbonell, J., Goldstein, J.: The use of MMR, diversity-based reranking for reordering documents and producing summaries. In: SIGIR, pp. 335–336. ACM (1998)
2. Chapelle, O., Metlzer, D., Zhang, Y., Grinspan, P.: Expected reciprocal rank for graded relevance. In CIKM, pp. 621–630. ACM (2009)
3. Clarke, C.L., Kolla, M., Cormack, G.V., Vechtomova, O., Ashkan, A., Büttcher, S., MacKinnon, I.: Novelty and diversity in information retrieval evaluation. In: SIGIR, pp. 659–666. ACM (2008)
4. Kharazmi, S., Sanderson, M., Scholer, F., Vallet, D.: Using score differences for search result diversification. In: SIGIR, pp. 1143–1146. ACM (2014)
5. Wang, J., Zhu, J.: Portfolio theory of information retrieval. In: SIGIR, pp. 115–122. ACM (2009)
6. Wu, Q., Burges, C.J., Svore, K.M., Gao, J.: Adapting boosting for information retrieval measures. Inf. Retr. **13**(3), 254–270 (2010)
7. Zhai, C.X., Cohen, W.W., Lafferty, J.: Beyond independent relevance: methods and evaluation metrics for subtopic retrieval. In: SIGIR, pp. 10–17. ACM (2003)
8. Zhang, B., Li, H., Liu, Y., Ji, L., Xi, W., Fan, W., Chen, Z., Ma, W.-Y.: Improving web search results using affinity graph. In: SIGIR, pp. 504–511. ACM (2005)
9. Zuccon, G., Azzopardi, L.: Using the quantum probability ranking principle to rank interdependent documents. In: Gurrin, C., He, Y., Kazai, G., Kruschwitz, U., Little, S., Roelleke, T., Rüger, S., Rijsbergen, K. (eds.) ECIR 2010. LNCS, vol. 5993, pp. 357–369. Springer, Heidelberg (2010). doi:10.1007/978-3-642-12275-0_32

An In-Depth Study of Implicit Search Result Diversification

Hai-Tao Yu[1(✉)], Adam Jatowt[2], Roi Blanco[3], Hideo Joho[1], Joemon Jose[4],
Long Chen[4], and Fajie Yuan[4]

[1] University of Tsukuba, Tsukuba, Japan
{yuhaitao,hideo}@slis.tsukuba.ac.jp
[2] Kyoto University, Kyoto, Japan
adam@dl.kuis.kyoto-u.ac.jp
[3] University of A Coruña, A Coruña, Spain
rblanco@udc.es
[4] University of Glasgow, Glasgow, UK
{joemon.jose,long.chen}@glasgow.ac.uk, f.yuan.1@research.gla.ac.uk

Abstract. In this paper, we present a novel Integer Linear Programming formulation (termed *ILP4ID*) for implicit *search result diversification* (SRD). The advantage is that the exact solution can be achieved, which enables us to investigate to what extent using the greedy strategy affects the performance of implicit SRD. Specifically, a series of experiments are conducted to empirically compare the state-of-the-art methods with the proposed approach. The experimental results show that: (1) The factors, such as different initial runs and the number of input documents, greatly affect the performance of diversification models. (2) *ILP4ID* can achieve substantially improved performance over the state-of-the-art methods in terms of standard diversity metrics.

Keywords: Search result diversification · ILP · Optimization

1 Introduction

Accurately and efficiently providing desired information to users is still difficult. A key problem is that users often submit short queries that are ambiguous and/or underspecified. As a remedy, one possible solution is to apply *search result diversification* (SRD) characterized as finding the optimally ranked list of documents which maximizes the overall relevance to multiple possible intents, while minimizing the redundancy among the returned documents. Depending on *whether the subtopics underlying a query are known beforehand*, the problem of SRD can be differentiated into *implicit SRD* and *explicit SRD*. In this work, we do not investigate supervised methods for SRD, but *we focus instead on implicit methods*, for which the possible subtopics underlying a query are *unknown*.

Despite the success achieved by the state-of-the-art methods, the key underlying drawback is that: the commonly used greedy strategy works well on the

© Springer International Publishing AG 2016
S. Ma et al. (Eds.): AIRS 2016, LNCS 9994, pp. 342–348, 2016.
DOI: 10.1007/978-3-319-48051-0_29

premise that the preceding choices are optimal or close to the optimal solution. However, in most cases, this strategy fails to guarantee the optimal solution. Moreover, the factors, such as the initial runs and the number of input documents, are not well investigated in most of the previous studies on implicit SRD.

In this paper, a novel Integer Linear Programming (ILP) formulation for implicit SRD is proposed. Based on this formulation, the exactly optimal solution can be obtained and validated. We then compare the effectiveness of the proposed method *ILP4ID* with the state-of-the-art algorithms using the standard TREC diversity collections. The experimental results prove that *ILP4ID* can achieve improved performance over the baseline methods.

In Sect. 2, we first survey the well-known approaches for implicit SRD. In Sect. 3, the method *ILP4ID* based on ILP is proposed. A series of experiments are then conducted and discussed in Sect. 4. Finally, we conclude the paper in Sect. 5.

2   Related Work

In this section, we first give a brief survey of the typical approaches for SRD. For a detailed review, please refer to the work [3]. We begin by introducing some notations used throughout this paper. For a given query q, $D = \{d_1, ..., d_m\}$ represents the top-m documents of an initial retrieval run. $r(q, d_i)$ denotes the relevance score of a document d_i w.r.t. q. The similarity between two documents d_i and d_j is denoted as $s(d_i, d_j)$.

For implicit SRD, some approaches, such as *MMR* [1] and MPT [4], rely on the *greedy best first strategy*. At each round, it involves examining each document that has not been selected, computing a gain using a specific heuristic criterion, and selecting the one with the maximum gain. To remove the need of manually tuning the trade-off parameter λ, Sanner et al. [2] propose to perform implicit SRD through the greedy optimization of *Exp-1-call@k*, where a latent subtopic model is used in the sequential selection process.

The Desirable Facility Placement (*DFP*) model [6] is formulated as:

$$S^* = \max_{S \subset D, |S| = k} \lambda \cdot \sum_{d \in S} r(d) + (1 - \lambda) \cdot \sum_{d' \in D \setminus S} \max_{d \in S} \{s(d, d')\} \tag{1}$$

where $\mathcal{R}(S) = \sum_{d \in S} r(d)$ denotes the overall relevance. $\mathcal{D}(S) = \sum_{d' \in D \setminus S} \max_{d \in S} \{s(d, d')\}$ denotes the diversity of the selected documents. For obtaining S^*, they initialize S with the k most relevant documents, and then iteratively refine S by swapping a document in S with another one in $D \setminus S$. At each round, interchanges are made only when the current solution can be improved. Finally, the selected documents are ordered according to the contribution to Eq. 1.

Instead of solving the target problem approximately, we formulate in this paper implicit SRD as an ILP problem. Moreover, the effects of different initial runs and the number of used documents on the diversification models are explored. This study is complementary to the work by Yu and Ren [5], where explicit subtopics are required.

3   ILP Formulation for Implicit SRD

We formulate implicit SRD as a process of selecting and ranking k exemplar documents from the top-m documents. We expect to maximize not only the overall relevance of the k exemplar documents w.r.t. a query, but also the *representativeness* of the exemplar documents w.r.t. the non-selected documents. The ILP formulation of selecting k exemplar documents is given as:

$$\max_{\mathbf{x}} \lambda \cdot (m\text{-}k) \cdot \sum_{i=1}^{m} x_{ii} \cdot r(q, d_i) + (1\text{-}\lambda) \cdot k \cdot \sum_{i=1}^{m} \sum_{j=1:j\neq i}^{m} x_{ij} \cdot s(d_i, d_j) \qquad (2)$$

$$s.t. \ x_{ij} \in \{0, 1\}, \ i \in \{1, ..., m\}, j \in \{1, ..., m\} \qquad (3)$$

$$\sum_{i=1}^{m} x_{ii} = k \qquad (4)$$

$$\sum_{j=1}^{m} x_{ij} = 1, i \in \{1, ..., m\} \qquad (5)$$

$$x_{jj} - x_{ij} \geq 0, i \in \{1, ..., m\}, j \in \{1, ..., m\} \qquad (6)$$

In particular, the binary square matrix $\mathbf{x} = [x_{ij}]_{m \times m}$ is defined as: $m = |D|$, x_{ii} indicates whether document d_i is selected, and $x_{ij:i\neq j}$ indicates whether document d_i chooses document d_j as its exemplar. Restriction by Eq. 4 guarantees that k documents are selected. Restriction by Eq. 5 means that each document must have one representative exemplar. The constraint given by Eq. 6 enforces that if there is one document d_i selecting d_j as its exemplar, then d_j must be an exemplar. $\mathcal{R}'(\mathbf{x}) = \sum_{i=1}^{m} x_{ii} \cdot r(q, d_i)$ depicts the overall relevance of the selected exemplar documents. $\mathcal{D}'(\mathbf{x}) = \sum_{i=1}^{m} \sum_{j=1:j\neq i}^{m} x_{ij} \cdot s(d_i, d_j)$ denotes diversity. In view of the fact that there are k numbers (each number is in $[0, 1]$) in the relevance part $\mathcal{R}'(\mathbf{x})$, and m-k numbers (each number is in $[0, 1]$) in the diversity part $\mathcal{D}'(\mathbf{x})$. The coefficients m-k and k are added in order to avoid possible skewness issues, especially when $m \gg k$. Finally, the two parts are combined through the parameter λ.

Although solving arbitrary ILPs is an NP-hard problem, modern ILP solvers can find the optimal solution for moderately large optimization problems in reasonable time. We use the free solver GLPK in this study. Once the k exemplar documents are selected, they are further ranked in a decreasing order of their respective contributions to objective function given by Eq. 2. We denote the proposed approach as *ILP4ID*, namely, *a novel Integer Linear Programming method for implicit SRD*.

Looking back at *DFP* given by Eq. 1, if we view S as the set of exemplar documents, and $D \setminus S$ as the complementary set of non-selected documents, the calculation of $\max_{d \in S} \{s(d, d')\}$ can be then interpreted as selecting the most representative exemplar $d \in S$ for $d' \in D \setminus S$. Thus $\mathcal{D}(S)$ is equivalent to $\mathcal{D}'(\mathbf{x})$. Therefore, *DFP* is a special case of *ILP4ID* when the coefficients m-k and k are not used. Since *ILP4ID* is able to obtain the exact solution w.r.t. the formulated objective function, and *DFP* relies on an approximate algorithm, thus *ILP4ID* can be regarded as the *theoretical upper-bound* of *DFP*.

Moreover, *MMR*, *MPT* and *QPRP* can be rewritten as different variants of *ILP4ID* since the study [6] has shown that they can be rewritten as different variants of *DFP*. However, *ILP4ID* is not the upper-bound of *MMR*, *MPT* and *QPRP*. This is because the space of feasible solutions for *ILP4ID* and *DFP* relying on a two-step diversification is different from the one for *MMR* or *MPT* or *QPRP*, which generates the ranked list of documents in a greedy manner.

4   Experiments

4.1   Experimental Setup

The four test collections released in the diversity tasks of TREC Web Track from 2009 to 2012 are adopted (50 queries per each year). Queries numbered 95 and 100 are discarded due to the lack of judgment data. The evaluation metrics we adopt are nERR-IA and α-nDCG, where nERR-IA is used as the main measure as in TREC Web Track. The metric scores are computed using the top-20 ranked documents and the officially released script *ndeval* with the default settings. The ClueWeb09-T09B is indexed via the Terrier 4.0 platform. The language model with Dirichlet smoothing (denoted as *DLM*) and *BM25* are deployed to generate the initial run.

In this study, the models *MMR* [1], *MPT* [4], 1-*call@k* [2] and *DFP* [6] introduced in Sect. 2 are used as baseline methods. In particular, for 1-*call@k*, we follow the setting as [2]. For *MPT*, the relevance variance between two documents is approximated by the variance with respect to their term occurrences. For *DFP*, the iteration threshold is set to 1000. For *MMR*, *MPT*, *DFP* and the proposed model *ILP4ID*, we calculate the similarity between a pair of documents based on the Jensen-Shannon Divergence. The relevance values returned by *DLM* and *BM25* are then normalized to the range $[0, 1]$ using the MinMax normalization.

4.2   Experimental Evaluation

Optimization Effectiveness. We first validate the superiority of *ILP4ID* over *DFP* in solving the formulated objective function. In particular, we set $\lambda = 0$ (for $\lambda \neq 0$, the results can be compared analogously), both *DFP* and *ILP4ID* work the same, namely selecting k exemplar documents. For a specific topic, we compute the representativeness (denoted as $\mathcal{D}$) of the subset S of k exemplar documents, which is defined as $\mathcal{D}'(\mathbf{x})$ in Sect. 3. The higher the representativeness is, the more effective the adopted algorithm is. Finally, for each topic, we

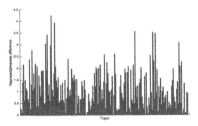

Fig. 1. Optimization effectiveness comparison.

compute the difference between $\mathcal{D}_{ILP4ID}$ and $\mathcal{D}_{DFP}$. As an illustration, we use the top-100 documents of the initial retrieval by *BM25*. Figure 1 shows the performance of *DFP* and *ILP4ID* in finding the best k exemplars, where the x-axis represents the 198 topics, and the y-axis represents the representativeness difference (i.e., $\mathcal{D}_{ILP4ID} - \mathcal{D}_{DFP}$).

From Fig. 1, we see that $\mathcal{D}_{ILP4ID} - \mathcal{D}_{DFP} \geq 0$ for all topics. Because *ILP4ID* always returns the exact solution for each topic, while *DFP* can not guarantee to find the optimal solution due to the adopted approximation algorithm. Since the process of selecting exemplar documents plays a fundamental role for implicit SRD, the effectiveness of DFP is therefore impacted, which is shown in the next section.

Implicit SRD Performance. We use 10-fold cross-validation to tune the trade-off parameters, namely b for *MPT* and λ for *MMR, DFP* and *ILP4ID*. Particularly, λ is tuned in the range $[0, 1]$ with a step of 0.1. For b, the range is $[-10, 10]$ with a step of 1. The metric nERRIA@20 is used to determine the best result. Table 1 shows how *MMR, MPT, DFP*, 1-*call@k* and *ILP4ID* vary when we change the initial runs (i.e., *BM25* and *DLM*) and the number of input documents (i.e., top-m documents of the initial run, where $m \in \{30, 50, 100\}$). Based on the Wilcoxon signed-rank test with $p < 0.05$, the superscripts $* \diamond \dagger$ indicate statistically significant difference to the best result of each setting, respectively.

At first glance, we see that *BM25* substantially outperforms *DLM*. Moreover, given the better initial run by *BM25*, all the models tend to show better performance than that based on the initial run with *DLM*.

A closer look at the results (columns 2–4) shows that *MPT* and 1-*call@k* exhibit poor performance, which even does not enhance the naive-baseline results with *BM25*. For *MMR, DFP* and *ILP4ID*, they show a positive effect of deploying a diversification model. Moreover, the proposed model *ILP4ID* outperforms all the other models in terms of both nERR-IA@20 and α-nDCG@20 across different cutoff-values of used documents. When using the top-100 documents, the improvements in terms of nERR-IA@20 over *BM25, MMR, MPT, DFP* and 1-*call@k* are 20.76 %, 15.03 %, 59.05 %, 5.18 % and 69.67 %, respectively.

Given a poor initial run with *DLM* (columns 6–8), for *MMR*, $\lambda = 0$ (i.e., using top-50 or top-100 documents) indicates that at each step *MMR* selects

Table 1. The performances of each model based on the initial run with *BM25* (columns 2-4) and the initial run with *DLM* (columns 6-8), respectively. The best result of each setting is indicted in bold.

m	Model	nERR-IA@20	α-nDCG@20	Model	nERR-IA@20	α-nDCG@20
	BM25	0.2168$^{◊†}$	0.2784$^{*◊†}$	*DLM*	0.1596$^{*◊†}$	0.2235$^{*◊}$
30	$MMR(\lambda=0.64)$	0.2257	0.2888	$MMR(\lambda=0.1)$	0.1595*	0.226*
	$MPT(b=10)$	0.2078*	0.2701*	$MPT(b=10)$	0.176*	0.2464
	$DFP(\lambda=0.65)$	0.2285*	0.2916*	$DFP(\lambda=0.4)$	**0.2177**	**0.2626**
	1-call@k	0.1918*	0.2632*	*1-call@k*	0.1873*	0.248*
	$ILP4ID(\lambda=0.79)$	**0.2387**	**0.2995**	$ILP4ID(\lambda=0.57)$	0.2107*	0.2578*
50	$MMR(\lambda=0.62)$	0.2247	0.288	$MMR(\lambda=0)$	0.1353$^{◊}$	0.1983$^{◊}$
	$MPT(b=10)$	0.1889$^{◊}$	0.2409$^{◊}$	$MPT(b=10)$	0.1823	**0.2542**
	$DFP(\lambda=0.65)$	0.2522	0.3111	$DFP(\lambda=0.4)$	0.197	0.2394$^{◊}$
	1-call@k	0.1783$^{◊}$	0.2458$^{◊}$	*1-call@k*	0.1663$^{◊}$	0.2233$^{◊}$
	$ILP4ID(\lambda=0.78)$	**0.2565**	**0.3112**	$ILP4ID(\lambda=0.57)$	0.2026	0.2445
100	$MMR(\lambda=0.67)$	0.2276†	0.2917	$MMR(\lambda=0)$	0.1107†	0.1515†
	$MPT(b=10)$	0.1646†	0.2059†	$MPT(b=10)$	0.161	**0.2227**
	$DFP(\lambda=0.68)$	0.2489	0.3094	$DFP(\lambda=0.4)$	**0.1836**	0.2181
	1-call@k	0.1543†	0.2109†	*1-call@k*	0.1535†	0.1988†
	$ILP4ID(\lambda=0.78)$	**0.2618**	**0.3157**	$ILP4ID(\lambda=0.56)$	0.1731†	0.2114

a document merely based on its similarity with the previously selected documents. When using only the top-30 documents, all models (except *MMR*) outperform *DLM* that does not take into account the feature of diversification. The improvements of *MPT*, *DFP*, *1-call@k* and *ILP4ID* over *DLM* in terms of nERR-IA@20 are 10.28 %, 36.4 %, 17.36 % and 32.02 %, respectively. However, when we increase the number of used documents of the initial retrieval, *MPT* shows a slightly improved performance when using the top-50 documents, but the other models consistently show decreased performance. For *MMR*, the results are even worse than *DLM*. These consistent variations imply that there are many *noisy documents* within the extended set of documents.

For *MPT*, $b = 10$ indicates that *MPT* performs a risk-aversion ranking, namely an unreliably-estimated document (with big variance) should be ranked at lower positions.

Given the above observations, we explain them as follows: Even though *1-call@k* requires no need to fine-tune the trade-off parameter λ, the experimental results show that *1-call@k* is not as competitive as the methods like *MPT*, *DFP* and *ILP4ID*, especially when more documents are used. The reason is that: for *1-call@k*, both relevant and non-relevant documents of the input are used to train a latent subtopic model, thus it greatly suffers from the noisy information. Both *MMR* and *MPT* rely on the best first strategy, the advantage of which is that it is simple and computationally efficient. However, at a particular round, the document with the maximum gain via a specific heuristic criterion may cause error propagation. For example, for *MMR*, a long and highly relevant document may also include some noisy information. Once noisy information is included, the diversity score of a document measured by its maximum similarity w.r.t. the previously selected documents would not be precise enough. This well explains why *MMR* and *MPT* commonly show an impacted performance with the increase of the number of used documents. *DFP* can alleviate the aforesaid problem based on the swapping process. Namely, it iteratively refines S by swapping a document in S with another unselected document whenever the current solution can be improved. However, *DFP* is based on the hill climbing algorithm. A potential problem is that hill climbing may not necessarily find the global maximum, but may instead converge to a local maximum. *ILP4ID* casts the implicit SRD task as an ILP problem. Thanks to this, *ILP4ID* is able to simultaneously consider all the candidate documents and globally identify the optimal subset. The aforementioned issues are then avoided, allowing *ILP4ID* to be more robust to the noisy documents.

To summarize, *ILP4ID* substantially outperforms the baseline methods in most reference comparisons. Furthermore, the factors like different initial runs and the number of input documents greatly affect the performance of a diversification model.

5   Conclusion and Future Work

In this paper, we present a novel method based on ILP to solve the problem of implicit SRD, which can achieve substantially improved performance when compared to state-of-the-art baseline methods. This also demonstrates the impact of optimization strategy on the performance of implicit SRD. In the future, besides examining the efficiency, we plan to investigate the potential effects of factors, such as query types and the ways of computing document similarity, on the performance of diversification models, in order to effectively solve the problem of implicit SRD.

References

1. Carbonell, J., Goldstein, J.: The use of MMR, diversity-based reranking for reordering documents and producing summaries. In: Proceedings of the 21st SIGIR, pp. 335–336 (1998)
2. Sanner, S., Guo, S., Graepel, T., Kharazmi, S., Karimi, S.: Diverse retrieval via greedy optimization of expected 1-call@k in a latent subtopic relevance model. In: Proceedings of the 20th CIKM, pp. 1977–1980 (2011)
3. Santos, R.L.T., Macdonald, C., Ounis, I.: Search result diversification. Found. Trends Inf. Retr. **9**(1), 1–90 (2015)
4. Wang, J., Zhu, J.: Portfolio theory of information retrieval. In: Proceedings of the 32nd SIGIR, pp. 115–122 (2009)
5. Yu, H., Ren, F.: Search result diversification via filling up multiple knapsacks. In: Proceedings of the 23rd CIKM, pp. 609–618 (2014)
6. Zuccon, G., Azzopardi, L., Zhang, D., Wang, J.: Top-k retrieval using facility location analysis. In: Baeza-Yates, R., de Vries, A.P., Zaragoza, H., Cambazoglu, B.B., Murdock, V., Lempel, R., Silvestri, F. (eds.) ECIR 2012. LNCS, vol. 7224, pp. 305–316. Springer, Heidelberg (2012)

Predicting Information Diffusion in Social Networks with Users' Social Roles and Topic Interests

Xiaoxuan Ren and Yan Zhang[(⊠)]

Department of Machine Intelligence, Peking University, Beijing 100871, China
{renxiaoxuan,zhy.cis}@pku.edu.cn

Abstract. In this paper, we propose an approach, Role and Topic aware Independent Cascade (RTIC), to uncover information diffusion in social networks, which extracts the opinion leaders and structural hole spanners and analyze the users' interests on specific topics. Results conducted on three real datasets show that our approach achieves substantial improvement with only limited features compared with previous methods.

Keywords: Information diffusion · Social role · Retweeting prediction

1 Introduction

There are a lot of studies on information diffusion in the Online Social Network (OSN), and many of them focus on retweeting, which is a main form of information diffusion. Suh *et al.* examine a number of features that might affect retweetability of tweets based on a Generalized Linear Model (GLM) [10]. Zhang *et al.* consider a diversity of personal attributes for retweeting behavior prediction, including followers, followees, bi-followers, longevity, gender and verification status [14].

Different from the above complicated consideration, we present a simple observation from the group level. During the process of information diffusion, users' social roles play an important part. Usually opinion leaders [8] will dominate the retweeting within a group. When information needs to spread across the groups, the users who span the structural holes [3] are more critical. Besides the social roles, users' topic interests are another important part. Users have significant interest differences in distinct topics, which indirectly affect whom they follow and whether they will retweet the tweet, thus affecting the whole information diffusion process. Therefore, based on users' social roles and topic interests, we propose a simple prediction approach for information diffusion.

2 Related Work

Social Role Analysis. For the opinion leaders extracting, there are some algorithms based on the original PageRank [2] algorithm. For instance, Haveliwala

© Springer International Publishing AG 2016
S. Ma et al. (Eds.): AIRS 2016, LNCS 9994, pp. 349–355, 2016.
DOI: 10.1007/978-3-319-48051-0_30

proposes the Topic-Sensitive PageRank [6], and Weng *et al.* propose the Twitterrank [11]. For structural hole spanners' finding, Granovetter proposes that nodes will benefit from linking different parts of the network together [5]. Based on this, Burt puts forward the conception of structural hole [3]. Lou an Tang propose an algorithm called HIS to identify structural hole spanners [7].

Topic Interest Analysis. Most previous studies attempt to mine the users' topic interests through their original tweets or the tweets retweeted from others, and mainly use the model like LDA. For instance, tag-LDA proposed by Si and Sun [9] and Twitter-LDA put forward by Zhao and Jiang [15]. In our work, we analyze the users' topic interests not only with the tweets, but the users' following relationship as well.

Information Diffusion Prediction. Zaman *et al.* train a probabilistic collaborative filter model to predict future retweets [13]. Yang *et al.* study the information diffusion via the retweeting behavior, and also propose a semisupervised framework to predict the retweeting [12].

3   Social Role and Topic Interest Analysis

Let $G = (V, E)$ denotes a social network, where V is a set of users, and $E \subseteq V \times V$ is a set of links between users, in which $e_{vu} \in E$ denotes the directed link from user v to user u $(v, u \in V)$.

3.1   Opinion Leader Finding

We use the PageRank algorithm [2] to calculate the users' influence.

Definition 1. Let i be a user(node) of the social network, $i \in V$. Let $F(i)$ be the set of user i's followers, and $R(i)$ is the set of users that i follows. $N(i) = |R(i)|$ is the number of user i's followees.

Let $PR(i)$ donates the influence value of user i, then the equation for calculating the influence value of each user is $PR(i) = d \sum_{j \in F_i} \frac{PR(j)}{N_j} + \frac{1-d}{n}$, where d is the damping coefficient, and n is the number of users in the social network.

However, a user with a high PR value may not be the authority figure for a specific topic. Considering this, we introduce the topic authoritativeness to evaluate the user's influence on specific topics. We use the 16 top-level categories from *Open Directory Project*[1] to classify the topics of the social network. For each topic, we give each opinion leader i a value $a_{i,t} \in [0, 10]$ to represent i's authoritativeness on topic t, then we can get a new PR value for each opinion leader which we denote as $APR(i) = a_{i,t}PR(i)$.

---

[1] http://www.dmoz.org.

3.2 Structural Hole Spanner Finding

We use the HIS algorithm [7] for the structural hole spanner mining over disjoint groups.

Definition 2. Giving a network $G = (V, E)$, denote that G is grouped into m groups $G = \{G_1, G_2, \ldots, G_m\}$, with $V = G_1 \cup \ldots \cup G_m$, and $G_i \cap G_j = \varnothing (0 < i < j \leq m)$; let $I(v, C_i) \in [0, 1]$ denotes the importance score of user v in community i; for each subset of communities $S(S \subseteq C$ and $|S| \geq 2)$, let $SH(v, S) \in [0, 1]$ denotes the structural hole spanner score of user v in S.

In order to split the social network G into m groups, we use the Louvain algorithm [1]. Then the two scores $I(v, C_i)$ and $SH(v, S)$ can be defined in terms of each other in a mutual recursion:

$$I(v, C_i) = \max_{e_{uv} \in E, S \subseteq C \wedge C_i \in S} \{I(v, C_i), \alpha_i I(u, C_i) + \beta_s SH(u, S)\} \tag{1}$$

$$SH(v, S) = I(v, C_i) \tag{2}$$

where α_i and β_S are two tunable parameters. For initialization, we use the PageRank algorithm to initialize the importance score $I(v, C_i)$:

$$\begin{aligned} I(v, C_i) &= PR(v), v \in C_i \\ I(v, C_i) &= 0, \qquad v \notin C_i \end{aligned} \tag{3}$$

3.3 Topic Interest Analysis

We classify the users into two types, the famous users and the normal users[2]. We first collect all the original tweets from the famous users, then we filter the original tweets of each user v, if he/she has sent the tweets relevant to the topic t, we can say the user v has interests in topic t. Then we get the normal users' topic interests from whom they follow. We collect all the famous users that a normal user u follows, then count the famous users who have interests on topic t that user u follows, and get the number of famous users related as c.

4 Information Diffusion Prediction

In our work, we extend the Independent Cascade (IC) model [4] with users' social roles and topic interests, and propose an approach called Role and Topic aware Independent Cascade (RTIC).

For each time of the retweeting behavior, we can consider it as a classification problem: given a tweet m, a topic t, a user v and its inactive followers set $U = \{u_1, u_2, \ldots, u_n\}$, where n is the number of v's inactive followers, and a timestamp t, the goal is to categorize user u's $(u \in U)$ status at time $t + 1$. We

---

[2] We define that users who have followed by at least 1000000 are the famous users, others are the normal users.

denote the classification outcome as $y_{v,u}^t$. If user u retweet the tweet m at time t from user v, then $y_{v,u}^t = 1$; otherwise $y_{v,u}^t = 0$. We consider three factors: x_1, x_2 and x_3. x_1 is the APR value of user v, x_2 is the SH score of user v, while x_3 indicates the number c of famous users who have interests on topic t that u follows. In this paper, we use the logistic regression model to predict the value of influence probability $P(y_{v,u}^t = 1|x_1, x_2, x_3)$, and the logistic regression classifier is as follows:

$$P(y_{v,u}^t = 1|x_1, x_2, x_3) = \frac{1}{1 + e^{-(\beta_0 + \beta_1 x_1 + \beta_2 x_2 + \beta_3 x_3)}} \tag{4}$$

where β_0 is a bias, and β_1, β_2, β_3 are weights of the features.

We define a threshold $\epsilon \in [0, 1]$, if $P \geq \epsilon$, then the user u will retweet the tweet from the user v, otherwise not. Let tw_u denote the number of tweets of user u, and tw_{avg} denotes the average number of tweets of all users, then we define the ϵ value as $\epsilon = \frac{1}{1 + e^{\frac{tw_u}{tw_{avg}} - 1}}$.

5   Experiment

5.1   Experimental Setup

Dataset. We evaluate our approach on three different real Weibo datasets from [14]. Three big events are extracted from the original datasets for further analysis, which are "Mo Yan won the Nobel Prize in Literature", "Liu Xiang quit the 110-metre hurdle in the 2012 Olympic Games" and "Fang Zhouzi criticizes Han Han's articles are ghostwritten"[3]. The statistics are shown in Table 1.

Table 1. Data statistics.

Data sets	Original tweets	Retweet users	Users	Follow-relationships
Mo Yan	6	3,419	956,163	3,688,435
Liu Xiang	5	14,422	1,201,762	6,021,954
Han Han	4	3,327	584,341	1,866,145

Baseline. We consider the following comparison methods in our experiments:

LRC-B. Proposed in [14], it uses the logistic regression model and uses a variety of features to train the logistic regression classifier.

No-Roles. This is a variant of the RTIC approach, which only considers the users' topic interests, and ignores the users' social roles.

---

[3] In the following pages, we call these three data sets as "Mo Yan data set", "Liu Xiang data set" and "Han Han data set.".

No-Interests. This is also a variant of the RTIC approach which only considers the users' social roles as opinion leaders or structural hole spanners.

Hyperparameter. Through extensive empirical studies, we set d as 0.85, $\alpha_i = 0.3$ and $\beta_S = 0.5 - 0.5^{|S|}$. We change the number of opinion leaders to see the performance variation of information diffusion prediction, and finally we set the top 0.01 % users as opinion leaders. Same thing for structural hole spanners.

5.2 Performance Analysis

Social Role Analysis. We calculate the influence coverage of each dataset, which means the percentage of the opinion leaders' followers in the all users of the data set. The results are 75.6 %, 68.6 %, 63.1 % respectively, which are quite strong. For the structural hole spanners mining, we count the number of groups that a structural hole spanner's followers can cover, and the results show that most structural hole spanners' followers can cover above 50 % of the whole groups, and some even can cover 100 % of the groups, which means that as long as a structural hole spanner retweet a tweet, the information can propagate to a large amount of users from the unrelated groups.

Information Diffusion Prediction. We use the Mo Yan dataset as the training set, and the Han Han dataset as the testing set. Table 2 shows the performance of the proposed approach RTIC and baselines of information diffusion prediction task.

Table 2. Performance of information diffusion prediction (%).

Method	Prec	Rec	F_1
RTIC	75.69	76.78	76.23
LRC-B	65.74	77.11	70.97
No-Roles	61.92	68.95	65.24
No-Interests	48.53	50.24	49.37

The proposed approach outperforms the baseline method LRC-B by 5.26 % in terms of F_1 Measure. We may conclude that the features considered in LRC-B are basically about the users' own profiles, but ignore the structure of the network and the semantic information, which is quite important in the real social networks. The performance of the both variant approaches is unsatisfactorily, however, the No-Roles is better than the No-Interests method.

5.3 Parameter Analysis

We use each data set as the training set to train the four parameters β_0, β_1, β_2 and β_3. The results are shown in Table 3. The quite similar parameters indicate that it's possible to use one dataset to train the parameters.

Table 3. Parameters of data set.

Data Set	β_0	β_1	β_2	β_3
Mo Yan	-8.508	4.518	3.159	0.055
Liu Xiang	-8.437	4.326	3.474	0.053
Han Han	-8.042	4.624	3.097	0.055

6   Conclusion and Future Work

In this paper, we propose an approach called Role and Topic aware Independent Cascade (RTIC), which aims to simplify the approach by reducing the number of features considered. We evaluate our approach on real Weibo data sets, and the results show it outperforms other baseline approaches with only a limited number of features considered. As a future work, we want to study how the influential nodes work in the whole information diffusion process.

Acknowledgment. This research is supported by NSFC with Grant No. 61532001 and No. 61370054, and MOE-RCOE with Grant No. 2016ZD201.

References

1. Blondel, V.D., Guillaume, J.L., Lambiotte, R., Lefebvre, E.: Fast unfolding of communities in large networks. J. Stat. Mech. Theory Exper. **2008**(10), P10008 (2008)
2. Brin, S., Page, L.: The anatomy of a large-scale hypertextual web search engine. In: Proceedings of WWW 1998. ACM (1998)
3. Burt, R.S.: Structural Holes: The Social Structure of Competition. Harvard University Press, Cambridge (2009)
4. Goldenberg, J., Libai, B., Muller, E.: Talk of the network: a complex systems look at the underlying process of word-of-mouth. Mark. Lett. **12**(3), 211–223 (2001)
5. Granovetter, M.S.: The strength of weak ties. Am. J. Sociol. **78**, 1360–1380 (1973)
6. Haveliwala, T.H.: Topic-sensitive PageRank. In: Proceedings of WWW 2002, pp. 517–526. ACM (2002)
7. Lou, T., Tang, J.: Mining structural hole spanners through information diffusion in social networks. In: Proceedings of WWW 2013, pp. 825–836. International World Wide Web Conferences Steering Committee (2013)
8. Rogers, E.M.: Diffusion of Innovations. Simon and Schuster, New York (2010)
9. Si, X., Sun, M.: Tag-LDA for scalable real-time tag recommendation. J. Comput. Inf. Syst. **6**(1), 23–31 (2009)
10. Suh, B., Hong, L., Pirolli, P., Chi, E.H.: Want to be retweeted? Large scale analytics on factors impacting retweet in Twitter network. In: 2010 IEEE Second International Conference on Social Computing (SocialCom), pp. 177–184. IEEE (2010)
11. Weng, J., Lim, E.P., Jiang, J., He, Q.: TwitterRank: finding topic-sensitive influential twitterers. In: Proceedings WDSM 2010, pp. 261–270. ACM (2010)

12. Yang, Z., Guo, J., Cai, K., Tang, J., Li, J., Zhang, L., Su, Z.: Understanding retweeting behaviors in social networks. In: Proceedings of CIKM 2010, pp. 1633–1636. ACM (2010)
13. Zaman, T.R., Herbrich, R., Van Gael, J., Stern, D.: Predicting information spreading in Twitter. In: Workshop on Computational Social Science and the Wisdom of Crowds, NIPs, vol. 104, pp. 599–601. Citeseer (2010)
14. Zhang, J., Tang, J., Li, J., Liu, Y., Xing, C.: Who influenced you? Predicting retweet via social influence locality. TKDD **9**(3), 25 (2015)
15. Zhao, X., Jiang, J.: An empirical comparison of topics in Twitter and traditional media. Singapore Management University School of Information Systems Technical Paper Series (2011). Accessed 10 Nov 2011

When MetaMap Meets Social Media in Healthcare: Are the Word Labels Correct?

Hongkui Tu[1][(✉)], Zongyang Ma[2], Aixin Sun[2], and Xiaodong Wang[1]

[1] National Laboratory of Parallel and Distributed Processing,
National University of Defense Technology, Changsha, China
`tuhkjet@foxmail.com, xdwang@nudt.edu.cn`
[2] School of Computer Science and Engineering,
Nanyang Technological University, Singapore, Singapore
`{zyma,axsun}@ntu.edu.sg`

Abstract. Health forums have gained attention from researchers for studying various topics on healthcare. In many of these studies, identifying biomedical words by using the MetaMap is often a pre-processing step. MetaMap is a popular tool for recognizing Unified Medical Language System (UMLS) concepts in free text. However, MetaMap favors identifying terminologies used by professionals rather than laymen terms by the common users. The word labels given by MetaMap on social media may not be accurate, and may adversely affect the next level studies. In this study, we manually annotate the correctness of medical words extracted by MetaMap from 100 posts in HealthBoards and get a precision of 43.75 %. We argue that directly applying MetaMap on social media data in healthcare may not be a good choice for identifying the medical words.

Keywords: Social media · Healthcare · MetaMap · UMLS

1 Introduction

Health forums like HealthBoards and MedHelp enable patients to learn and communicate on health issues online. A major advantage of such medical forums is that patients are able to get advices from peers, and sometimes professionals. Compared with traditional medical data (*e.g.,* electronic health records), online social platforms are able to provide a large amount of streaming data contributed by millions of users. Mining such large scale user generated content (UGC) helps us better understanding of users and patients on many health related topics. In many studies conducted on the medical forum data [3,4], Unified Medical Language System (UMLS) has been introduced to identify the medical words,

This work was done when the first author visiting School of Computer Science and Engineering, Nanyang Technological University, supported by Chinese Scholarship Council (CSC) scholarship.

© Springer International Publishing AG 2016
S. Ma et al. (Eds.): AIRS 2016, LNCS 9994, pp. 356–362, 2016.
DOI: 10.1007/978-3-319-48051-0_31

as a pre-processing. The most popular tool for recognizing UMLS concepts in text is MetaMap.[1]

MetaMap favors domain-specific terminologies and is not designed to use on UGC in medical forum. Therefore, the correctness of the word labels by directly applying MetaMap on forum data may not be high enough to support the next level studies. We further illustrate this point by using two example sentences taken from a medial forum and the corresponding word labels assigned by MetaMap.

- Welcome <u>back</u> [Body Location or Region], <u>thanks</u> [Gene or Genome] for your <u>advice</u> [Health Care Activity].
- <u>I</u> [Inorganic Chemical] have exactly the <u>same</u> [Qualitative Concept] <u>thing</u> [Entity], <u>calf pain</u> [Sign or Symptom] when <u>sitting</u> [Physiologic Function], especially <u>driving</u> [Daily or Recreational Activity]!

In the example sentences, the words or phrases that are identified by MetaMap are underlined, and the assigned semantic types are in square brackets. Observe that, words like "I", "thanks" are commonly used in online medical forums. However, these words are often wrongly assigned to incorrect semantic types by MetaMap. The words that are closely related to medical domain like "calf pain" are assigned to correct semantic types. For the professional textual documents, it is very rare to have phrases like "welcome back". On the contrast, in the UGC from social media, such informal phrases frequently appear. Here, we focus on the evaluation of the words and phrases that are "recognized" by MetaMap, and *recall* is not considered in this study, because manually labeling a large number of words and phrases from social medical data is not practical.

More specifically, we evaluate the accuracy of the word labels by MetaMap on sentences from a popular medical forum HealthBoards,[2] which was launched in 1998 and it now consists of over 280 message boards for patients. The anonymized dataset is released by the authors in another study [5].[3] We manually annotate the correctness of the word labels by MetaMap on 100 randomly selected posts covering different message boards. These posts consist of 665 sentences and MetaMap returns 3758 word labels.[4] After manually annotating the correctness of the word labels assigned by MetaMap, we make five observations in Sect. 2.

There are several works [1,6,7] aiming to quantify the results of MetaMap on various types of free text. Stewart *et al.* [7] compare the precision of MetaMap and MGrep for medical lexicons mapping on a medical mailing list dataset. Denecke *et al.* [1] do a qualitative study with MetaMap and cTakes on "Health Day News" and blog postings from "WebMD". Compared with social forums, the language of News and Blogs is relatively more formal. Most relevant to our study,

---

[1] https://metamap.nlm.nih.gov/.

[2] http://www.healthboards.com/.

[3] http://www.mpi-inf.mpg.de/departments/databases-and-information-systems/research/impact/peopleondrugs/.

[4] Note that MetaMap is able to label phrases. We use "words" in our discussion for simplicity.

Park *et al.* [6] characterize three types of failures when applying MetaMap to online community text related to breast cancer. The authors further incorporate Stanford POS tagger and parser with some handcraft rules to revise the outputs. Our study complements the study reported in [6] as we focus on a different dataset (*i.e.*, posts from HealthBoards) and the data covers a wide range of topics. We make observations on the correctness of labels by MetaMap based on their semantic types. We also show that the labels of several semantic types have relative high precision, which could benefit next stage study, depending on the task.

2   MetaMap Meets HealthBoards

Data Annotation and Observations. We randomly selected 100 posts from the data set and split them into sentences with NLTK.[5] We get 665 sentences in total. Then all words in the sentences are labeled with all the semantic types using the Java API of MetaMap 2014 Windows version.[6] In using the API, the word sense disambiguation option was selected and the other options/parameters were set to their default values. As the result, 3758 words are labeled by MetaMap and there are 1383 unique words. We recruit 2 graduated students to annotate the 665 sentences and compute the Kappa coefficient ($\kappa = 0.889$), which suggests that annotators can reach a high agreement.

Observation 1. *The precision of the word labels by MetaMap with all semantic types is fairly low at 43.75 % only.*

With all the 135 semantic types in consideration, the precision we obtained on the 3758 word labels is 43.75 %, which is not promising. The low precision suggests that MetaMap is very unlikely to assign correct semantic types to words obtained from social medical data. We also observe that some semantic types (*e.g.*, Reptile, Cell Function) are not used to label any words. It is understood that because of the extremely wide coverage of the UMLS concept space, not all concepts are discussed in medical forums. Either the semantic types are not of interests of the patients (*e.g.*, Reptile) or the semantic types are too domain-specific for discussions among non-experts (*e.g.*, Cell Function).

Observation 2. *The precision on the semantic types of general concepts is relatively high and much lower for the word labels of concepts in (narrow) domain-specific semantic types.*

As not all semantic types are covered in forum discussion, we now study the largest 38 semantic types ranked by the number of instances in the annotated data. Each of the 38 semantic types, listed in Table 1, has at least 20 word labels. Observe that the semantic types obtaining high precisions (*e.g.*, greater than 85 %) are mostly general types such as [Entity] (*e.g.*, things), [Population Group]

---

[5] http://www.nltk.org.

[6] https://metamap.nlm.nih.gov/download/public_mm_win32_main_2014.zip.

Table 1. Precisions (denoted by Pr) on the largest 38 semantic types by the number of word labels (or instances) in each semantic type. The semantic types are ranked by their precisions in descending order. The semantic types with more than 50 instances are in boldface.

Semantic Type	Pr (%)	#Instance	Semantic Type	Pr (%)	#Instance
Entity	96.30	27	**Social Behavior**	55.77	52
Population Group	96.00	25	**Qualitative Concept**	45.18	436
Daily or Recreational Activity	94.12	34	**Finding**	43.70	135
Body Part, Organ, or Organ Component	94.03	67	Medical Device	42.31	26
Sign or Symptom	89.58	48	Disease or Syndrome	38.46	65
Mental Process	86.74	181	Organism Attribute	37.93	29
Food	84.62	39	Conceptual Entity	37.50	48
Organism Function	80.49	41	**Intellectual Product**	37.23	94
Mental or Behavioral Dysfunction	78.38	37	Occupational Activity	33.33	21
Professional or Occupational Group	77.27	22	Health Care Activity	32.50	40
Pharmacologic Substance	75.32	77	Activity	31.25	48
Family Group	75.00	20	**Idea or Concept**	26.27	118
Quantitative Concept	70.62	177	**Spatial Concept**	19.42	103
Organic Chemical	66.67	36	**Functional Concept**	14.39	271
Temporal Concept	63.83	282	**Amino Acid, Peptide, or Protein**	5.13	78
Manufactured Object	60.71	56	**Immunologic Factor**	3.08	65
Biomedical or Dental Material	60.00	20	**Geographic Area**	2.65	113
Therapeutic or Preventive Procedure	59.46	37	**Inorganic Chemical**	2.36	339
Body Location or Region	57.14	21	**Gene or Genome**	0.00	91

(*e.g.,* people, women), [Daily or Recreational Activity] (*e.g.,* read, speaking), [Body Part, Organ, or Organ Component] (*e.g.,* ears, hair). Concepts in these semantic types, however, are less related to healthcare in many cases. However, [Sign or Symptom] (*e.g.,* tired, sleepless) got a precision at 89.58 %. The words assigned with this type are seemingly of less ambiguity. Among the 6 semantic types that have higher precision than 85 % and with lots of instances is [Mental Process]. This type is more likely to be assigned to words like "think" and "hope", which are words that are widely used in online discussions and may not be specifically for communicating medical related issues. There are five semantic types where the word labels are of extremely low precisions, below 10 %. They are [Amino Acid, Peptide, or Protein], [Immunologic Factor], [Geographic Area], [Inorganic Chemical] and [Gene or Genome]. As the names suggest, the concepts in these semantic types are very domain-specific. Without considering the context of the words, the word labels assigned by MetaMap are of very low quality. In particular, the word "I" is assigned to [Inorganic Chemical], which is also highlighted in [6]. It is understood that this word may not appear many times in medical literature or medical records; when dealing with content directly contributed by patients, the usage of this word could be extremely high, but not referring to inorganic chemical.

Observation 3. *Most semantic types with a large number of instances are not medical or healthcare related.*

In Table 1, the semantic types each has more than 50 instances are in boldface. Other than [Inorganic Chemical] which contains most appearance of the wrongly labeled word "I" as discussed above, the semantic types with a large number of instances are mostly general concepts, *e.g.,* [Qualitative Concept], [Temporal Concept], [Functional Concept], [Mental Process] and [Quantitative Concept]. The names of these semantic types are self-explanatory, and the words are obviously not specifically related to healthcare or medical. The other large semantic types include [Idea or Concept] containing words like "life", and [Spatial Concept] containing words like "outside" and "in". Both semantic types have low precision at 30 % or below.

Observation 4. *Limiting to a selected subset of (more medical related) semantic types does not necessarily lead to much higher quality word labels on medical forums.*

As mentioned before, the semantic type [Sign or Symptom], which is clearly related to medical and healthcare, has a relatively large number of instances with a precision of 89.58 %. Example words assigned to this semantic type include "sleepless", "insomnia" and "tiredness". In [2], the authors selected 14 semantic types which are more related to medical and health topics in our understanding. These 14 semantic types are [Antibiotic] (3), [Clinical Drug] (1), [Congenital Abnormality] (9), [Disease or Syndrome] (65), [Drug Delivery Device] (0), [Experimental Model of Disease] (0), [Finding] (135). [Injury or Poisoning] (16), [Mental or Behavioral Dysfunction] (29), [Pharmacologic Substance] (77), [Sign or Symptom] (48), [Steroid] (9), [Therapeutic or Preventive Procedure] (37), and [Vitamin] (0). The numbers following each semantic type are the number of word labels in the annotated data. The total number of instances under these 14 semantic types is 437. Observe that, there are three semantic types have no words assigned to them. Nevertheless, in the following discussion, we will simply use *the selected 14 semantic types* to refer to this set of semantic types without taking out the three with no instances.

The precision of the word labels assigned to these 14 semantic types is 56.75 %, which is better than the overall precision computed on all the 135 semantic types, but remains low. In other words, the quality of word labels on the selected set of more medical related semantic types is not promising.

Observation 5. *Removing the high-frequent words may not necessarily be a good choice in pre-processing.*

In [2], the authors filter out the popular words from their data which are ranked among the top-2000 in Google N-Grams.[7] Following this idea, we filter from the 3758 labeled words the top-2000 Google N-Grams, resulting in 1688 word labels. Removing more than half of the identified word labels suggests that

---

[7] https://books.google.com/ngrams.

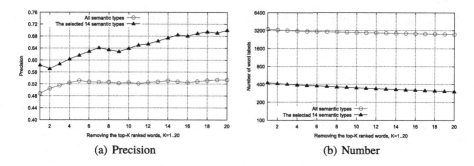

(a) Precision   (b) Number

Fig. 1. Precision and number of word labels after removing the top-K ranked words, $K = 1...20$

Table 2. The top 20 most frequent words in *all* and the *selected 14* semantic types respectively

All semantic types	i to not my m good time help now out hope so take ve one feel go well then think
14 semantic types	meds pain find said little hi read water worse brand used shot but ready able cortisone dh tired depression others

there exists a big overlap between the set of words identified by MetaMap and the popular words in Google N-Grams. The precision computed on the remaining 1688 word labels is 58.77 %. Although this is a much better value compared to 43.75 % before the filtering, there remain about two fifth incorrect word labels after the filtering. We argue that, simply removing the most frequent words based on Google N-Gram may not be a good choice in pre-processing.

From our data, it is observed that some frequent words are repeatedly wrongly labeled, *e.g.,* "I" labeled to [Inorganic Chemical]. It is therefore interesting to evaluate the quality of the word labels after removing some most frequent words. The precisions computed after removing the top-K most frequent words ($1 \leq K \leq 20$) are plotted in Fig. 1(a). For all semantic types, removing the most frequent words does help to improve the precision to the maximum of 53.38 %. The plot of the precision values on the 14 selected semantic types, however, is not very smooth. As shown in Table 2, some frequent words are indeed medical related, *e.g.,* pain, tired, and depression. Removing such words leads to decrease in precision because of their large frequencies.

3   Summary

We evaluate the quality of word labels by MetaMap on text from a medical forum HealthBoards. Our study shows that directly applying MetaMap on social media data on healthcare leads to low quality word labels. One of the main reasons is that MetaMap is designed for processing medical data written by professionals rather than UGC in online forums. We hope the negative results obtained in our study will motivate more research on the effective application of MetaMap on social media data in healthcare.

Acknowledgment. This research was supported by Singapore Ministry of Education Academic Research Fund Tier-1 Grant RG142/14, National Natural Science Foundation of China (No. 61070203, No. 61202484, No. 61472434 and No. 61572512) and a grand from the National Key Lab. of High Performance Computing (No. 20124307120033).

References

1. Denecke, K.: Extracting medical concepts from medical social media with clinical NLP tools: a qualitative study. In: Proceedings of the 4th Workshop on Building and Evaluation Resources for Health and Biomedical Text Processing (2014)
2. Gupta, S., MacLean, D.L., Heer, J., Manning, C.D.: Research and applications: induced lexico-syntactic patterns improve information extraction from online medical forums. JAMIA **21**(5), 902–909 (2014)
3. Liu, X., Chen, H.: Identifying adverse drug events from patient social media: a case study for diabetes. IEEE Intell. Syst. **30**(3), 44–51 (2015)
4. Metke-Jimenez, A., Karimi, S., Paris, C.: Evaluation of text-processing algorithms for adverse drug event extraction from social media. In: Proceedings of ACM Workshop on Social Media Retrieval and Analysis (SoMeRA), pp. 15–20 (2014)
5. Mukherjee, S., Weikum, G., Danescu-Niculescu-Mizil, C.: People on drugs: credibility of user statements in health communities. In: Proceedings KDD, pp. 65–74. ACM (2014)
6. Park, A., Hartzler, A.L., Huh, J., McDonald, D.W., Pratt, W.: Automatically detecting failures in natural language processing tools for online community text. JMIR **17**(8), e212 (2015)
7. Stewart, S.A., Von Maltzahn, M.E., Abidi, S.R.: Comparing metamap to MGrep as a tool for mapping free text to formal medical lexicons. In: Proceedings of the 1st International Workshop on Knowledge Extraction & Consolidation from Social-Media, pp. 63–77 (2012)

Evaluation with Confusable Ground Truth

Jiyi Li[(⊠)] and Masatoshi Yoshikawa

Graduate School of Informatics, Kyoto University,
Yoshida-Honmachi, Sakyo-ku, Kyoto 606-8501, Japan
{jyli,yoshikawa}@i.kyoto-u.ac.jp

Abstract. Subjective judgment with human rating has been an important way of constructing ground truth for the evaluation in the research areas including information retrieval. Researchers aggregate the ratings of an instance into a single score by statistical measures or label aggregation methods to evaluate the proposed approaches and baselines. However, the rating distributions of instances are diverse even if the aggregated scores are same. We define a term of confusability which represents how confusable the reviewers are on the instances. We find that confusability has prominent influence on the evaluation results with a exploration study. We thus propose a novel evaluation solution with several effective confusability measures and confusability aware evaluation methods. They can be used as a supplementary to existing rating aggregation methods and evaluation methods.

Keywords: Human rating · Ground truth · Evaluation · Confusability

1 Introduction

Subjective judgment by human beings has been an important way for constructing ground truth because in many cases the gold standard cannot be collected or computed automatically and need to be judged manually. In various styles of subjective judgment, human ratings is a frequently used manner. When constructing the ground truth, researchers aggregate the ratings from multiple reviewers into a single score, which can be mean, majority voting and so on, and use the aggregated scores as the ground truth in the experiments, for example, [1,2] use mean scores of multiple ratings on the similarity of document pairs.

However, the rating distributions of instances are diverse, even if the aggregated scores are same. For example, assuming that there are two rating sets with five ratings, $\{3, 3, 3, 3, 3\}$ and $\{2, 2, 3, 4, 4\}$, they have different rating distributions but the aggregated mean scores are same. The aggregation loses the information of the differences of rating distributions. We make an exploration study with approaches of document similarity computation on a document collection. It shows that these differences have important influences on the evaluation results.

Because the ground truth are always constructed with quality control, we can assume that the human ratings by reviewers in this scenario have sufficiently high

© Springer International Publishing AG 2016
S. Ma et al. (Eds.): AIRS 2016, LNCS 9994, pp. 363–369, 2016.
DOI: 10.1007/978-3-319-48051-0_32

quality and these differences of rating distributions as an inherent property of a judged instance or its rating set. This property represents how confusable the reviewers are when judging this instance. We denote this property as *confusability*. We need to consider the confusability information in the ground truth when we carry out experiments and evaluations, while existing evaluation methods which only use aggregated scores of human ratings do not consider it. Our confusability metrics and confusability aware evaluation methods are not proposed to instead traditional ones. They are used with the existing ones together to supplement the quality of the evaluation results.

There are some existing work which also consider the inconsistency of human ratings. On one hand, Cohen's kappa [3] is used to measure inter-rater agreement of two reviewers on a set of instances. In our scenario, we need to measure the disagreement of a set of reviewers on one instance. On the other hand, label aggregation methods in crowdsourcing such as [4] with a probabilistic model can be used to generate more rational aggregated rating than statistical measures like mean and majority voting. However, they still output a unique rating for an instance and the confusability information of the dataset is still lost when using them for evaluation. Our work concentrates on how to include the confusability in the evaluation beyond the use of single aggregation score.

Our contributions are as follows. First, we make an exploration study which analyzes the *influences of confusability* in the evaluation results. Second, we propose several *confusability metrics and confusability aware evaluation methods* for different purposes. We utilize a public document dataset for illustration.

2   Exportation Study on Confusability in Human Ratings

In a given dataset $\mathcal{D}$, we denote d_i as an *instance*. In the ground truth, we define the person who provides the ratings as *reviewer*. $\mathbf{r}_i = \{r_{ij}\}_j$ is the set of ratings of d_i, and the number of ratings of instance d_i is $|\mathbf{r}_i| = n_i$. $\mathcal{R}$ is the space of rating values and $r_{ij} \in \mathcal{R}$. The ratings are numerical values such as $\mathcal{R} = \{1, 2, 3\}$, or categorical values such as $\{-, +\}$ mapped to $\mathcal{R} = \{0, 1\}$. The result of instance d_i generated by an approach k is s_i^k.

Dataset LP50: This public document collection contains 50 news articles selected from the Australian Broadcasting Corporation's news mail service [1]. It not only provides the data of human ratings with aggregated measures, but also the non-aggregated ratings of all reviewers. There are 1225 document pairs and 83 reviewers. The semantic similarity of each document pair is labeled by around 10 reviewers with ratings from 1 (highly unrelated) to 5 (highly related).

2.1   Influences of Confusability in Evaluation

The confusability can be measured in various ways in our topic, while in this exploration study we measure it with rating difference, which is the difference between maximum and minimum rating of an instance d_i, $difr_i = max\{\mathbf{r}_i\} - min\{\mathbf{r}_i\}$. When $difr_i$ is lower, it means that the ratings of the reviewers are

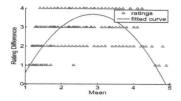

Fig. 1. Rating Difference and Mean

Table 1. Approach performance on entire LP50 and subset of each rating difference

$difr_i$	LSA	KBG	$difr_i$	LSA	KBG
0	**0.859**	0.714	3	0.466	**0.483**
1	**0.755**	0.622	4	0.512	**0.562**
2	**0.597**	0.567	Entire	0.575	0.570

more consistent, the rating of d_i is easier to be judged by human beings. Figure 1 shows the distribution of the rating differences $difr_i$ on the mean ratings $mear_i$. It shows that the document pairs that are easy to be judged ($difr_i = 0$) are the document pairs that are very similar ($mear_i = 5$) or very dissimilar ($mear_i = 1$). When $mear_i$ is around 3, it is possible that the document semantic similarity of d_i is not easy to be judged by human beings ($difr_i \geq 2$). In such condition, it may be also difficult for an approach k to generate consistent similarity results with the human ratings which are used as the ground truth.

We implement two approaches to analyze the relationships between approach performance and the confusability. One is Latent Semantic Analysis (**LSA**) which represents documents with vectors of latent topics using singular value decomposition. The other one is Knowledge Based Graph (**KBG**) [2] which represents document with a entity sub-graph extracted from a knowledge base.

Following the existing work such as [2], we use Pearson correlation coefficient which is a traditional metric without considering confusability for evaluation. The correlation between the result set $\mathbf{s}^k = \{s_i^k\}_i$ of approach k (here k is LSA or KBG) and mean rating set $mear = \{mear_i\}_i$ on a set of document pairs is

$$Corr(\mathbf{s}^k, mear) = \frac{\sum_{i=1}^{\mathcal{N}}(s_i^k - \mu_{\mathbf{s}^k})(mear_i - \mu_{mear})}{\sqrt{\sum_{j=1}^{\mathcal{N}}(s_j^k - \mu_{\mathbf{s}^k})^2}\sqrt{\sum_{j=1}^{\mathcal{N}}(mear_j - \mu_{mear})^2}}.$$

$\mathcal{N}$ is the number of document pairs. $\mu_{\mathbf{s}^k}$ and μ_{mear} are the mean scores of the results and human ratings. The range of this metric is $[-1, 1]$.

In Table 1, besides the performance on the entire dataset (row "entire"), we also divide the dataset into several subsets based on rating difference and list the performance on each subset. It shows two observations. First, generally when the rating difference increases, the difficulty of document similarity judgment increases and the performance of both two models decrease. Second, each approach performs differently on different subsets with different confusability.

We conclude two statements in this exploration study. First, confusability in ground truth has prominent influences on the evaluation results; second, the credibility of the evaluation results changes in different confusability. Therefore, we need to consider the confusability in ground truth and evaluation.

3   Evaluation with Confusable Ground Truth

3.1   Confusability Measures

For analyzing human ratings in ground truth, the statistical measures used in traditional methods which aggregate the human ratings of a given instance can be mean ($mear_i$), median ($medr_i$), majority voting ($majr_i$) and so on. They do not consider the confusability information. In contrast, we propose three candidate confusability measures.

1. *Rating Difference*: It is the largest difference of two ratings in the rating set and has been used in the study in Sect. 2.1, $difr_i = max\{\mathbf{r}_i\} - min\{\mathbf{r}_i\}$.
2. *Standard Deviation*: It is to qualify the amount of variation or dispersion of the ratings set of an instance. We use the sample standard deviation and define it as $stdr_i = \sqrt{\sum_{j=1}^{n_i}(r_{ij} - mear_i)^2/(n_i - 1)}$.
3. *Entropy*: It is to describe the uncertainty in the ratings set of an instance. When this uncertainty is low, it means that more reviewers can reach same ratings and the confusability is low. We define it as $entr_i = -\sum_k p_i(k)lnp_i(k)$, $p_i(k) = n_i(k)/n_i$, where $n_i(k)$ is the number of ratings that are equal to k.

We plot these three confusability measures in Fig. 2. It shows that these confusability measures are linearly dependent to each other in a certain degree. When all reviewers of d_i reach consensus, $difr_i = stdr_i = entr_i = 0$. Only one measure cannot describe the confusability completely, and we need several measures to describe it from multiple aspects. For example, for two rating sets with five ratings, $\{5, 5, 5, 5, 1\}$ and $\{4, 3, 3, 2, 2\}$. The first set has a lower $entr_i$ (=0.72), but higher $difr_i$ (=4) and $stdr_i$ (=1.79); the second set has a higher $entr_i$ (=1.52), but lower $difr_i$ (=2) and $stdr_i$ (=0.84). Furthermore, as shown in Fig. 1 in Sect. 2.1, the distribution of a confusability measure on a statistical measure has some patterns of distribution, in contrast to distributing randomly.

In addition, even relatively low values of confusability measures can be significantly different from zero. Statistical significance for confusability measures makes no claim on how important is a sufficient magnitude in a given scenario or what value can be regarded as absolutely low or high confusability.

3.2   Confusability Aware Evaluation

The confusability in the ground truth of a given dataset actually depends on the instance selection process of constructing this dataset. The dataset creators can follow some rules in the instance selection process. For example, only selecting the instances with low confusability or keeping a balance on the number of low-confusability and high-confusability instances. Then the dataset users can provide confusability information with the experimental results.

We propose four candidate methods for confusability aware evaluation, which leverage the confusability information in two manners. One is selecting the

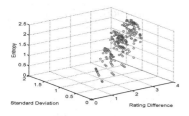

Fig. 2. Relations of Rating Difference, Standard Deviation and Entropy

Table 2. Evaluation results with united confusability aware metrics

	Metrics	LSA	KBG
Entire dataset	difr-mear	**0.309**	0.257
	stdr-mear	**0.345**	0.306
	entr-mear	**0.318**	0.282
$difr_i \leq 2$	difr-mear	**0.303**	0.204
	stdr-mear	**0.387**	0.311
	entr-mear	**0.360**	0.287

instances according to the confusabilty (instance selection based methods), i.e., the following method 1 and 2. The other is combining the confusability measures with existing evaluation metrics (confusability aware metric based methods), i.e., the following method 3 and 4. We use the LP50 dataset and same approaches in Sect. 2.1 to illustrate experimental results by using these evaluation methods.

1. Evaluation instances with different confusability respectively:

This method has been used in the exploration study in Sect. 2.1. In addition to evaluate the performance on the entire dataset, we also divide the dataset into several subsets according to the confusability. The evaluation results by this method is shown in Table 1.

2. Evaluation instances with low confusability only:

We can only use the instances of which the confusability satisfy a given condition. For example, if the condition is $difr_i = 0$, which means we only use the instances that all reviewers reach consensus, as shown in Table 1, the performance of LSA and KBG are 0.859 and 0.714; if the condition is $difr_i \leq 2$, the performance of LSA and KBG are 0.645 and 0.595. They show that when confusability of instances is low, the performance of LSA is better.

In contrast to method 1, the advantage of this method is that it ensures high credibility of the experimental results. The disadvantages of this method is that it omits some kinds of instances. For example, for LP50 dataset, it uses the instances with low or high similarities but ignores the instances with middle similarities. Our proposals do not strongly accept or reject the instances with high confusability. This decision is made according to the detailed cases.

3. Separated confusability aware metrics:

A simple manner of combining confusability information with existing evaluation metrics is to provide additional confusability measures without modifying existing metrics. We use the average of confusability values on all instances in the evaluated set for this purpose. For example, the performance on the entire dataset of LSA and KBG are showed in Table 1. In addition, on the entire dataset, the average rating difference is 1.89; the average standard deviation is 0.67; the average entropy is 0.99. These information of the dataset are independent with the evaluated approaches. The advantage of this method is that it is easy to be integrated with existing experimental methods and results. The disadvantage of this method is that it only provides the information, but the confusability does not influence the evaluation results.

4. United confusability aware metrics:
In this method, we integrate the confusability information into the existing evaluation metrics. We use the Pearson correlation coefficient defined in Sect. 2.1 as the example of evaluation metric. We modify the formula by adding the confusability information as the weights of instances in the following format. We consider that the instances with higher confusability has lower credibility and should have lower influences in the evaluation results. We thus assign lower weights to the instances with higher confusability. The confusability aware formats for other evaluation metrics average MAP or nDCG on instances in the dataset can be formulated in same way:

$$caCorr(\mathbf{s}^k, \mathbf{st}) = \frac{\sum_{i=1}^{\mathcal{N}} w_i(s_i^k - \mu_{\mathbf{s}^k})(st_i - \mu_{\mathbf{st}})}{\sqrt{\sum_{j=1}^{\mathcal{N}}(s_j^k - \mu_{\mathbf{s}^k})^2}\sqrt{\sum_{j=1}^{\mathcal{N}}(st_j - \mu_{\mathbf{st}})^2}}, \quad w_i = 1 - norm(ca_i).$$

st_i can be $mear_i$, $medr_i$ or $majr_i$; ca_i can be $difr_i$, $stdr_i$ or $entr_i$, it is normalized into the range of $[0, 1]$. For example, if we use $mear_i$ as st_i and $entr_i$ as ca_i, we can define this metric as entropy-confusability-aware-mean correlation.

The performance on the entire dataset of LSA and KBG are illustrated in Table 2. It shows that LSA has prominently better performance than KBG when considering confusability in the evaluation metrics. In contrast, without considering confusability, LSA and KBG have similar performance on the entire dataset in Table 1. *Although the performance of LSA and KBG are not distinguishable when using traditional evaluation method, they have prominent difference when using our confusability aware evaluation method. It shows the significance of confusability aware evaluation and the effectiveness of our evaluation method.*

In an evaluation task, an instance selection based method and a confusability aware metric based method can be combined to construct an evaluation solution. For example, if using method 2 and 4, we can evaluate the performance of LSA and KBG on the subset which contains the instances with low confusability $difr_i \leq 2$. The results are shown in Table 2. It is consistent with the observation that LSA performs better when the confusability is lower.

All these confusability aware evaluation metrics and methods do not raise the problem on scalability. They can also be used to save the cost in the ground truth construction by setting less reviewers to instances with low confusability.

4   Conclusion

In this paper, we discuss the influence of confusability on evaluation results. We propose several confusability aware evaluation metrics and methods. In future work, we will analyze more characteristics of the confusability aware evaluations.

Acknowledgments. Thanks to Dr. Yasuhito Asano and Dr. Toshiyuki Shimizu for your kind comments on this topic.

References

1. Lee, M.D., Welsh, M.: An empirical evaluation of models of text document similarity. In: Proceedings of CogSci 2005, pp. 1254–1259 (2005)
2. Schuhmacher, M., Ponzetto, S.P.: Knowledge-based graph document modeling. In: Proceedings of WSDM 2014, pp. 543–552 (2014)
3. McHugh, M.L.: Interrater reliability: the kappa statistic. Biochem. Med. **22**(3), 276–282 (2012). http://www.ncbi.nlm.nih.gov/pmc/articles/PMC3900052/
4. Whitehill, J., Wu, T.F., Bergsma, J., Movellan, J.R., Ruvolo, P.L.: Whose vote should count more: optimal integration of labels from labelers of unknown expertise. In: Proceedings of NIPS 2009, pp. 2035–2043 (2009)

Author Index

Printed in the United States
By Bookmasters